Stem Cell Biology and Regenerative Medicine

Volume 77

Our understanding of stem cells has grown rapidly over the last decade. While the apparently tremendous therapeutic potential of stem cells has not yet been realized, their routine use in regeneration and restoration of tissue and organ function is greatly anticipated. To this end, many investigators continue to push the boundaries in areas such as the reprogramming, the stem cell niche, nanotechnology, biomimetics and 3D bioprinting, to name just a few. The objective of the volumes in the Stem Cell Biology and Regenerative Medicine series is to capture and consolidate these developments in a timely way. Each volume is thought-provoking in identifying problems, offering solutions, and providing ideas to excite further innovation in the stem cell and regenerative medicine fields.

Raymond C. B. Wong • Yohei Hayashi
Editors

Advances in Using Stem Cells to Treat Neural Diseases

Springer

Editors
Raymond C. B. Wong
Centre for Eye Research Australia
The University of Melbourne
East Melbourne, VIC, Australia

Yohei Hayashi
Research and Development Center
CiRA Foundation
Osaka, Japan

ISSN 2196-8985 ISSN 2196-8993 (electronic)
Stem Cell Biology and Regenerative Medicine
ISBN 978-3-031-94100-9 ISBN 978-3-031-94101-6 (eBook)
https://doi.org/10.1007/978-3-031-94101-6

This Springer imprint is published by the registered company Springer Nature Switzerland AG
The registered company address is: Gewerbestrasse 11, 6330 Cham, Switzerland

Preface

This book, *Advances in Using Stem Cells to Treat Neural Diseases*, dives into the forefront of stem cell research, uncovering the tremendous potential of stem cell technology for neural regeneration. The goal of this book is to offer valuable insights into the diverse strategies for using stem cells to study neurodegenerative and neurodevelopmental disorders, as well as advances in research of neural regeneration strategies.

This comprehensive volume provides a collection of chapters carefully curated by international experts to discuss advances in various aspects of tissue engineering, stem cell biology, developmental biology, and neurosciences. These chapters explore the progress made in stem cell research and highlight recent advances in cell reprogramming and tissue engineering to develop disease modeling and regenerative therapy for neural disorders. This book offers an unparalleled snapshot of the current landscape of stem cell research and an overview of up-and-coming technologies that could revolutionize the field. We hope that this unique resource will provide an important reference for stem cell biologists, neuroscientists, tissue engineers, molecular biologists, as well as students, academics, and clinicians with an interest in neural regeneration.

Most importantly, we would like to sincerely express our heartfelt appreciation to the book chapter authors for their dedicated effort in providing insightful discussions within their respective research areas. In addition, we are grateful for the funding support of the RIKEN-University of Melbourne Collaborative project scheme and the JSPS open partnership joint research projects co-funding schemes (RCB Wong and Y Hayashi). Also we extend our thanks to the editorial team of Life and Biomedical Sciences Books at Springer Nature, Dr. Gonazolo Cordova, Merry Stuber, and Ilse Kooijman, for their valuable assistance and guidance in bringing this book together.

East Melbourne, VIC, Australia — Raymond C. B. Wong
Osaka, Japan — Yohei Hayashi

Preface

Editorial: Advances in Stem Cell and Regenerative Approaches to Study and Treat Neural Diseases

The remarkable advances in stem cell research highlighted a paradigm shift in neuroscience research, offering a valuable opportunity to develop new methods to study and treat a range of neural diseases.

The development of induced pluripotent stem cell (iPSC) technology provides a feasible approach to generate patient-specific iPSCs and genome-edited iPSCs as an unlimited cellular source for disease modeling and cell therapy. **Hayashi** introduces the general approaches and strategies for generating, using, and editing iPSCs for utilizing these cells for disease modeling and drug discovery. As a prominent example, **Akaba et al.** introduce that patient-derived, disease-specific iPSCs for MECP2-related neurodevelopmental disorders have become indispensable for modeling disease phenotypes and exploring therapeutic strategies. Also, **Polanco et al.** provide a comprehensive overview of the wide assortment of gene editing tools available to iPS cells, aiming to serve as a guide for both the Parkinson's disease (PD) research community and the broader regenerative medicine field.

To model neurodegenerative diseases from patient cells, it is critical to develop tissue engineering and cell processing technologies. **Harati et al.** explore the role of cell-cell interactions in regulating stem cell behavior and discuss the potential of advanced bioengineering techniques using bioengineered materials and organ-in-a-chip to facilitate development of in vitro disease models and cell therapy in the brain. **Inzunza et al.** provide an overview of the latest advancements in 3D brain in vitro models, especially for Alzheimer's disease (AD) research, and their advantages and limitations. Also, **Maeda** introduces a new approach using induced neurons (iNs), which are converted directly from somatic cells for studying AD and various diseases. **Frederiksen et al.** extend the discussion of iPSC application for disease modeling for amyotrophic lateral sclerosis (ALS) and frontotemporal dementia (FTD), and discuss the current cell therapy strategies for these neurodegenerative diseases affecting the central nervous system.

Similarly, iPSC has been used to successfully model a number of retinal degenerative diseases, providing novel insights into the mechanism driving disease pathogenesis. **Sharma et al.** provide an in-depth discussion on the use of iPSC-derived retinal organoids for modeling a range of inherited retinal diseases including retinitis pigmentosa and Leber's congenital amaurosis, as well as its application for cell replacement therapy. Expanding on this topic, **Liu et al.** provide a comprehensive review of the strategies to generate retinal organoids, and highlight the current challenges and future development to improve the retinal organoid technology.

Lastly, for developing therapeutics using stem cells, **Morizane** describes the overview of cell therapy for PD using pluripotent stem cells and discusses practical points to improve in stem cell-based therapies for PD. Also, **Banno et al.** highlight the current situations and future perspectives of drug discovery and development for neurodegenerative diseases using iPSCs. **Urrutia-Cabrera et al.** discuss the use of cellular reprogramming to stimulate endogenous regeneration in the retina, and highlight the potential of this approach as a novel gene therapy strategy to treat retinal degeneration.

The presented chapters in this book cover comprehensive topics to offer profound insight into the development of cell and gene therapy for neurodevelopment and neurodegenerative diseases. Altogether the chapters provide a timely overview of the dynamic and evolving field of stem cell research, highlighting the recent cutting-edge research in neural regeneration. We hope that this book will be of high interest to many students and specialists in the fields of stem cell biology, regenerative medicine, and neuroscience.

Centre for Eye Research Australia
Royal Victorian Eye and Ear Hospital
East Melbourne, VIC, Australia
Ophthalmology, Department of Surgery
University of Melbourne
East Melbourne, VIC, Australia

Raymond C. B. Wong

Research and Development Center
CiRA Foundation
Osaka, Japan

Yohei Hayashi

Contents

Chapter 1
Disease Modeling and Drug Development Using Patient-Derived and Genome-Edited Pluripotent Stem Cells

Yohei Hayashi

Abstract Patient-derived induced pluripotent stem cells (hiPSCs) and genome-edited hiPSCs are useful in biomedical studies for various diseases. Since hiPSCs maintain self-renewal and pluripotency, once hiPSCs are generated from somatic cells of a donor patient, they can be used indefinitely as valuable bioresources. For this reason, there are hiPSC banks in many countries to utilize them in basic research and drug development. In addition, by developing genome-edited hiPSC lines carrying mutations responsible for genetic diseases generated from standard or healthy-donor hiPSCs or repaired mutations from patient-derived hiPSCs, it is possible to perform relatively accurate research studies using these isogenic hiPSC lines. For disease modeling, hiPSCs are differentiated into specific cell lineages or organoids. Since some of the differentiated cell types are highly heterogeneous, visualizing particular cell types and cell status with fluorescent reporters is valuable. Disease models are developed based on recapitulating abnormal phenotypes of cell characteristics and behavior, abnormal expression and abnormal localization of genes and proteins, and other metabolites. These models contribute to elucidating the molecular mechanisms of diseases. Furthermore, high-throughput assays based on hiPSC-derived models can be used to screen molecular targets and compounds. These systematic models are widely applicable as alternative methods which evade animal experiments for intractable diseases.

Keywords Human-induced pluripotent stem cells (hiPSCs) · Disease-specific hiPSCs · Characterization · Self-renewal · Pluripotency · Genome editing · Organoid · Banking · Disease model · Abnormal phenotypes · High-throughput assays · Drug development · Fluorescent reporters

---

Y. Hayashi (✉)
Research and Development Center, CiRA Foundation, Kyoto, Japan
e-mail: yohei.hayashi@cira-foundation.or.jp

R. C. B. Wong and Y. Hayashi (eds.), *Advances in Using Stem Cells to Treat Neural Diseases*, Stem Cell Biology and Regenerative Medicine 77,
https://doi.org/10.1007/978-3-031-94101-6_1

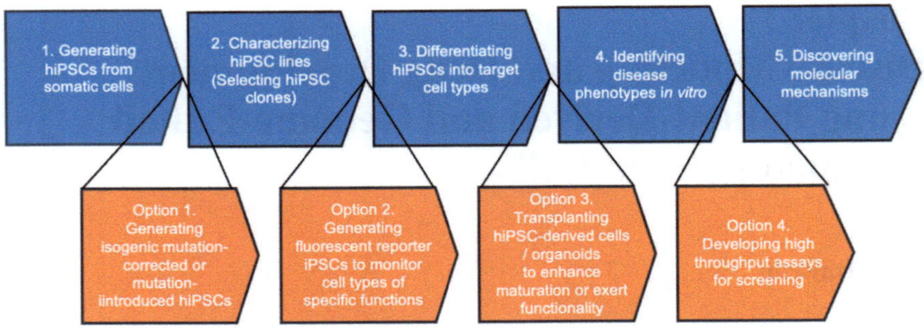

Fig. 1.1 The general workflow of disease-specific hiPSC study

1.1 Introduction

In 2007, human-induced pluripotent stem cells (hiPSCs) were established for the first time [1, 2], and their utilization was expected to recapitulate pathological conditions of intractable diseases by creating hiPSCs from somatic cells of patients with intractable diseases. In the next year, 2008, hiPSCs derived from patients with intractable diseases such as ADA deficiency, Schwachmann-Diamond syndrome, Gaucher disease, Duchenne and Becker muscular dystrophy, Parkinson's disease, Huntington's disease, juvenile type one diabetes, Down syndrome, and Lesch-Nyhan syndrome were generated [3, 4]. In the next year, 2009, a study was reported in which hiPSCs derived from patients with familial autonomic neuropathy were used to induce differentiation into cell types degenerated by the disease, successfully reproducing abnormal phenotypes at the cellular level, and analyzing their gene expression patterns [5]. Starting with these studies, various disease-specific hiPSCs have been generated to date and used for modeling abnormal phenotypes at the cellular level that reflect pathological conditions. Furthermore, these hiPSCs have been utilized in research to elucidate the pathogenic mechanisms and in the development of therapeutic methods. Also, through advances in genome editing technologies, genome-edited hiPSCs have been complementarily used for disease modeling. The general workflow of patient-derived hiPSC study is shown in Fig. 1.1.

1.2 Generation of Patient-Derived hiPSCs

With advances in hiPSC generation technology, it is now possible to produce hiPSCs from several milliliters of peripheral blood or tiny skin samples from a patient suffered from an intractable disease. The reprogramming methods used for hiPSC generation were initially retroviral or lentiviral vectors, which had to be inserted into the genome; however, in recent years, episomal plasmid vectors, Sendai virus vectors, and mRNA transfection, which are not inserted, are commonly used.

Usually, several single-cell-derived clones should be picked up and expanded from a donor to use further studies.

1.3 Characterization of Patient-Derived hiPSCs

After generating hiPSCs, these hiPSC lines need to be characterized for self-renewal, pluripotency, genomic integrity, no expression of reprogramming vectors, and no pathogen contamination. For example, a criteria of successful patient-derived hiPSCs is listed in the scientific guidelines for Lab Resources section in Stem Cell Research journal [6]. This characterization process includes the selection of good clones using these aspects. As examples of the generation and characterization of hiPSCs, I have contributed to the establishment of patient-derived hiPSCs for fibrodysplasia ossificans progressiva (FOP) [7, 8], ring chromosome [9], juvenile nephronophthisis [10], adrenoleukodystrophy (ALD) [11], 22q11.2 deletion syndrome [12], Wilson's disease [13], Rett syndrome [14], glucose transporter one deficiency syndrome [15], and biliary atresia [16].

1.4 Using Deposited Disease-Specific hiPSCs Cells in Cell Banks

As hiPSCs maintain pluripotency and self-renewal, they can proliferate indefinitely as immortalized cells. Therefore, once patient-derived hiPSCs have been generated and cryopreserved, they can be used indefinitely in principle. These hiPSCs can be deposited in cell banks as valuable bioresources and made available for permanent use by a wide range of research communities. Here I introduce the current status of hiPSC banks.

1.4.1 Global hiPSC Banking for Research

By 2021, there were more than 6000 disease-specific pluripotent stem cell lines derived from patients with more than 1000 diseases in 36 countries worldwide [17]. In addition to RIKEN Cell Bank in Japan, other particularly large cell banks include the CIRM iPSC Repository (Cellular Dynamics/Fujifilm) in California, USA, WiCell in Wisconsin, USA, and EBiSC in Europe. Other regions have also established their own iPSC service consortia, including the National Stem Cell Bank of Korea and the Taiwan Human Disease iPSC Service Consortium in Taiwan (Table 1.1).

Table 1.1 The list of major disease-specific hiPSC banks worldwide for academic research

Bank name	Region	Web URL
RIKEN BRC Cell Engineering Division (Cell Bank)	Japan	https://cell.brc.riken.jp/en/
Coriell Institute for Medical Research	USA	https://www.coriell.org/
WiCell Research Institute	USA	https://www.wicell.org/
European Bank for induced pluripotent Stem Cells (EBiSC)	Europe	https://ebisc.org/
National Stem Cell Resource Center	China	https://nsctrc.tongji.edu.cn/about_english.html
Korean National Stem Cell Bank (KSCB)	Korea	http://kscr.nih.go.kr/
Human Disease iPSC Consortium Resource Center (Taiwan Human Disease iPSC Consortium)	Taiwan	https://ncfb.nycu.edu.tw/en/ipsc.html

1.4.2 Banking of Patient-Derived hiPSCs in Japan

In Japan, disease-specific hiPSCs for 231 diseases, 753 patients, and 3110 hiPSC lines have been deposited in the RIKEN Cell Bank (as of February 6, 2023). In addition, 718 hiPSC lines from 130 healthy Japanese individuals have been deposited in the RIKEN Cell Bank for comparison. Most of these disease-specific hiPSC lines were established by the "Establishment Center (Director: Shinya Yamanaka)" as a part of the AMED (Japan Agency for Medical Research and Development) project [18]. In this project, hiPSCs have been generated from many patients who are suffered from intractable diseases as designated in Japan. Most of the cell lines do not require "consent for donation" or "collaborative research" and can be used by profit-making companies. However, review and approval by the ethics committee of the research institution are required.

1.5 Application of Genome Editing Technologies

1.5.1 Generation of Isogenic hiPSCs for Insertion or Repair of Specific Mutations

Genome editing technology, represented by CRISPR-Cas9, which has been remarkably developed in recent years, makes it possible to generate hiPSCs with the same genetic background (i.e., isogenic) as disease-specific hiPSCs or, conversely, to generate hiPSCs carrying an isogenic mutation from healthy donor-derived hiPSCs [19–21]. This method is particularly applicable for single-nucleotide mutation or polymorphism and is important for conducting high-quality research that can withstand statistical analysis, especially for rare diseases that are caused by a single gene and for which it is difficult to obtain a sufficient number of patient-derived hiPSCs [22, 23].

1.5.2 Development of Fluorescent-Reporter hiPSCs Using Genome Editing Technology

Using a fluorescent reporter to visualize specific cell types and their status is highly beneficial. By employing genome editing techniques, we can precisely monitor and select differentiated cell types by inserting a gene that encodes a fluorescent protein associated with marker genes in hiPSCs. Despite advances in genome editing technology with homology-directed repair, which allow for the insertion of desired reporter genes into the human genome, the efficiency remains low because the donor vectors randomly integrate into the host genome. To concentrate the knocked-in cells, we have developed a method utilizing a "double-tk donor vector system," where the expression units of the herpes simplex virus-thymidine kinase (HSV-tk) are positioned on both outer sides of the homology arms. This method is more effective for selecting knocked-in hiPSCs than conventional donor vector systems that lack an HSV-tk cassette or contain only a single HSV-tk cassette [24]. With this approach, we have created fluorescent-reporter hiPSC lines and made them available to the research community through RIKEN Cell Bank. Here are some examples: EEF1A1-iRFP670 [24], H2B-EGFP line [24], H2B-tdTomato line [24], PAX6-tdTomato line [25], SOX17-tdTomato line [25], ISL1-tdTomato line [26], and MBP-tdTomato expressing line [27].

1.6 Disease Modeling Using Disease-Specific hiPSCs

According to a survey report commissioned by Japan Agency for Medical research and Development (AMED) to a consulting company, the most common types of diseases for which hiPSCs are produced are neurological diseases, followed by various genetic diseases, endocrine system diseases, cardiovascular diseases, and psychiatric diseases [28]. In terms of the number of studies selected for Grants-in-Aid for Scientific Research in Japan, the most common disease areas of iPSC-related studies are psychiatric and nervous system, congenital diseases, hematology/immunology, ophthalmology, and musculoskeletal system, in that order. Also, in the number of scientific papers used for drug screening using hiPSCs from 2016 to 2018, "Neurology" was by far the most common. These results indicate that there are unmet medical needs, which could be solved by the research and development utilizing hiPSCs, especially in neurological diseases. On the other hand, at the basic research level, with the development of differentiation technologies, research in various areas such as the urinary system including kidney, respiratory system including lung and trachea, and digestive system including stomach and intestines has been reported one after another. Expansion into research in these and other unexplored areas is expected in the future.

1.7 Drug Discovery and Development Using Disease-Specific hiPSCs

Disease-specific hiPSCs are utilized in many drug discovery and development processes. First, the discovery of abnormal phenotypes, abnormal gene/protein expression, and abnormal metabolites at the cellular level can lead to identify drug targets. The high throughput of these assay systems is expected to be applied to phenotypic screening. Since drug development using disease-specific hiPSCs is not an established method, it is important to determine how to integrate them into the entire drug development process. In addition, various types of cells differentiated from hiPSCs are useful for various safety tests. Because hiPSCs are immortalized human cells derived from patients and retain self-renewal and pluripotency, platforms of drug discovery and development using disease-specific hiPSCs have the following features and advantages.

First, compared to tests using experimental animals, hiPSCs are superior as an alternative method to animal experiments in terms of their replacement and reduction. Furthermore, since human cells are used, there is no need to consider species differences. However, it is difficult to evaluate the effects of higher-complexed functions and pharmacokinetics in individuals since it is an in vitro experimental system. Second, compared to conventional immortalized human cell lines, the ability to differentiation into various cell types while maintaining a normal genome set enables to address a wider range of diseases. However, it should be noted that differentiation into specific cell types from hiPSCs is costly and time-consuming, and the efficiency of the differentiation is limited. Third, compared to primary cultured human cells, hiPSC-derived cell types can be supplied stably. Furthermore, some cell types, such as the central nervous system and cardiac muscles, can be practically obtained only from pluripotent stem cells for research purposes. However, it should be noted that most hiPSC-derived cell types are immature and fetal. Fourth, disease-specific hiPSCs are also useful to conduct research to predict the effects of gene therapy, such as the transduction of normal genes with repair effects into diseased hiPSCs using vectors or other methods, or the use of nucleic acids as a therapeutic method. By incorporating such gene modification technologies and gene therapy methods, drug discovery and development using disease-specific hiPSCs can demonstrate their safety and effectiveness in various therapeutic modalities.

1.8 Examples of Disease Studies Using Disease-Specific hiPSCs

Here I show examples of our studies using patient-derived and genome-edited hiPSCs for disease modeling and potential drug discovery.

1.8.1 Wilson's Disease

Wilson's disease is an intractable disease caused by the mutation of the ATP7B gene, which functions to expel copper from cells, resulting in the accumulation of unwanted copper in the body, which causes systemic damage mainly in the liver and brain [29–32]. Conventional treatment with dietary restrictions and drugs that promote copper discharge may cause side effects, and the development of new treatment methods is desired [33–36]. Immortalized cell lines and disease model mice have been conventionally used to study the pathology of Wilson's disease, but there are problems of interspecies differences between humans and experimental animals, differences in the types of mutations between people, and large gaps in the genetic background of individuals [37–44]. Thus, we used hiPSCs that carried the mutation of ATP7B, the causative gene, derived from multiple Wilson's disease patients and ATP7B-edited hiPSCs [13] (schematics in Fig. 1.2).

In this study, hiPSCs were generated from four patients of Wilson's disease. These Wilson's disease-patient-derived hiPSCs maintained self-renewal capacity, pluripotency, genomic integrity, and the compound heterozygous mutations in the ATP7B gene. We then examined the expression and subcellular localization of proteins produced from the mutant ATP7B gene in these hiPSCs and found that, depending on the type of mutation, either the expression level or subcellular localization, or both, were aberrant. The protein ceruloplasmin (Cp), synthesized in hepatocytes, is a copper-containing ferroxidase that is secreted into the blood after binding to copper incorporated by the ATP7B protein in the trans-Golgi network [45–48]. Low levels of Cp in the blood are commonly used as a diagnostic criterion for Wilson's disease. These results indicate that hepatocytes differentiated from hiPSCs derived from Wilson's disease patients can recapitulate the decreased expression or abnormal localization of ATP7B protein and decreased secretion of Cp.

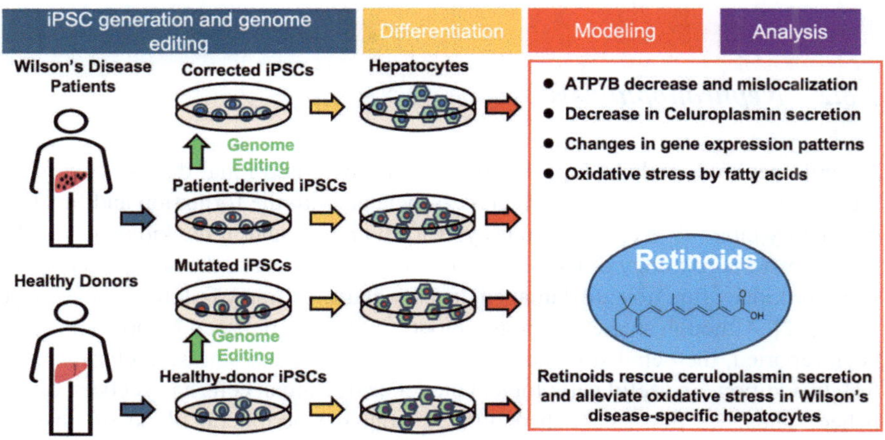

Fig. 1.2 The schematic overview of our study using Wilson's disease-specific hiPSCs

Furthermore, two hiPSC lines with two different ATP7B gene mutations were generated from healthy-donor-derived hiPSC lines, respectively, by genome editing. We also generated a hiPSC line with a corrected mutation in a known Wilson disease patient-derived hiPSC line (ATP7B R778L homozygote). Using these Wilson disease patient-derived hiPSC lines and ATP7B-edited hiPSC lines, we explored the correlation between disease phenotype and genotype to find new therapeutic drug candidates. Transcriptome analysis identified genes associated with abnormalities in the retinoid signaling pathway and lipid metabolism in Wilson disease hepatocytes. By searching for chemicals that increase Cp secretion in hiPSC-derived hepatocytes, we identified all-trans retinoic acid (ATRA) and retinoids, which have been shown to be clinically safe, as promising candidates. ATRA also alleviated the production of the reactive oxygen species (ROS) induced by lipid accumulation in oleic acid-treated Wilson's disease hepatocytes.

In this study, we have developed a versatile research platform for developing therapeutic techniques and investigating molecular pathogenesis in Wilson's disease and other liver diseases, as this disease model faithfully reproduces the initial characteristics of Wilson's disease hepatocytes. Furthermore, since the detection methods are clear and easy to apply, these assays will be utilized to create a high-throughput screening system for the identification of ideal therapeutic agents. We also demonstrated for the first time that retinoids restore Cp secretion and alleviate oxidative stress in Wilson's disease hepatocytes. Since decreased Cp secretion and hepatic lipidosis are early symptoms of Wilson's disease, our results indicate that retinoids may prevent or delay the progression of these symptoms. The results indicate that retinoids may prevent or delay the progression of these symptoms. On the other hand, the molecular mechanisms by which abnormalities in the ATP7B gene lead to abnormalities in Cp secretion and lipid metabolism remain elusive. Future studies should investigate the relationship between copper and lipid metabolism. We hope that the results of this research will lead to the development of new treatments for Wilson's disease and related liver diseases.

1.8.2 Nephronophthisis

Juvenile nephronophthisis type 1 (NPH1) is an inherited renal disease caused by mutations in the NPHP1 gene, which is associated with the formation and function of primary cilia [49–51]. Numerous cysts form and expand in the kidneys, and the overall histomorphology of the kidneys collapses. As a result, the nephrons responsible for kidney function are damaged, usually causing kidney failure in the teenage years [52]. Although studies on mouse models with the abnormal Nphp1 gene have been reported, the detailed pathogenesis of the disease has not been clarified [53, 54]. In addition, no useful pathological model using human cells has yet been established. Thus, we first established hiPSCs from the peripheral blood of juvenile nephronophthisis patients with a total deletion of the NPHP1 gene [10]. Next, we aimed to generate kidney organoids from these hiPSCs and compared them with

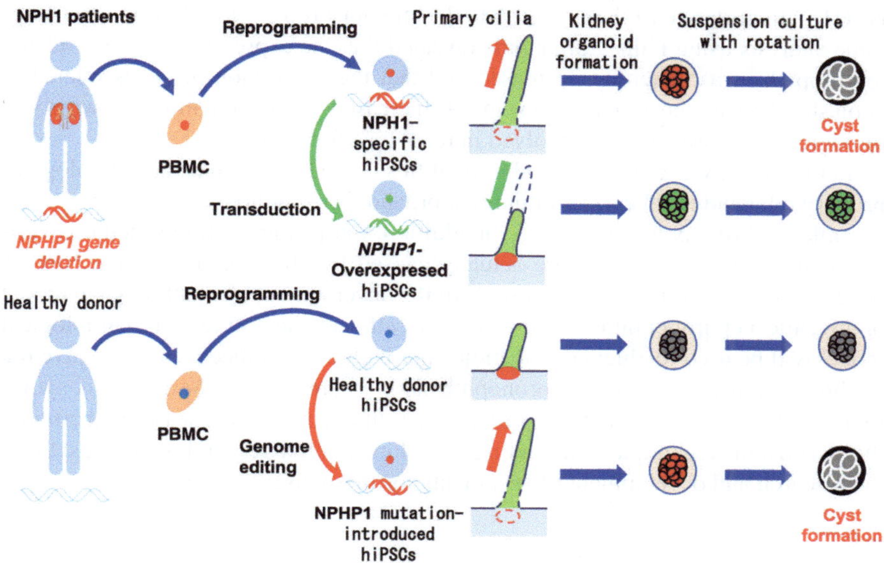

Fig. 1.3 Overview schematics of this study

hiPSCs derived from healthy individuals and patient-derived hiPSCs transfected with the NPHP1 gene [55] (schematics in Fig. 1.3).

We aimed to create a pathological model to analyze the pathogenic mechanism using NPH1-patient-derived hiPSCs and NPHP1-edited hiPSCs generated from healthy-donor hiPSCs using genome editing technology. Also, a plasmid DNA vector expressing the human NPHP1 gene was transduced into these patient-derived hiPSCs. These transduced hiPSCs were collected and evaluated for expression of the nephrocystin-1 protein produced by the NPHP1 gene, which was completely lost in patient-derived hiPSCs, while expression was confirmed in NPHP1-transfected hiPSCs. Immunofluorescence staining of the nephrocystin-1 protein confirmed its expression in NPHP1-transgenic hiPSCs. In addition, hiPSCs with a frameshift mutation in the NPHP1 gene were generated from healthy donor-derived hiPSCs by genome editing technology, and using CRISPR-Cas9 targeting the NPHP1 gene, deletion or insertion of a base was confirmed at the target site. The expression of nephrocystin-1 protein was lost in these NPHP1-edited hiPSCs. Nephrocystin-1 protein is localized in the transition zone of primary cilia. Therefore, we visualized primary cilia and nephrocystin-1 protein by immunofluorescence staining and compared the length of primary cilia and the percentage of cells forming primary cilia. NPH1-patient-derived and NPHP1-edited hiPSCs abnormally exhibited longer primary cilia and decreased the proportion of cells that formed primary cilia. Furthermore, kidney organoids were produced from these hiPSCs and examined. To reproduce the shear stress caused by the flow of primary urine (liquid filtered from blood) in the kidney, kidney organoids were cultured in suspension with constant rotation. Cysts were formed in kidney organoids generated from

NPH1-patient-derived hiPSCs and NPHP1 mutant hiPSCs. In addition, transcriptome analysis using kidney organoids revealed that the expression of genes related to ciliopathy was generally decreased and that the expression of genes related to cholesterol biosynthesis and cell cycle was increased. These signaling pathways and biological processes may be involved in renal cyst formation.

In this study, we recapitulated abnormalities in the formation and morphology of primary cilia and renal cysts within renal organoids using hiPSCs established from juvenile nephronophthisis patients. In addition, the genome editing methods verified that these abnormalities are actually caused by abnormalities in the NPHP1 gene. These results are the first example of the development of a pathological model of juvenile nephronophthisis using human cells. In the future, this pathological model will be used to elucidate the molecular pathological mechanisms of juvenile nephronophthisis. Juvenile nephronophthisis is a rare disease; however, its symptoms and pathophysiology share similarities with other kidney diseases and ciliopathies. Thus, the techniques and findings developed in this research will be useful in the research and development of treatments for these diseases as well.

1.9 Current Issues and Future Perspectives for Disease Modeling Using hiPSCs

As described above, disease-specific hiPSCs can be used to recapitulate abnormal phenotypes under culture conditions. In addition, other pioneering examples of research have developed systems that can evaluate the efficacy of drugs and conduct phenotypic screening through drug repositioning and other assays. In the future, the following issues will need to be overcome to apply for drug discovery and development utilizing disease-specific hiPSCs. First, methods to generate homogeneous disease-specific hiPSCs in a short period of time are required. Reprogrammed hiPSCs are highly heterogeneous. Even if general characterization as basic quality tests is performed, the entire genome and epigenetic status of clonal lines cannot be completely understood. As a result, differentiation efficiency and cell functions vary among clones, and even cell lines derived from the same patient may vary in properties among clones or among somatic cells of different origin. In such cases, differences between individuals or between diseased patients and healthy individuals, which should be detectable, can become undetectable. Higher quality and more homogeneous methods for the generation, selection, and characterization of hiPSCs are required. Furthermore, methods to homogenize already generated disease-specific hiPSCs would be effective. Second, methods to produce homogeneous differentiated cells and organoids are required. Organoid technology is a very good method to create tissues of a certain size and to demonstrate their tissue functions, but it is often difficult to ensure homogeneity for highly accurate drug discovery research because the control of differentiation state depends on cell autonomy. It is necessary to develop differentiation and organoid production methods that can be applied uniformly to a wider range of cell types. Third, advancement using MPS (MicroPhysiological Systems), such as organ-on-a-chip, which incorporates organoids on a chip, has been attracting attentions in recent years, and it aims to reproduce the interactions between organs in the human body in vitro. The MPS technology is also used to reproduce the interactions between organs in the human body in vitro. The development of massively parallel systems and the reproduction of pharmacokinetics through multiple organs are also goals of the MPS technology. It is expected to be applied to drug discovery processes. Fourth, the expansion of application to more diseases other than single genetic disorders, such as genetic polymorphisms, are required. Although hiPSCs inherit patient-derived genomes, the "epigenetic memory" of the original cells is essentially erased, and thus their effectiveness in treating diseases caused by multiple factors other than single genetic disorders has been considered low. Recent developments in research have shown that it is also effective as a model for infectious diseases. It is possible to approach the area of diseases involving multiple factors, such as immunology. Furthermore, it is expected to contribute not only in the research phase but also in drug discovery toward the realization of tailor-made medicine that takes individual genetic polymorphisms into account, such as evaluating the efficacy of therapeutic drugs in the development field.

1.10 Conclusion

In conclusion, the development and application of hiPSCs have significantly advanced our understanding of intractable diseases and opened new avenues for research and therapeutic development. Since their establishment in 2007, hiPSCs have been generated from patients with various intractable diseases, enabling the modeling of disease phenotypes at the cellular level and the study of pathogenic mechanisms. Advances in genome editing technologies have further enhanced the utility of hiPSCs in research. The generation and characterization of patient-derived hiPSCs have become more efficient, allowing for the production of hiPSCs from minimal samples and the use of non-integrating vectors. These hiPSCs can be cryopreserved and deposited in cell banks, making them valuable resources for the global research community. Genome editing technologies, such as CRISPR-Cas9, have enabled the creation of isogenic hiPSCs and fluorescent-reporter hiPSCs, facilitating high-quality research and the visualization of specific cell types. Disease-specific hiPSCs have been instrumental in disease modeling, particularly for neurological and genetic diseases, and have shown promise in drug discovery and development. Despite these advancements, challenges remain, including the need for more homogeneous hiPSC generation and differentiation methods, the development of organoids and MPS, and the expansion of hiPSC applications to a broader range of diseases. Addressing these challenges will further enhance the potential of hiPSCs in advancing personalized medicine and therapeutic development.

Acknowledgements This work was funded by a grant from Japan Agency for Medical Research and Development (AMED) (JP23bm1423010), a grant from Japan Society for the Promotion of Science (JSPS) Grants-in-Aid for Scientific Research (25K22926), and a collaborative research grant from Kaneka corporation. The funders had no role in study design, data collection and interpretation, or the decision to submit the work for publication.

Conflict of Interest The author declares that no competing interests exist.

References

1. Takahashi K et al (2007) Induction of pluripotent stem cells from adult human fibroblasts by defined factors. Cell 131(5):861–872
2. Yu J et al (2007) Induced pluripotent stem cell lines derived from human somatic cells. Science 318(5858):1917–1920
3. Dimos JT et al (2008) Induced pluripotent stem cells generated from patients with ALS can be differentiated into motor neurons. Science 321(5893):1218–1221
4. Park IH et al (2008) Disease-specific induced pluripotent stem cells. Cell 134(5):877–886
5. Lee G et al (2009) Modelling pathogenesis and treatment of familial dysautonomia using patient-specific iPSCs. Nature 461(7262):402–406
6. Scientific guidelines for Lab Resources (2017) https://www.sciencedirect.com/journal/stem-cell-research/about/lab-resources/scientific-guidelines-for-lab-resources
7. Matsumoto Y et al (2013) Induced pluripotent stem cells from patients with human fibrodysplasia ossificans progressiva show increased mineralization and cartilage formation. Orphanet J Rare Dis 8:190
8. Hayashi Y et al (2016) BMP-SMAD-ID promotes reprogramming to pluripotency by inhibiting p16/INK4A-dependent senescence. Proc Natl Acad Sci USA 113(46):13057–13062
9. Bershteyn M et al (2014) Cell-autonomous correction of ring chromosomes in human induced pluripotent stem cells. Nature 507(7490):99–103
10. Arai Y et al (2020) Generation of two human induced pluripotent stem cell lines derived from two juvenile nephronophthisis patients with NPHP1 deletion. Stem Cell Res 45:101815
11. Kuramochi Y et al (2021) Generation of two human induced pluripotent stem cell lines derived from two X-linked adrenoleukodystrophy patients with ABCD1 mutations. Stem Cell Res 53:102337
12. Shimizu T et al (2022) Generation of human induced pluripotent stem cell lines derived from four DiGeorge syndrome patients with 22q11.2 deletion. Stem Cell Res 61:102744
13. Song D et al (2022) Retinoids rescue ceruloplasmin secretion and alleviate oxidative stress in Wilson's disease-specific hepatocytes. Hum Mol Genet 31(21):3652–3671
14. Mori M et al (2024) Generation of human induced pluripotent stem cell lines derived from four Rett syndrome patients with MECP2 mutations. Stem Cell Res 77:103432
15. Li R et al (2024) Generation of human induced pluripotent stem cell lines derived from two glucose transporter 1 deficiency syndrome patients. Stem Cell Res 81:103584
16. Ge N et al (2024) Generation of human induced pluripotent stem cell lines derived from patients of cystic biliary atresia. Hum Cell 38(1):18
17. Chen Y et al (2021) Integrated collection of stem cell bank data, a data portal for standardized stem cell information. Stem Cell Reports 16(4):997–1005
18. Saito MK et al (2023) A disease-specific iPS cell resource for studying rare and intractable diseases. Inflamm Regen 43(1):43
19. Mandegar MA et al (2016) CRISPR interference efficiently induces specific and reversible gene silencing in human iPSCs. Cell Stem Cell 18(4):541–553

20. Ran FA et al (2013) Double nicking by RNA-guided CRISPR Cas9 for enhanced genome editing specificity. Cell 154(6):1380–1389
21. Ran FA et al (2013) Genome engineering using the CRISPR-Cas9 system. Nat Protoc 8(11):2281–2308
22. Miyaoka Y et al (2014) Isolation of single-base genome-edited human iPS cells without antibiotic selection. Nat Methods 11(3):291–U345
23. Miyaoka Y et al (2016) Systematic quantification of HDR and NHEJ reveals effects of locus, nuclease, and cell type on genome-editing. Sci Rep 6:23459
24. Nakade K et al (2023) Efficient selection of knocked-in pluripotent stem cells using a dual cassette cellular elimination system. Cell Rep Methods 3(12):100662
25. Matsuo-Takasaki M et al (2024) Complete suspension culture of human induced pluripotent stem cells supplemented with suppressors of spontaneous differentiation. Elife 12:RP89724
26. Tsukamoto S et al (2021) Generation of two ISL1-tdTomato reporter human induced pluripotent stem cell lines using CRISPR-Cas9 genome editing. Stem Cell Res 53:102363
27. Takagi D et al (2024) Generation of MBP-tdTomato reporter human induced pluripotent stem cell line for live myelin visualization. Stem Cell Res 79:103493
28. Final Report of Japan Agency for Medical Research and Development (AMED) Commissioned Study FY2020 on Utilization of Disease-Specific iPS Cell Bank Project (in Japanese). (2021). https://www.amed.go.jp/content/000079225.pdf
29. Ala A et al (2007) Wilson's disease. Lancet 369(9559):397–408
30. Arioz C, Li Y, Wittung-Stafshede P (2017) The six metal binding domains in human copper transporter, ATP7B: Molecular biophysics and disease-causing mutations. Biometals 30(6):823–840
31. Huster D et al (2012) Diverse functional properties of Wilson disease ATP7B variants. Gastroenterology 142(4):947–956 e5
32. Gomes A, Dedoussis GV (2016) Geographic distribution of ATP7B mutations in Wilson disease. Ann Hum Biol 43(1):1–8
33. Bandmann O, Weiss KH, Kaler SG (2015) Wilson's disease and other neurological copper disorders. Lancet Neurol 14(1):103–113
34. Brewer GJ (2009) Zinc and tetrathiomolybdate for the treatment of Wilson's disease and the potential efficacy of anticopper therapy in a wide variety of diseases. Metallomics 1(3):199–206
35. Huster D et al (2007) High copper selectively alters lipid metabolism and cell cycle machinery in the mouse model of Wilson disease. J Biol Chem 282(11):8343–8355
36. Kusuda Y et al (2000) Novel mutations of the ATP7B gene in Japanese patients with Wilson disease. J Hum Genet 45(2):86–91
37. Roberts EA, Schilsky ML, American Association for Study of Liver Diseases (2008) Diagnosis and treatment of Wilson disease: an update. Hepatology 47(6):2089–2111
38. Ferenci P et al (2003) Diagnosis and phenotypic classification of Wilson disease. Liver Int 23(3):139–142
39. Polishchuk EV et al (2014) Wilson disease protein ATP7B utilizes lysosomal exocytosis to maintain copper homeostasis. Dev Cell 29(6):686–700
40. van den Berghe PV et al (2009) Reduced expression of ATP7B affected by Wilson disease-causing mutations is rescued by pharmacological folding chaperones 4-phenylbutyrate and curcumin. Hepatology 50(6):1783–1795
41. Rauch H (1983) Toxic milk, a new mutation affecting copper-metabolism in the mouse. J Hered 74(3):141–144
42. Yoshida MC et al (1987) New mutation causing hereditary hepatitis in the laboratory rat. J Hered 78(6):361–365
43. Terada K, Sugiyama T (1999) The Long-Evans Cinnamon rat: an animal model for Wilson's disease. Pediatr Int 41(4):414–418
44. Buiakova OI et al (1999) Null mutation of the murine ATP7B (Wilson disease) gene results in intracellular copper accumulation and late-onset hepatic nodular transformation. Hum Mol Genet 8(9):1665–1671

45. Ayton S et al (2013) Ceruloplasmin dysfunction and therapeutic potential for Parkinson disease. Ann Neurol 73(4):554–559
46. Czaja MJ et al (1987) Molecular studies of ceruloplasmin deficiency in Wilsons-disease. J Clin Invest 80(4):1200–1204
47. Hellman NE, Gitlin JD (2002) Ceruloplasmin metabolism and function. Annu Rev Nutr 22:439–458
48. Sato M, Gitlin JD (1991) Mechanisms of copper incorporation during the biosynthesis of human ceruloplasmin. J Biol Chem 266(8):5128–5134
49. Antignac C et al (1993) A gene for familial juvenile nephronophthisis (recessive medullary cystic kidney disease) maps to chromosome 2p. Nat Genet 3(4):342–345
50. Saunier S et al (1997) A novel gene that encodes a protein with a putative SRC homology 3 domain is a candidate gene for familial juvenile nephronophthisis. Hum Mol Genet 6(13):2317–2323
51. Saunier S et al (2000) Characterization of the NPHP1 locus: mutational mechanism involved in deletions in familial juvenile nephronophthisis. Am J Hum Genet 66(3):778–789
52. Caridi G et al (2001) Juvenile nephronophthisis and related variants: clinical features and molecular approach. Contrib Nephrol 136:57–67
53. Jiang ST et al (2008) Targeted disruption of Nphp1 causes male infertility due to defects in the later steps of sperm morphogenesis in mice. Hum Mol Genet 17(21):3368–3379
54. Li D et al (2021) An Nphp1 knockout mouse model targeting exon 2-20 demonstrates characteristic phenotypes of human nephronophthisis. Hum Mol Genet 31(2):232–243
55. Arai Y et al (2024) Patient-derived and gene-edited pluripotent stem cells lacking NPHP1 recapitulate juvenile nephronophthisis in abnormalities of primary cilia and renal cyst formation. Front Cell Dev Biol 12:1370723

Chapter 2
Advances in Stem Cell Research for *MECP2*-Related Disorders

Yuichi Akaba, Satoru Takahashi, and Keita Tsujimura

Abstract Rett syndrome (RTT) and *MECP2* duplication syndrome (MDS), which are categorized as *MECP2*-related disorders, are severe neurodevelopmental conditions caused by mutations or abnormalities in the dosage of the *MECP2* gene. These diseases share the common clinical features of severe intellectual disability and motor developmental deficits and are characterized by diverse cellular and molecular phenotypes, including synaptic dysfunction and abnormal neuronal growth. Advances in stem cell research, particularly the emergence of induced pluripotent stem cell (iPSC) technology, have provided groundbreaking tools for investigating these diseases. Patient-derived disease-specific iPSCs for *MECP2*-related disorders have become indispensable for modeling disease phenotypes and exploring therapeutic strategies. Two-dimensional (2D) neuronal culture systems using iPSCs enable the study of disease-specific neuronal characteristics and synaptic abnormalities, offering critical insights into the cellular and molecular underpinnings of RTT and MDS. Furthermore, the development of a culture system for three-dimensional (3D) iPSC-derived brain organoids has made it possible to mimic the complex architecture and developmental processes of the human brain and better understand the neurodevelopmental aspects of *MECP2*-related diseases. The integration of iPSC-based modeling systems with multi-omics technologies has accelerated the identification of disease mechanisms and therapeutic targets. These innovative methodologies have also enabled the evaluation of novel compounds for the treatment of these diseases and the implementation of personalized treatment strategies to bridge the gap between basic research and clinical applications. This chapter highlights the transformative role of stem cell research in elucidating the

Y. Akaba · S. Takahashi
Department of Pediatrics, Asahikawa Medical University, Asahikawa, Hokkaido, Japan
e-mail: satoru5p@asahikawa-med.ac.jp

K. Tsujimura (✉)
Rett Syndrome Organization Japan, Hirakata, Osaka, Japan

MECP2 Duplication Syndrome Japan, Suita, Osaka, Japan
e-mail: t.keita@npo-rett.jp

R. C. B. Wong and Y. Hayashi (eds.), *Advances in Using Stem Cells to Treat Neural Diseases*, Stem Cell Biology and Regenerative Medicine 77,
https://doi.org/10.1007/978-3-031-94101-6_2

pathophysiology of *MECP2*-related disorders and in developing therapeutic avenues for these diseases. In this section, we outline how the use of stem cell-based model systems has greatly advanced our understanding of the pathophysiologies underlying RTT and MDS, thus paving the way for the development of novel therapeutic approaches.

Keywords MECP2 · Rett syndrome · RTT · MECP2 duplication syndrome · MDS · iPS cell · Neuron · Organoid · microRNA · miRNA · Developmental disorder · Disease modeling · Neurological disease · Neurological disorder

2.1 Introduction

Stem cells were first identified in the 1960s [1, 2]. Since then, our understanding of their properties has significantly advanced. Stem cells can self-renew and differentiate into multiple cell types, including embryonic stem cells (ESCs) and adult tissue stem cells. ESCs, derived from the inner cell mass of blastocysts, are pluripotent and proliferate extensively, making them highly valuable in clinical applications. However, their use raises ethical concerns and carries risks such as transplant rejection by the recipient's immune system and tumorigenesis [3, 4]. By contrast, adult tissue stem cells can be harvested from the same patient for autologous therapy, thereby avoiding ethical issues and eliminating the risk of transplant rejection. However, their limited differentiation capacities restrict their versatility in therapeutic applications. Induced pluripotent stem cells (iPSCs) have recently emerged as an innovative technology. Discovered by Yamanaka and Takahashi in 2006 [5], iPSCs are generated by reprogramming somatic cells into a pluripotent state, enabling differentiation into nearly any cell type. iPSCs bypass the ethical concerns of ESCs and have the potential to significantly reduce the risk of immune rejection owing to their autologous origin [6].

Developing treatments and elucidating the pathophysiology of neurological diseases remain challenging because of the difficulty in accessing the affected regions of the human brain [6]. As a result, direct investigation of disease mechanisms is often not feasible. Thus, previous studies have primarily relied on studying gene expression and regulation in non-neuronal cell lines or animal models to analyze cellular and biochemical mechanisms. In addition, many studies have proposed the use of bone marrow-derived mesenchymal stem cells (MSCs) for neural regeneration as MSCs are relatively accessible because of their presence in the bone marrow [7, 8]. However, in certain cases, animal and cellular models fail to accurately recapitulate the human pathology. It is important to recognize that the structural and functional complexities of the frontal cortex are quite different between humans and nonhuman mammals [9]. Moreover, a consensus now exists that naïve MSCs do not differentiate into neural cells [10]. Advancements in iPSC technology have

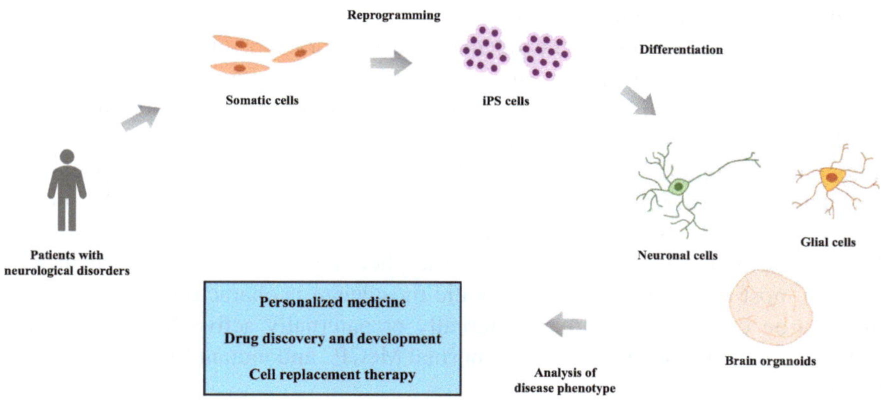

Fig. 2.1 Overview of the potential applications of disease-specific iPSCs in neurological disease research. iPSCs have the unique ability to self-renew and differentiate into various cell types as they can be reprogrammed from somatic cells. These properties make iPSCs a valuable tool for studying neurological diseases with promising applications in disease modeling, drug discovery, and cell replacement therapy

revolutionized academic fields by enabling the generation of disease-specific iPSCs. This approach is particularly valuable for studying neurological diseases because it offers the unique ability to reproduce patient-specific pathological phenomena in vitro in inaccessible nervous system tissues. Thus, iPSC-based modeling systems provide an indispensable platform for studying disease mechanisms and developing novel therapeutic strategies (Fig. 2.1). The following section outlines the achievements of disease-specific iPSC research focusing on methyl-CpG–binding protein 2 (*MECP2*)-related disorders.

2.2 Rett Syndrome and *MECP2* Duplication Syndrome

Gene dosage abnormalities, caused by either an excess or deficiency of gene dosage, are strongly associated with neurological symptoms such as psychiatric disorders, cognitive impairments, behavioral abnormalities, and neurodevelopmental disorders [11, 12]. A prominent example of a gene dosage disorder is a condition resulting from mutations or abnormalities in the X-linked *MECP2*. MeCP2 was originally identified as a DNA methylation-mediated transcriptional repressor [13, 14]. Mutations or duplications in *MECP2* are associated with two major neurodevelopmental disorders: Rett syndrome (RTT; OMIM ID 312750) and *MECP2* duplication syndrome (MDS; OMIM ID 300260) [15].

RTT is a progressive neurodevelopmental disorder that primarily affects females, with an estimated prevalence of 1 in 10,000–1 in 15,000 live female births. Although RTT is primarily diagnosed based on clinical criteria, more than 90% of cases are attributed to mutations in the *MECP2* gene [16]. Patients with RTT typically develop normally during early infancy. However, from 6 to 18 months of age, they experience a regression in developmental milestones, characterized by a regression in purposeful hand use and spoken language, with the development of gait abnormalities and hand stereotypies [17]. Other clinical features include autism, intellectual disability, epilepsy, feeding difficulties, growth retardation, sleep disturbances, and autonomic dysfunction [17]. Several studies have identified genotype-phenotype correlations, providing crucial insights into the mutation-specific molecular mechanisms of *MECP2* [18–20]. Random X-chromosome inactivation (XCI) is also a key factor contributing to the variability in the phenotypic severity observed in RTT [21]. In most cases, the resulting somatic mosaicism is characterized by an equal distribution of cells with either maternally or paternally active X-chromosomes. This leads to a random distribution of normal MeCP2 and mutant MeCP2 within the brain [22].

MDS, another *MECP2*-related neurodevelopmental disorder, is caused by genomic duplications or extra copies of the Xq28 region, which includes *MECP2* gene. Primary clinical features predominantly observed in males, include infantile hypotonia, severe intellectual and developmental disabilities, progressive spasticity, frequent respiratory infections, refractory epilepsy, and autistic traits [23, 24]. The complexity of MDS is shaped by diverse genetic variations, with duplication sizes ranging from 0.355 to 11.2 Mb [25]. The minimally shared region included only *MECP2* and *IRAK1* [25]. This phenotype is largely driven by an increased MeCP2 dosage, as shown in mouse models, where twofold overexpression of MeCP2 reproduced key symptoms of MDS, including stereotypic behaviors, epilepsy, and spasticity [26].

2.3 *MECP2* Gene and Protein Functions

The *MECP2* gene is located on the X-chromosome in the Xq28 region and encodes a protein that plays a critical role in neural development and the regulation of gene expression. Structurally, the *MECP2* gene consists of four exons and produces two major isoforms, MeCP2_e1 and MeCP2_e2, through alternative splicing [27–29] (Fig. 2.2). These isoforms differ in their N-terminal sequences owing to the use of distinct initiation codons and regulatory mechanisms, while sharing most functional domains. Although the two isoforms have overlapping functions, their expression patterns vary depending on tissue type and developmental stage [30]. Notably, MeCP2_e1 isoform-specific knockout mice exhibit motor and metabolic abnormalities resembling RTT, indicating that the e1 isoform plays a critical role in disease pathogenesis [31].

MeCP2 is a multifunctional protein that recognizes DNA methylation and plays a critical role in epigenetic regulation of gene expression (Fig. 2.3). Two major domains, the methyl-CpG binding domain (MBD) and the transcriptional repression domain (TRD), have been considered important for MeCP2 protein function. MeCP2 represses gene expression by recruiting a corepressor complex that includes

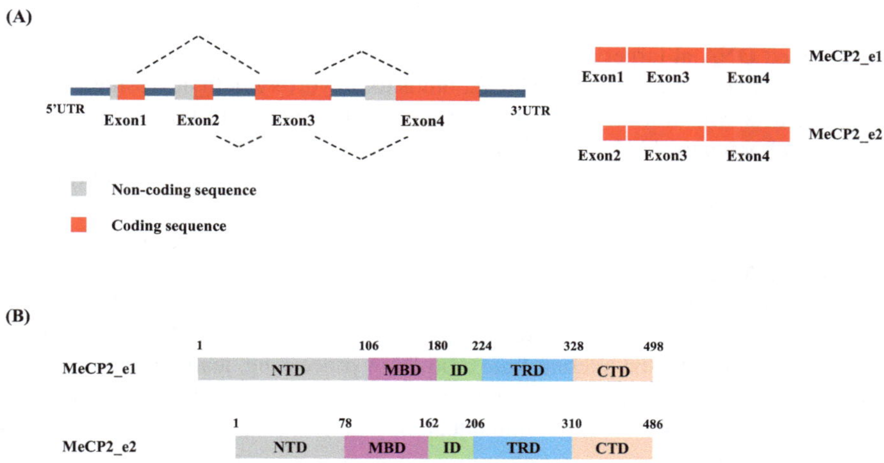

Fig. 2.2 Isoforms and structural domains of the *MECP2* gene and protein. (**a**) Schematic representation of the *MECP2* gene exons and the alternative splicing pattern leading to the generation of two protein isoforms, MeCP2_e1 and MeCP2_e2. (**b**) Major structural domains of MeCP2_e1 and MeCP2_e2. *NTD* N-terminal domain, *MBD* methylated DNA-binding domain, *ID* intervening domain, *TRD* transcription repression domain, *CTD* C-terminal domain

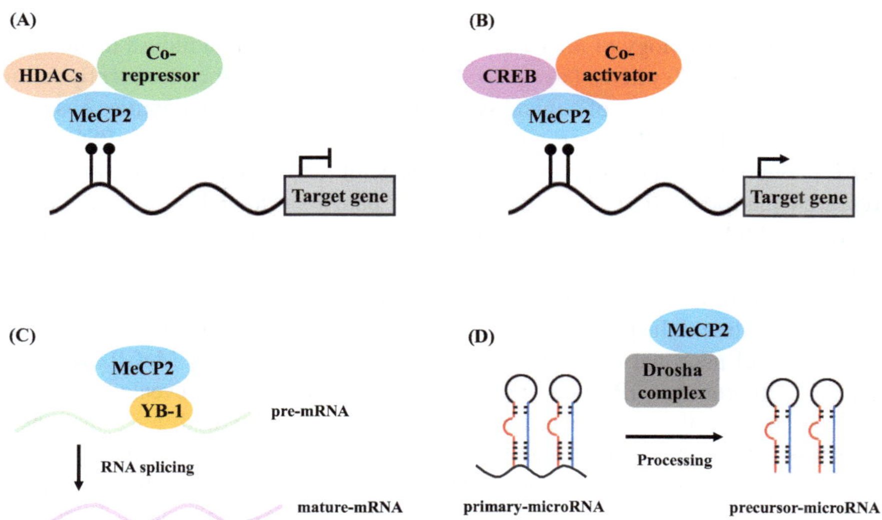

Fig. 2.3 Epigenetic mechanisms of MeCP2 in gene expression regulation. (**a**) MeCP2 binds to methylated CpG sites and recruits repressor complexes, such as HDACs, to suppress transcription. (**b**) MeCP2 interacts with CREB to activate transcription and recruit co-activators. (**c**) MeCP2 regulates RNA splicing through interaction with YB-1, facilitating pre-mRNA maturation. (**d**) MeCP2 modulates primary-microRNA processing to precursor-microRNA through interaction with the Drosha complex, influencing microRNA expression and gene regulation. *HDACs* histone deacetylases, *CREB* cAMP response element-binding protein, *YB-1* Y-box binding protein 1

histone deacetylases (HDACs), Sin3A, NCoR/SMRT [14, 32]. MeCP2 binds to methylated DNA at target genes through its MBD and recruits corepressors via its TRD. Under certain conditions, MeCP2 binds to 5-hydroxymethylcytosine (5 hmC) and interacts with transcription factors to relax chromatin and function as a transcriptional activator [33]. MeCP2 interacts with cAMP response element-binding protein (CREB) to facilitate the transcription of specific genes [34]. Additionally, MeCP2 functions as a chromatin linker by interacting with nucleosomes to stabilize higher-order chromatin structures, thereby supporting the widespread regulation of genes [35]. MeCP2 also regulates post-transcriptional processes, including alternative splicing, by interacting with Y-box binding protein 1 (YB-1), thereby promoting diverse cell-specific gene expression [36, 37]. Furthermore, MeCP2 collaborates with the enzyme Drosha to contribute to miRNA processing, leading to modulating intracellular signaling and gene regulatory networks [38, 39]. Through these functions, MeCP2 serves as a central hub integrating diverse pathways critical for maintaining brain function.

2.4 Human iPSC (hiPSC)-Based Modeling for RTT and MDS

The establishment of disease-specific iPSCs has marked a breakthrough in elucidating the pathophysiology of *MECP2*-related disorders. Although mouse models and other experimental systems have provided valuable insights, they frequently fail to recapitulate the spectrum of clinical symptoms and structural complexities observed in the human brain. However, since the first successful derivation of iPSCs from RTT patients in 2009 [40], these disease-specific cells have emerged as pivotal research platforms. iPSC lines derived from patients with RTT have effectively reproduced RTT-specific neuronal abnormalities, such as impaired synapse formation, shortened dendrites, reduced soma size, and disrupted calcium signaling during neural differentiation [41]. Furthermore, studies using RTT iPSC-based models have demonstrated the potential reversibility of neuronal abnormalities, with partial restoration of neuronal morphology and function observed after MeCP2 reactivation [42]. In contrast, iPSCs derived from MDS patients have revealed excessive neuronal excitability and aberrant excitatory synaptic activity in differentiated neurons [43]. These abnormalities are considered contributing factors to the pathology, and therapeutic strategies targeting *MECP2* overexpression are being actively explored. Additionally, compound screening using disease-specific iPSCs identified several compounds that can ameliorate synaptic dysfunction and restore neuronal function [44]. Recent advancements in iPSC-based models have demonstrated their utility in the development of therapies for *MECP2*-related disorders. Bajikar et al. highlighted the potential of iPSC-derived neurons for modeling antisense oligonucleotide therapy, emphasizing their role in drug discovery for MDS treatment [45]. Thus, disease-specific iPSCs provide a robust platform for modeling the pathophysiology of RTT and MDS, thereby linking basic research with therapeutic

Table 2.1 Rett syndrome patient-specific iPSC lines

hiPS cell line	Sex	Age	Ethics	Source cell type	Variant in the *MECP2* gene	Data source
BRC-i024A (HiPS-RTT_CUH01)	Female	9 years	Japanese	PBMC	c.806delG, p.(Gly269Alafs*20)	BRC
BRC-i025A (HiPS-RTT_CUH02)	Female	15 years	Japanese	PBMC	c.763C>T, p.(Arg255*)	BRC
BRC-i026A (HiPS-RTT_CUH05)	Female	7 years	Japanese	PBMC	c.916C>T, p.(Arg306Cys)	BRC
BRC-i027A (HiPS-RTT_CUH06)	Female	23 years	Japanese	PBMC	c.316C>T, p.(Arg106Trp)	BRC
HPS1156, HPS1157, HPS1158, HPS1159, HPS1160, HPS1161	Female	20s	Unknown	PBMC	Unknown	BRC
HPS3036, HPS3037, HPS3038, HPS3039, HPS3040, HPS3041	Female	20s	Japanese	PBMC	Unknown	BRC
HPS3042, HPS3043, HPS3044, HPS3045, HPS3046, HPS3047	Female	20s	Japanese	PBMC	Ser134Cys	BRC
HPS3048, HPS3049, HPS3050, HPS3051, HPS3052, HPS3053	Female	Under 10 years	Japanese	PBMC	Arg306His	BRC
HPS3084, HPS3085, HPS3086, HPS3087, HPS3088, HPS3089	Female	30s	Japanese	PBMC	Cys880Thr	BRC
HPS3090, HPS3091, HPS3092, HPS3093, HPS3094, HPS3095	Female	30s	Japanese	PBMC	Unknown	BRC

(continued)

Table 2.1 (continued)

hiPS cell line	Sex	Age	Ethics	Source cell type	Variant in the *MECP2* gene	Data source
GM27437	Female	3 years	Jewish	Fibroblast	1155del32	CIMR
FJMUUHi002-A	Female	12 years	Chinese	PBMC	c.397C>T, p.(Arg133Cys)	hPSCreg
WIC03i-02-11E	Female	11 years	Caucasian	Fibroblast	Arg294*	WiCell
WIC05i-127-325	Female	8 years	Caucasian	Fibroblast	Arg306Cys	WiCell
WIC07i-07982-4	Female	25 years	Caucasian	Fibroblast	Val247*	WiCell
QQ0167	Female	26 years	Unknown	PBMC	Unknown	HipSci
QQ0112	Female	7 years	Unknown	PBMC	Unknown	HipSci
QQ0113	Female	38 years	Unknown	PBMC	Unknown	HipSci
R306C-iPS	Female	8 years	Caucasian	Fibroblast	c.916C>T, p.(Arg306Cys)	SKIP
RTTe1-hiPSC #27, #39, #48, #96	Female	Unknown	Caucasian	Fibroblast	c.47_57del; p.(Gly16Glufs*22)	SKIP
RS1-52M, RS1-61P	Female	Unknown	Japanese	Fibroblast	c.806delG, p.(Gly269Alafs*288)	SKIP
RS2-65M, RS2-62P	Female	Unknown	Japanese	Fibroblast	c.806delG, p.(Gly269Alafs*288)	SKIP

PBMC peripheral blood mononuclear cells, *BRC* BioResource Center, *CIMR* Cambridge Institute for Medical Research, *hPSCreg* Human Pluripotent Stem Cell Registry, *WiCell* WiCell Research Institute, *HipSci* Human Induced Pluripotent Stem Cell Initiative, *SKIP* Stemcell Knowledge and Information Portal

development. The iPSC lines established from RTT and MDS patients are presented in Tables 2.1 and 2.2 [43, 45, 46], respectively.

2.4.1 2D Neuronal Differentiation Models

Neuronal differentiation protocols involve a stepwise process for generating mature neurons [47]. iPSCs are first induced to form neuroectoderm, a process facilitated by dual SMAD inhibition, which is downstream of the bone morphogenetic protein (BMP) and Activin/TGFβ signaling pathways [48, 49]. This induction drives the transition to neural stem cells (NSCs), and guides them toward specific neural lineages. The final stage of terminal differentiation into mature neurons is achieved through the application of molecular cues, including small molecules and growth/differentiation factors, in the cell culture environment.

Marchetto et al. first developed a neuronal model of RTT using hiPSCs with four distinct *MECP2* mutations (Thr158Met, Arg306Cys, Gln244*, and 1155del32) [41]. They initially generated embryoid bodies (EBs), manually collected and isolated neural rosettes, and re-plated them to derive neural progenitor cells (NPCs). These progenitors are subsequently induced to differentiate into mature neurons.

Table 2.2 *MECP2* duplication syndrome patient-specific iPSC lines

hiPS cell line	Sex	Age	Ethics	Source cell type	Duplication region in the Xq28 locus	Reference
Patient 1	Male	Unknown	Unknown	Fibroblast	300 kb (152.73—153.02 Mb) (NCBI36/hg18)	Nageshappa S, et al. [43]
Patient 2	Male	Unknown	Unknown	Fibroblast	15.25 Mb (139.33—154.58 Mb) (NCBI36/hg18)	Nageshappa S, et al. [43]
Patient 3	Male	Unknown	Unknown	Fibroblast	500 kb (152.66—153.15 Mb) (NCBI36/hg18)	Nageshappa S, et al. [43]
BCMi003-A	Male	13 years	Non-Hispanic	Fibroblast	1.2 Mb (152.62—153.83 Mb) (GRCh37/hg19)	Mendonca D, et al. [46]
BCMi004-A	Male	17 years	Non-Hispanic	Fibroblast	680 kb (152.85—153.53 Mb) (GRCh37/hg19)	Mendonca D, et al. [46]
BCMi005-A	Male	21 years	Non-Hispanic	Fibroblast	11.2 Mb (144.05—155.25 Mb) (GRCh37/hg19)	Mendonca D, et al. [46]
BCMi006-A	Male	29 years	Non-Hispanic	Fibroblast	408 kb (153.15—153.56 Mb) (GRCh37/hg19)	Mendonca D, et al. [46]
BCMi007-A	Male	4 years	Non-Hispanic	Fibroblast	355 kb (153.06—153.42 Mb) (GRCh37/hg19)	Mendonca D, et al. [46]
Proband 1	Male	24 years	Unknown	Fibroblast	439 kb (153.18—153.62 Mb) (GRCh37/hg19)	Bajikar SS, et al. [45]
Proband 2	Male	22 years	Unknown	Fibroblast	439 kb (153.18—153.62 Mb) (GRCh37/hg19)	Bajikar SS, et al. [45]
Proband 3	Male	7 years	Unknown	Fibroblast	2.4 Mb (152.83—155.25 Mb) (GRCh37/hg19)	Bajikar SS, et al. [45]
Proband 4	Male	6 years	Unknown	Fibroblast	465 kb (153.09—153.55 Mb) (GRCh37/hg19)	Bajikar SS, et al. [45]

Their findings revealed that RTT-derived neurons exhibit impaired maturation, characterized by a reduction in synapse number, decreased soma size, altered calcium signaling, functional defects in firing activity, and an imbalance in excitatory and inhibitory signaling (Fig. 2.4). Subsequent studies delved deeper into specific aspects of neuronal dysfunction in RTT. For instance, Cheung et al. utilized XCI to isolate *MECP2*-null hiPSCs alongside isogenic controls, enabling the direct comparison of mutant and wild-type *MECP2* in terms of neuronal phenotypes [50]. Their research not only corroborated the phenotypes reported by Marchetto et al.

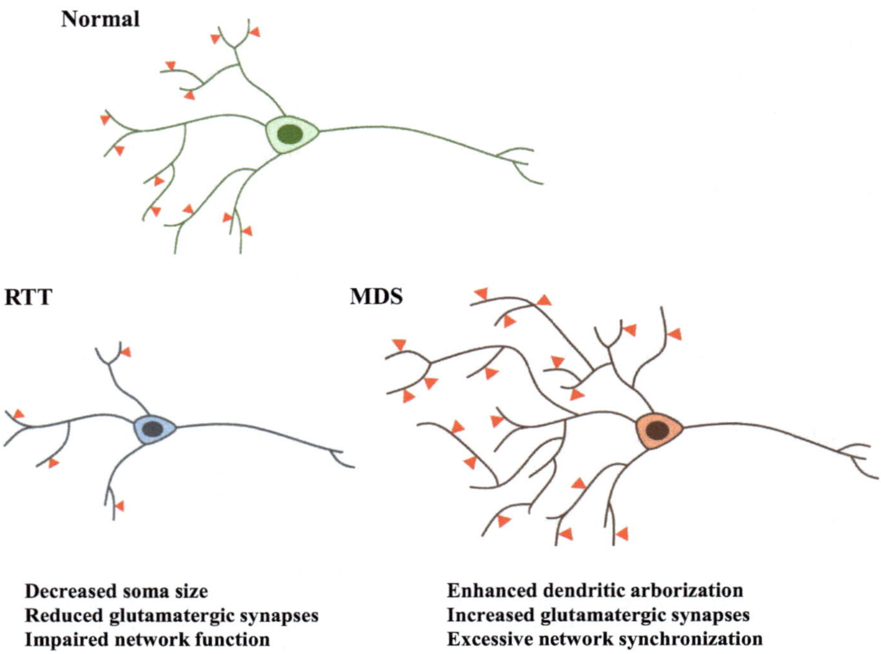

Fig. 2.4 Phenotypic characteristics of neurons differentiated from iPSCs generated from patients with Rett syndrome (RTT) and *MECP2* duplication syndrome (MDS). Neurons derived from RTT-iPSCs exhibit reduced soma size, fewer synapses, and impaired synaptic function and activity. In contrast, neurons derived from MDS-iPSCs showed increased dendritic branching, more synapses, and enhanced synchronized burst activity

but also provided a refined model for studying RTT-specific neuronal abnormalities at both the morphological and functional levels. Synaptic functions have been extensively evaluated in neurons derived from RTT-iPSCs carrying various *MECP2* mutations. Studies have consistently shown a marked reduction in the number of excitatory synapses in RTT neurons compared to that in control iPSC-derived neurons, as evidenced by decreased VGLUT1 puncta, highlighting impairments in glutamate transport within RTT-differentiated neurons [41]. Synaptic and electrophysiological abnormalities have been observed in neurons with diverse *MECP2* mutations. Regardless of the experimental approach employed, RTT-iPSC-derived neurons consistently exhibited reduced postsynaptic activity frequency compared to controls, underscoring the pivotal role of *MECP2* mutations in driving synaptic transmission deficits [51, 52]. Additionally, three RTT patient-derived iPSC lines (Val247*, Arg294*, and Arg306Cys) differentiated into glial cells, including astrocytes and microglia [53]. To investigate whether human MeCP2-deficient astrocytes exert non-cell-autonomous effects on neurons, RTT patient-derived astrocytes were co-cultured with mouse primary hippocampal neurons.

Mutant astrocytes exhibit non-cell-autonomous detrimental effects on the morphology and function of mouse neurons. In another study using iPSCs derived from a patient with an *MECP2* frameshift mutation (Gly269Alafs∗), differentiated mutant astrocytes demonstrated abnormal growth patterns and gene expression compared to isogenic controls. Importantly, the glial marker glial fibrillary acidic protein (GFAP) is dysregulated in MeCP2-deficient neural cells, suggesting that abnormal astrocytic differentiation contributes to disease pathogenesis [54]. Transcriptomic and proteomic analyzes aimed at identifying molecular pathways involved in RTT have provided valuable insights into neurodevelopment. In a study utilizing neurons differentiated from iPSCs derived from patients with RTT, RNA-seq analysis revealed specific gene expression changes, including the induction of a senescence-associated secretory phenotype (SASP). Furthermore, the activation of the p53 pathway has been shown to negatively affect dendritic branching in RTT neurons, whereas treatment with a p53 inhibitor partially restores dendritic complexity [55]. Another study investigating exosomes from the culture supernatant of MeCP2-deficient hiPSC-derived neurons used proteomic analysis to reveal that neurodevelopmental signaling proteins associated with neuronal maturation, axon guidance, and synapse formation were downregulated compared with exosomes from isogenic controls. Treatment of MeCP2-deficient neuronal cultures with control exosomes improved neurodevelopmental deficits and enhanced neurogenesis, synapse formation, and network activity [56].

Studies on iPSC-derived neurons from affected individuals with MDS have revealed distinct molecular and functional abnormalities, advancing our understanding of the pathophysiology of this condition. Previous studies in mice have demonstrated that MeCP2 dosage significantly affects excitatory synapse formation, with increased MeCP2 protein levels promoting higher spine density during the early stages of neurogenesis [57]. The gene expression profiles of iPSC-derived neurons from MDS patients show a tendency for elevated expression levels of the synaptic genes *SLC17A7* (encoding VGLUT1) and *DLG4* (encoding PSD95). Quantification of the colocalized presynaptic and postsynaptic marker puncta further confirmed an increase in the number of glutamatergic synapses. Moreover, human *MECP2* duplication neurons exhibited a significant increase in dendritic branching and a notable enhancement in synchronized burst events within neuronal networks [43] (Fig. 2.5). In another study, Bajikar et al. compared dysregulated genes in MDS iPSC-derived neurons using RNA sequencing datasets generated from RTT iPSC-derived neurons. MDS neurons shared 102 dysregulated genes with RTT neurons, and the direction of dysregulation was significantly inversely correlated. Among the genes downregulated in MDS, four—*FOXH1*, *ZNF691*, *MAFK*, and *SOX1*—were identified as significantly exceeded the false discovery rate threshold [45]. *SOX1* and *MAFK* have been directly implicated in neurogenesis [58, 59], whereas *FOXH1* plays a critical role in embryonic development and may exert indirect effects [60, 61]. In contrast, the role of *ZNF691* in neurogenesis has yet to be elucidated. Further studies are necessary to determine whether these genes act downstream of MeCP2 and contribute to the pathophysiological mechanisms underlying the development of MDS.

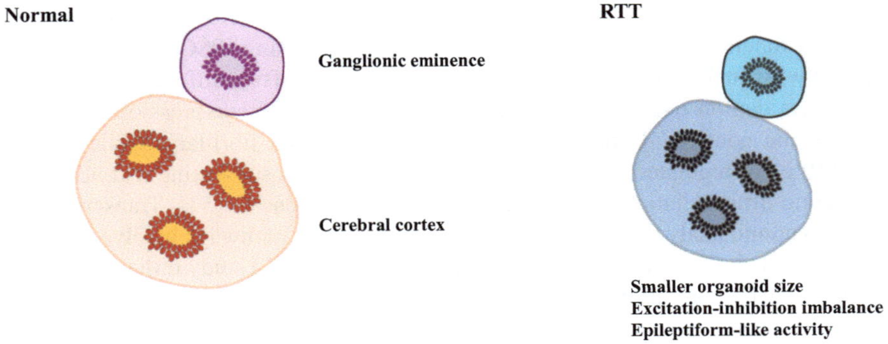

Fig. 2.5 Pathological phenotypes in brain organoids derived from patients with Rett syndrome (RTT). Brain organoids derived from patients with RTT exhibit smaller than those derived from healthy controls. In addition, assembloids formed by the fusion of the cortical and ganglionic eminence regions recapitulate pathological phenotypes, including excitation–inhibition imbalance and epileptiform-like activity

2.4.2 3D Brain Organoids

Brain organoids derived from hiPSCs are self-organizing tissues that have garnered attention as advanced models for brain research. These organoids comprise major cell types observed in the fetal human brain, including NPCs, neurons, and glial cells [62]. Furthermore, they replicate certain structural features of the developing brain, the formation of regions where early neurogenesis occurs, such as the ventricular and subventricular zones [63]. The use of morphogens and inhibitors enables the directed patterning of specific brain regions, including the cerebral cortex, forebrain [64], and midbrain [65]. The 3D architecture of brain organoids allows for a more comprehensive investigation of spatial and functional interactions among cells, providing a unique platform for studying the mechanisms underlying human brain development and neurological disorders and the evaluation of therapeutic interventions.

Mellios et al. generated brain organoid models using two RTT patient-derived hiPSC lines (316C > T and 705delG) and hiPSC-derived NPCs with a MeCP2 shRNA knockdown [66]. The RTT organoids exhibited enlarged ventricular areas and reduced ventricular wall thickness compared to control organoids. Furthermore, RTT organoids showed a decrease in the expression of the neuronal markers MAP2 and DCX at both the protein and mRNA levels. In contrast, the expression of the NPC marker PAX6 was enriched in RTT-derived neuronal cells. These findings suggested that RTT organoids exhibited impaired neurogenesis and neuronal differentiation. A recent study demonstrated that enhanced BMP-SMAD signaling contributes to the developmental abnormalities observed in brain organoids derived from patients with RTT [67]. Studies have found elevated levels of SMAD1 and phosphorylated SMAD1/5 (p-SMAD1/5) in RTT organoids, along with a significant reduction in circumference during the later stages of development. Treatment with BMP signaling inhibitors such as dorsomorphin and noggin partially restored these developmental abnormalities. Moreover, RTT organoids displayed increased astrocytic differentiation, which was reduced by the inhibition of BMP signaling. In advanced research, fusion organoids (also referred to as assembloids) that combine the cerebral cortex and ganglionic eminence have been developed to investigate network activity in RTT brain models [68]. These organoids functionally integrate the excitatory and inhibitory neurons. Consistent with clinical observations, *MECP2*-mutant cerebral cortex-ganglionic eminence fusion organoids/assembloids displayed high-amplitude, spontaneously synchronized calcium transients among neurons, a hallmark of epileptiform discharges. These results suggest that mutations in the *MECP2* gene lead to dysfunction of GABAergic neurons in cortical organoids, disrupting the balance between excitation and inhibition in neural circuits.

2.5 Clinical Prospects of hiPSC Technology

The use of iPSCs has enabled the generation of patient-specific cellular models and has played a crucial role in bridging the gap between basic research and clinical applications. This is especially true in the study of neurodevelopmental disorders, such as RTT and MDS, where its usefulness has been remarkably demonstrated. Furthermore, studies using hiPSC-based models have elucidated the mechanisms of action of several drugs, and these findings are expected to influence the design of clinical trials and the development of therapeutic strategies.

One of the therapeutic targets in the development of RTT treatments is brain-derived neurotrophic factor (BDNF), which plays an essential role in neuronal and synaptic maturation. MeCP2 binds to the promoter region of the *BDNF* gene and regulates its transcription in an activity-dependent manner. This mechanism is hypothesized to be dysfunctional in RTT patients [69]. Research on heterozygous female RTT mouse models and patients with RTT has demonstrated that the functional loss of MeCP2 leads to the downregulation of BDNF transcripts [70]. However, the clinical application of BDNF as a therapeutic agent is challenging because of its poor blood–brain barrier (BBB) permeability. Insulin-like growth factor 1 (IGF1) has garnered attention as an alternative. IGF1 exerts similar effects on neuronal survival and maturation by overlapping signaling pathways with BDNF. Importantly, the IGF1 metabolite glycine-proline-glutamate (GPE) can cross the BBB, making it a promising candidate [71–73]. Evidence from in vitro studies has shown that treating hiPSC- and hESC-derived RTT neurons harboring various *MECP2* mutations with IGF1 can rescue pathological phenotypes, including an increase in the number of glutamatergic synapses, restoration of soma size and dendritic complexity, and activation of the AKT/mTOR signaling pathway [41, 74]. Based on these findings, trofinetide, a proline-methylated analog of GPE, has successfully completed Phase III clinical trials and met the primary endpoints. Consequently, trofinetide was approved by the U.S. Food and Drug Administration (FDA) in 2023 as the first drug for the treatment of RTT. In addition to BDNF and

IGF1, other pharmacological treatment strategies for RTT are being actively explored. Abnormalities in the cholinergic system have been implicated in the motor and cognitive impairments observed in RTT mouse models [75]. Using iPSC-derived RTT neurons, one study investigated whether choline supplementation alleviated RTT-related deficits. The findings revealed that choline supplementation reduced synaptic dysfunction in RTT-iPSC-derived neurons in vitro [51]. The effects of HDAC6 inhibitors were evaluated using RTT-iPSCs. HDAC6 is a cytoplasmic enzyme that promotes the deacetylation of α-tubulin, thereby regulating microtubule dynamics [76]. A study demonstrated that HDAC6 inhibitors, including ACY-1215, significantly increased acetylated α-tubulin levels in neurons derived from RTT-iPSCs carrying *MECP2* mutations (Thr158Met and Arg306Cys) [77]. HDAC6 inhibitors are thought to contribute to the stabilization of the cytoskeleton and the improvement of axonal transport, potentially aiding the restoration of neuronal function [78]. These findings suggest that HDAC6 inhibitors can be explored as potential therapeutic agents for RTT. Approaches to directly restore MeCP2 function have focused on increasing the levels of functional MeCP2 protein. Recently, gene therapies for RTT have been developed using adeno-associated virus (AAV) vectors to mediate *MECP2* gene delivery or editing [79]. AAV, particularly AAV9, are suitable vectors for the central nervous system (CNS) because of their ability to cross the BBB [80]. Combining AAV9 with CRISPR/Cas9, a method targeting forkhead box G1 (*FOXG1*) mutations known to cause Rett-like syndrome, has been shown to effectively correct these mutations in hiPSC-derived neurons, achieving correction rates of 20–35% with high precision and minimal off-target effects [81]. *MECP2* overexpression leads to neurotoxicity and neuronal cell death [26]. To address this issue, a vector containing an *MECP2* expression cassette was developed to safely regulate *MECP2* expression. This cassette includes a modified endogenous *MECP2* promoter to restrict transcription and a 3' UTR with microRNA binding sites known to regulate *MECP2* expression [79, 82–84]. These design features ensure the safety of transgene expression in brain cells. Phase I/II open-label trials have been initiated to evaluate the safety and tolerability of this vector system.

Current strategies for developing MDS therapeutics, in contrast to those for RTT, focus on reducing MeCP2 levels. One potential approach utilizes antisense oligonucleotides (ASOs), which are chemically modified synthetic single-stranded DNA molecules engineered to regulate mRNA expression and consequently adjust protein levels [85]. In a study utilizing a humanized mouse model carrying two copies of the human *MECP2* gene and lacking the endogenous murine *Mecp2* gene, intracerebroventricular administration of *MECP2*-specific ASOs led to a reduction in MeCP2 expression throughout the brain. ASO treatment improved several behavioral abnormalities and dose-dependent restoration of specific gene expression regulated by MeCP2 [86]. A recent study investigated the efficacy of ASO therapy in neurons derived from iPSCs from patients with MDS. This study demonstrated that ASO treatment normalized MeCP2 expression levels and partially corrected abnormal gene expression patterns [45]. Additionally, ASO therapy improved certain abnormal morphological features observed in the MDS neurons. As previously noted, MeCP2 dysregulation affects neuronal function, with both overexpression and underexpression leading to detrimental effects. A critical challenge for ASO therapy is determining how precisely MeCP2 expression can be modulated within the physiological range, and how this modulation may vary among individual patients. Addressing these inter-patient variations is essential for advancing personalized medicine for the treatment of MeCP2-related disorders.

2.6 Limitations and Future Directions

Conventional in vitro cellular and mouse models are limited in their ability to fully replicate human diseases such as patient-specific genetic variations, human-specific epigenetic regulation, and complex neuronal connectivity. Neurons and brain organoids derived from disease-specific iPSCs for *MECP2*-related disorders have emerged as effective models for capturing patient-specific genetic abnormalities, recapitulating epigenetic dysregulation, and enabling the study of individualized disease mechanisms in a human cellular context. While 2D differentiation models derived from iPSCs allow for the generation of homogenous neural populations resembling specific brain regions, they fail to recreate complex neural network formation and cell-cell interactions. In contrast, 3D brain organoids can mimic the neurodevelopment and maturation of the human fetal brain, potentially addressing these limitations. Advances in organoid technologies, including the development of organoids resembling specific brain regions (e.g., the cerebral cortex, midbrain, cerebellum, and spinal cord) [87–89] and assembloids for modeling dynamic processes such as circuit formation and cell migration [90, 91], have paved the way for deeper insights into the mechanisms of *MECP2*-related disorders and therapeutic strategies. However, significant challenges remain for the development of 3D organoids. The lack of vascular systems limits oxygen and nutrient diffusion, leading to central necrosis, whereas batch-to-batch variability hinders reproducibility [92]. To address these issues, bioreactor systems have been introduced to improve organoid culture conditions [93], and coculture techniques that promote vascularization are under development [94]. Furthermore, strategies, such as the generation of mosaic organoids [44], have enabled the creation of more precise and clinically relevant models. As a disease-modeling platform, brain organoids are particularly valuable for studying neurodevelopmental disorders. When combined with multi-omics approaches, such as transcriptomics and proteomics, these models provide valuable insight into diseases-related molecular alterations and support drug discovery and personalized treatment development. Future advancements in stem cell research are expected to enable the generation of models that closely resemble the human brain, thereby playing crucial roles in elucidating disease mechanisms and developing novel therapies.

References

1. Becker AJ, McCulloch EA, Till JE (1963) Cytological demonstration of the clonal nature of spleen colonies derived from transplanted mouse marrow cells. Nature 197:452–454. https://doi.org/10.1038/197452a0
2. Siminovitch L, McCulloch EA, Till JE (1963) The distribution of colony-forming cells among spleen colonies. J Cell Comp Physiol 62:327–336. https://doi.org/10.1002/jcp.1030620313
3. Juengst E, Fossel M (2000) The ethics of embryonic stem cells—now and forever, cells without end. JAMA 284:3180–3184. https://doi.org/10.1001/jama.284.24.3180
4. Li J-Y, Christophersen NS, Hall V, Soulet D, Brundin P (2008) Critical issues of clinical human embryonic stem cell therapy for brain repair. Trends Neurosci 31:146–153. https://doi.org/10.1016/j.tins.2007.12.001
5. Takahashi K, Yamanaka S (2006) Induction of pluripotent stem cells from mouse embryonic and adult fibroblast cultures by defined factors. Cell 126:663–676. https://doi.org/10.1016/j.cell.2006.07.024
6. Okano H, Yamanaka S (2014) iPS cell technologies: significance and applications to CNS regeneration and disease. Mol Brain 7:22. https://doi.org/10.1186/1756-6606-7-22
7. Tropel P, Platet N, Platel J-C, Noël D, Albrieux M, Benabid A-L, Berger F (2006) Functional neuronal differentiation of bone marrow-derived mesenchymal stem cells. Stem Cells 24:2868–2876. https://doi.org/10.1634/stemcells.2005-0636
8. Montzka K, Lassonczyk N, Tschöke B, Neuss S, Führmann T, Franzen R, Smeets R, Brook GA, Wöltje M (2009) Neural differentiation potential of human bone marrow-derived mesenchymal stromal cells: misleading marker gene expression. BMC Neurosci 10:16. https://doi.org/10.1186/1471-2202-10-16
9. Preuss TM, Wise SP (2022) Evolution of prefrontal cortex. Neuropsychopharmacology 47:3–19. https://doi.org/10.1038/s41386-021-01076-5
10. Neirinckx V, Coste C, Rogister B, Wislet-Gendebien S (2013) Concise review: adult mesenchymal stem cells, adult neural crest stem cells, and therapy of neurological pathologies: a state of play. Stem Cells Transl Med 2:284–296. https://doi.org/10.5966/sctm.2012-0147
11. Genovese A, Butler MG (2023) The autism spectrum: behavioral, psychiatric and genetic associations. Genes (Basel) 14:677. https://doi.org/10.3390/genes14030677
12. Levitis E, Liu S, Whitman ET, Warling A, Torres E, Clasen LS, Lalonde FM, Sarlls J, Alexander DC, Raznahan A (2024) The variegation of human brain vulnerability to rare genetic disorders and convergence with behaviorally defined disorders. Biol Psychiatry 95:136–146. https://doi.org/10.1016/j.biopsych.2023.07.008
13. Nan X, Ng HH, Johnson CA, Laherty CD, Turner BM, Eisenman RN, Bird A (1998) Transcriptional repression by the methyl-CpG-binding protein MeCP2 involves a histone deacetylase complex. Nature 393:386–389. https://doi.org/10.1038/30764
14. Jones PL, Veenstra GJ, Wade PA, Vermaak D, Kass SU, Landsberger N, Strouboulis J, Wolffe AP (1998) Methylated DNA and MeCP2 recruit histone deacetylase to repress transcription. Nat Genet 19:187–191. https://doi.org/10.1038/561
15. Collins BE, Neul JL (2022) Rett syndrome and MECP2 duplication syndrome: disorders of MeCP2 dosage. Neuropsychiatr Dis Treat 18:2813–2835. https://doi.org/10.2147/NDT.S371483
16. Liyanage VRB, Rastegar M (2014) Rett syndrome and MeCP2. Neuromolecular Med 16:231–264. https://doi.org/10.1007/s12017-014-8295-9
17. Neul JL, Kaufmann WE, Glaze DG, Christodoulou J, Clarke AJ, Bahi-Buisson N, Leonard H, Bailey MES, Schanen NC, Zappella M et al (2010) Rett syndrome: revised diagnostic criteria and nomenclature. Ann Neurol 68:944–950. https://doi.org/10.1002/ana.22124
18. Neul JL, Fang P, Barrish J, Lane J, Caeg EB, Smith EO, Zoghbi H, Percy A, Glaze DG (2008) Specific mutations in methyl-CpG-binding protein 2 confer different severity in Rett syndrome. Neurology 70:1313–1321. https://doi.org/10.1212/01.wnl.0000291011.54508.aa

19. Caffarelli C, Gonnelli S, Pitinca MDT, Camarri S, Al Refaie A, Hayek J, Nuti R (2020) Methyl-CpG-binding protein 2 (MECP2) mutation type is associated with bone disease severity in Rett syndrome. BMC Med Genet 21:21. https://doi.org/10.1186/s12881-020-0960-2
20. Cuddapah VA, Pillai RB, Shekar KV, Lane JB, Motil KJ, Skinner SA, Tarquinio DC, Glaze DG, McGwin G, Kaufmann WE et al (2014) Methyl-CpG-binding protein 2 (MECP2) mutation type is associated with disease severity in Rett syndrome. J Med Genet 51:152–158. https://doi.org/10.1136/jmedgenet-2013-102113
21. Ishii T, Makita Y, Ogawa A, Amamiya S, Yamamoto M, Miyamoto A, Oki J (2001) The role of different X-inactivation pattern on the variable clinical phenotype with Rett syndrome. Brain Dev 23(Suppl 1):S161–S164. https://doi.org/10.1016/s0387-7604(01)00344-8
22. Hoffbuhr KC, Moses LM, Jerdonek MA, Naidu S, Hoffman EP (2002) Associations between MeCP2 mutations, X-chromosome inactivation, and phenotype. Ment Retard Dev Disabil Res Rev 8:99–105. https://doi.org/10.1002/mrdd.10026
23. Takeguchi R, Takahashi S, Akaba Y, Tanaka R, Nabatame S, Kurosawa K, Matsuishi T, Itoh M (2021) Early diagnosis of MECP2 duplication syndrome: insights from a nationwide survey in Japan. J Neurol Sci 422:117321. https://doi.org/10.1016/j.jns.2021.117321
24. Ta D, Downs J, Baynam G, Wilson A, Richmond P, Leonard H (2022) A brief history of MECP2 duplication syndrome: 20-years of clinical understanding. Orphanet J Rare Dis 17:131. https://doi.org/10.1186/s13023-022-02278-w
25. Carvalho CMB, Zhang F, Liu P, Patel A, Sahoo T, Bacino CA, Shaw C, Peacock S, Pursley A, Tavyev YJ et al (2009) Complex rearrangements in patients with duplications of MECP2 can occur by fork stalling and template switching. Hum Mol Genet 18:2188–2203. https://doi.org/10.1093/hmg/ddp151
26. Collins AL, Levenson JM, Vilaythong AP, Richman R, Armstrong DL, Noebels JL, David Sweatt J, Zoghbi HY (2004) Mild overexpression of MeCP2 causes a progressive neurological disorder in mice. Hum Mol Genet 13:2679–2689. https://doi.org/10.1093/hmg/ddh282
27. Kriaucionis S, Bird A (2004) The major form of MeCP2 has a novel N-terminus generated by alternative splicing. Nucleic Acids Res 32:1818–1823. https://doi.org/10.1093/nar/gkh349
28. Mnatzakanian GN, Lohi H, Munteanu I, Alfred SE, Yamada T, MacLeod PJM, Jones JR, Scherer SW, Schanen NC, Friez MJ et al (2004) A previously unidentified MECP2 open reading frame defines a new protein isoform relevant to Rett syndrome. Nat Genet 36:339–341. https://doi.org/10.1038/ng1327
29. Dragich JM, Kim Y-H, Arnold AP, Schanen NC (2007) Differential distribution of the MeCP2 splice variants in the postnatal mouse brain. J Comp Neurol 501:526–542. https://doi.org/10.1002/cne.21264
30. Olson CO, Zachariah RM, Ezeonwuka CD, Liyanage VRB, Rastegar M (2014) Brain region-specific expression of MeCP2 isoforms correlates with DNA methylation within Mecp2 regulatory elements. PLoS One 9:e90645. https://doi.org/10.1371/journal.pone.0090645
31. Vogel Ciernia A, Yasui DH, Pride MC, Durbin-Johnson B, Noronha AB, Chang A, Knotts TA, Rutkowsky JR, Ramsey JJ, Crawley JN et al (2018) MeCP2 isoform e1 mutant mice recapitulate motor and metabolic phenotypes of Rett syndrome. Hum Mol Genet 27:4077–4093. https://doi.org/10.1093/hmg/ddy301
32. Lyst MJ, Ekiert R, Ebert DH, Merusi C, Nowak J, Selfridge J, Guy J, Kastan NR, Robinson ND, de Lima Alves F et al (2013) Rett syndrome mutations abolish the interaction of MeCP2 with the NCoR/SMRT co-repressor. Nat Neurosci 16:898–902. https://doi.org/10.1038/nn.3434
33. Mellén M, Ayata P, Dewell S, Kriaucionis S, Heintz N (2012) MeCP2 binds to 5hmC enriched within active genes and accessible chromatin in the nervous system. Cell 151:1417–1430. https://doi.org/10.1016/j.cell.2012.11.022
34. Chahrour M, Jung SY, Shaw C, Zhou X, Wong STC, Qin J, Zoghbi HY (2008) MeCP2, a key contributor to neurological disease, activates and represses transcription. Science 320:1224–1229. https://doi.org/10.1126/science.1153252

35. Della Ragione F, Vacca M, Fioriniello S, Pepe G, D'Esposito M (2016) MECP2, a multi-talented modulator of chromatin architecture. Brief Funct Genomics 15:420–431. https://doi.org/10.1093/bfgp/elw023
36. Young JI, Hong EP, Castle JC, Crespo-Barreto J, Bowman AB, Rose MF, Kang D, Richman R, Johnson JM, Berget S et al (2005) Regulation of RNA splicing by the methylation-dependent transcriptional repressor methyl-CpG binding protein 2. Proc Natl Acad Sci U S A 102:17551–17558. https://doi.org/10.1073/pnas.0507856102
37. Maunakea AK, Chepelev I, Cui K, Zhao K (2013) Intragenic DNA methylation modulates alternative splicing by recruiting MeCP2 to promote exon recognition. Cell Res 23:1256–1269. https://doi.org/10.1038/cr.2013.110
38. Cheng T-L, Wang Z, Liao Q, Zhu Y, Zhou W-H, Xu W, Qiu Z (2014) MeCP2 suppresses nuclear microRNA processing and dendritic growth by regulating the DGCR8/Drosha complex. Dev Cell 28:547–560. https://doi.org/10.1016/j.devcel.2014.01.032
39. Tsujimura K, Irie K, Nakashima H, Egashira Y, Fukao Y, Fujiwara M, Itoh M, Uesaka M, Imamura T, Nakahata Y et al (2015) miR-199a links MeCP2 with mTOR signaling and its dysregulation leads to Rett syndrome phenotypes. Cell Rep 12:1887–1901. https://doi.org/10.1016/j.celrep.2015.08.028
40. Hotta A, Cheung AYL, Farra N, Vijayaragavan K, Séguin CA, Draper JS, Pasceri P, Maksakova IA, Mager DL, Rossant J et al (2009) Isolation of human iPS cells using EOS lentiviral vectors to select for pluripotency. Nat Methods 6:370–376. https://doi.org/10.1038/nmeth.1325
41. Marchetto MCN, Carromeu C, Acab A, Yu D, Yeo GW, Mu Y, Chen G, Gage FH, Muotri AR (2010) A model for neural development and treatment of Rett syndrome using human induced pluripotent stem cells. Cell 143:527–539. https://doi.org/10.1016/j.cell.2010.10.016
42. Chen X, Han X, Blanchi B, Guan W, Ge W, Yu Y-C, Sun YE (2021) Graded and pan-neural disease phenotypes of Rett syndrome linked with dosage of functional MeCP2. Protein Cell 12:639–652. https://doi.org/10.1007/s13238-020-00773-z
43. Nageshappa S, Carromeu C, Trujillo CA, Mesci P, Espuny-Camacho I, Pasciuto E, Vanderhaeghen P, Verfaillie CM, Raitano S, Kumar A et al (2016) Altered neuronal network and rescue in a human MECP2 duplication model. Mol Psychiatry 21:178–188. https://doi.org/10.1038/mp.2015.128
44. Trujillo CA, Adams JW, Negraes PD, Carromeu C, Tejwani L, Acab A, Tsuda B, Thomas CA, Sodhi N, Fichter KM et al (2021) Pharmacological reversal of synaptic and network pathology in human MECP2-KO neurons and cortical organoids. EMBO Mol Med 13:e12523. https://doi.org/10.15252/emmm.202012523
45. Bajikar SS, Sztainberg Y, Trostle AJ, Tirumala HP, Wan Y-W, Harrop CL, Bengtsson JD, Carvalho CMB, Pehlivan D, Suter B et al (2024) Modeling antisense oligonucleotide therapy in MECP2 duplication syndrome human iPSC-derived neurons reveals gene expression programs responsive to MeCP2 levels. Hum Mol Genet 33:1986–2001. https://doi.org/10.1093/hmg/ddae135
46. Mendonca D, Cappuccio G, Sheppard J, Delacruz M, Bengtsson J, Carvalho CMB, Bajic A, Park H, Kim JJ, Jafar-Nejad P et al (2024) Generation of five induced pluripotent stem cell lines from patients with MECP2 duplication syndrome. Stem Cell Res 74:103292. https://doi.org/10.1016/j.scr.2023.103292
47. Hu B-Y, Weick JP, Yu J, Ma L-X, Zhang X-Q, Thomson JA, Zhang S-C (2010) Neural differentiation of human induced pluripotent stem cells follows developmental principles but with variable potency. Proc Natl Acad Sci U S A 107:4335–4340. https://doi.org/10.1073/pnas.0910012107
48. Smith JR, Vallier L, Lupo G, Alexander M, Harris WA, Pedersen RA (2008) Inhibition of Activin/Nodal signaling promotes specification of human embryonic stem cells into neuroectoderm. Dev Biol 313:107–117. https://doi.org/10.1016/j.ydbio.2007.10.003
49. Chambers SM, Fasano CA, Papapetrou EP, Tomishima M, Sadelain M, Studer L (2009) Highly efficient neural conversion of human ES and iPS cells by dual inhibition of SMAD signaling. Nat Biotechnol 27:275–280. https://doi.org/10.1038/nbt.1529

50. Cheung AYL, Horvath LM, Grafodatskaya D, Pasceri P, Weksberg R, Hotta A, Carrel L, Ellis J (2011) Isolation of MECP2-null Rett syndrome patient hiPS cells and isogenic controls through X-chromosome inactivation. Hum Mol Genet 20:2103–2115. https://doi.org/10.1093/hmg/ddr093
51. Chin EWM, Marcy G, Yoon S-I, Ma D, Rosales FJ, Augustine GJ, Goh ELK (2016) Choline ameliorates disease phenotypes in human iPSC models of Rett syndrome. Neuromolecular Med 18:364–377. https://doi.org/10.1007/s12017-016-8421-y
52. Djuric U, Cheung AYL, Zhang W, Mok RS, Lai W, Piekna A, Hendry JA, Ross PJ, Pasceri P, Kim D-S et al (2015) MECP2e1 isoform mutation affects the form and function of neurons derived from Rett syndrome patient iPS cells. Neurobiol Dis 76:37–45. https://doi.org/10.1016/j.nbd.2015.01.001
53. Williams EC, Zhong X, Mohamed A, Li R, Liu Y, Dong Q, Ananiev GE, Mok JCC, Lin BR, Lu J et al (2014) Mutant astrocytes differentiated from Rett syndrome patients-specific iPSCs have adverse effects on wild-type neurons. Hum Mol Genet 23:2968–2980. https://doi.org/10.1093/hmg/ddu008
54. Andoh-Noda T, Akamatsu W, Miyake K, Matsumoto T, Yamaguchi R, Sanosaka T, Okada Y, Kobayashi T, Ohyama M, Nakashima K et al (2015) Differentiation of multipotent neural stem cells derived from Rett syndrome patients is biased toward the astrocytic lineage. Mol Brain 8:31. https://doi.org/10.1186/s13041-015-0121-2
55. Ohashi M, Korsakova E, Allen D, Lee P, Fu K, Vargas BS, Cinkornpumin J, Salas C, Park JC, Germanguz I et al (2018) Loss of MECP2 leads to activation of P53 and neuronal senescence. Stem Cell Reports 10:1453–1463. https://doi.org/10.1016/j.stemcr.2018.04.001
56. Sharma P, Mesci P, Carromeu C, McClatchy DR, Schiapparelli L, Yates JR, Muotri AR, Cline HT (2019) Exosomes regulate neurogenesis and circuit assembly. Proc Natl Acad Sci U S A 116:16086–16094. https://doi.org/10.1073/pnas.1902513116
57. Jiang M, Ash RT, Baker SA, Suter B, Ferguson A, Park J, Rudy J, Torsky SP, Chao H-T, Zoghbi HY et al (2013) Dendritic arborization and spine dynamics are abnormal in the mouse model of MECP2 duplication syndrome. J Neurosci 33:19518–19533. https://doi.org/10.1523/JNEUROSCI.1745-13.2013
58. Pevny LH, Sockanathan S, Placzek M, Lovell-Badge R (1998) A role for SOX1 in neural determination. Development 125:1967–1978. https://doi.org/10.1242/dev.125.10.1967
59. Töröcsik B, Angelastro JM, Greene LA (2002) The basic region and leucine zipper transcription factor MafK is a new nerve growth factor-responsive immediate early gene that regulates neurite outgrowth. J Neurosci 22:8971–8980. https://doi.org/10.1523/JNEUROSCI.22-20-08971.2002
60. Fischer P, Chen H, Pacho F, Rieder D, Kimmel RA, Meyer D (2019) FoxH1 represses miR-430 during early embryonic development of zebrafish via non-canonical regulation. BMC Biol 17:61. https://doi.org/10.1186/s12915-019-0683-z
61. Kofron M, Puck H, Standley H, Wylie C, Old R, Whitman M, Heasman J (2004) New roles for FoxH1 in patterning the early embryo. Development 131:5065–5078. https://doi.org/10.1242/dev.01396
62. Paşca SP (2018) The rise of three-dimensional human brain cultures. Nature 553:437–445. https://doi.org/10.1038/nature25032
63. Lancaster MA, Knoblich JA (2014) Generation of cerebral organoids from human pluripotent stem cells. Nat Protoc 9:2329–2340. https://doi.org/10.1038/nprot.2014.158
64. Qian X, Nguyen HN, Song MM, Hadiono C, Ogden SC, Hammack C, Yao B, Hamersky GR, Jacob F, Zhong C et al (2016) Brain-region-specific organoids using mini-bioreactors for modeling ZIKV exposure. Cell 165:1238–1254. https://doi.org/10.1016/j.cell.2016.04.032
65. Jo J, Xiao Y, Sun AX, Cukuroglu E, Tran H-D, Göke J, Tan ZY, Saw TY, Tan C-P, Lokman H et al (2016) Midbrain-like organoids from human pluripotent stem cells contain functional dopaminergic and neuromelanin-producing neurons. Cell Stem Cell 19:248–257. https://doi.org/10.1016/j.stem.2016.07.005

66. Mellios N, Feldman DA, Sheridan SD, Ip JPK, Kwok S, Amoah SK, Rosen B, Rodriguez BA, Crawford B, Swaminathan R et al (2018) MeCP2-regulated miRNAs control early human neurogenesis through differential effects on ERK and AKT signaling. Mol Psychiatry 23:1051–1065. https://doi.org/10.1038/mp.2017.86
67. Nakashima H, Tsujimura K, Irie K, Imamura T, Trujillo CA, Ishizu M, Uesaka M, Pan M, Noguchi H, Okada K et al (2021) MeCP2 controls neural stem cell fate specification through miR-199a-mediated inhibition of BMP-Smad signaling. Cell Rep 35:109124. https://doi.org/10.1016/j.celrep.2021.109124
68. Samarasinghe RA, Miranda OA, Buth JE, Mitchell S, Ferando I, Watanabe M, Allison TF, Kurdian A, Fotion NN, Gandal MJ et al (2021) Identification of neural oscillations and epileptiform changes in human brain organoids. Nat Neurosci 24:1488–1500. https://doi.org/10.1038/s41593-021-00906-5
69. Martinowich K, Hattori D, Wu H, Fouse S, He F, Hu Y, Fan G, Sun YE (2003) DNA methylation-related chromatin remodeling in activity-dependent BDNF gene regulation. Science 302:890–893. https://doi.org/10.1126/science.1090842
70. Renthal W, Boxer LD, Hrvatin S, Li E, Silberfeld A, Nagy MA, Griffith EC, Vierbuchen T, Greenberg ME (2018) Characterization of human mosaic Rett syndrome brain tissue by single-nucleus RNA sequencing. Nat Neurosci 21:1670–1679. https://doi.org/10.1038/s41593-018-0270-6
71. D'Ercole AJ, Ye P, Calikoglu AS, Gutierrez-Ospina G (1996) The role of the insulin-like growth factors in the central nervous system. Mol Neurobiol 13:227–255. https://doi.org/10.1007/BF02740625
72. Zheng W-H, Quirion R (2004) Comparative signaling pathways of insulin-like growth factor-1 and brain-derived neurotrophic factor in hippocampal neurons and the role of the PI3 kinase pathway in cell survival. J Neurochem 89:844–852. https://doi.org/10.1111/j.1471-4159.2004.02350.x
73. O'Kusky JR, Ye P, D'Ercole AJ (2000) Insulin-like growth factor-I promotes neurogenesis and synaptogenesis in the hippocampal dentate gyrus during postnatal development. J Neurosci 20:8435–8442. https://doi.org/10.1523/JNEUROSCI.20-22-08435.2000
74. Li Y, Wang H, Muffat J, Cheng AW, Orlando DA, Lovén J, Kwok S-M, Feldman DA, Bateup HS, Gao Q et al (2013) Global transcriptional and translational repression in human-embryonic-stem-cell-derived Rett syndrome neurons. Cell Stem Cell 13:446–458. https://doi.org/10.1016/j.stem.2013.09.001
75. Nag N, Berger-Sweeney JE (2007) Postnatal dietary choline supplementation alters behavior in a mouse model of Rett syndrome. Neurobiol Dis 26:473–480. https://doi.org/10.1016/j.nbd.2007.02.003
76. Hubbert C, Guardiola A, Shao R, Kawaguchi Y, Ito A, Nixon A, Yoshida M, Wang X-F, Yao T-P (2002) HDAC6 is a microtubule-associated deacetylase. Nature 417:455–458. https://doi.org/10.1038/417455a
77. Landucci E, Brindisi M, Bianciardi L, Catania LM, Daga S, Croci S, Frullanti E, Fallerini C, Butini S, Brogi S et al (2018) iPSC-derived neurons profiling reveals GABAergic circuit disruption and acetylated α-tubulin defect which improves after iHDAC6 treatment in Rett syndrome. Exp Cell Res 368:225–235. https://doi.org/10.1016/j.yexcr.2018.05.001
78. Guo W, Naujock M, Fumagalli L, Vandoorne T, Baatsen P, Boon R, Ordovás L, Patel A, Welters M, Vanwelden T et al (2017) HDAC6 inhibition reverses axonal transport defects in motor neurons derived from FUS-ALS patients. Nat Commun 8:861. https://doi.org/10.1038/s41467-017-00911-y
79. Gadalla KKE, Vudhironarit T, Hector RD, Sinnett S, Bahey NG, Bailey MES, Gray SJ, Cobb SR (2017) Development of a novel AAV gene therapy cassette with improved safety features and efficacy in a mouse model of Rett syndrome. Mol Ther Methods Clin Dev 5:180–190. https://doi.org/10.1016/j.omtm.2017.04.007

80. Lykken EA, Shyng C, Edwards RJ, Rozenberg A, Gray SJ (2018) Recent progress and considerations for AAV gene therapies targeting the central nervous system. J Neurodev Disord 10:16. https://doi.org/10.1186/s11689-018-9234-0
81. Croci S, Carriero ML, Capitani K, Daga S, Donati F, Papa FT, Frullanti E, Lopergolo D, Lamacchia V, Tita R et al (2020) AAV-mediated FOXG1 gene editing in human Rett primary cells. Eur J Hum Genet 28:1446–1458. https://doi.org/10.1038/s41431-020-0652-6
82. Garg SK, Lioy DT, Cheval H, McGann JC, Bissonnette JM, Murtha MJ, Foust KD, Kaspar BK, Bird A, Mandel G (2013) Systemic delivery of MeCP2 rescues behavioral and cellular deficits in female mouse models of Rett syndrome. J Neurosci 33:13612–13620. https://doi.org/10.1523/JNEUROSCI.1854-13.2013
83. Klein ME, Lioy DT, Ma L, Impey S, Mandel G, Goodman RH (2007) Homeostatic regulation of MeCP2 expression by a CREB-induced microRNA. Nat Neurosci 10:1513–1514. https://doi.org/10.1038/nn2010
84. Oliveira AC, Bovolenta LA, Alves L, Figueiredo L, Ribeiro AO, Campos VF, Lemke N, Pinhal D (2019) Understanding the modus operandi of MicroRNA regulatory clusters. Cells 8:1103. https://doi.org/10.3390/cells8091103
85. Rinaldi C, Wood MJA (2018) Antisense oligonucleotides: the next frontier for treatment of neurological disorders. Nat Rev Neurol 14:9–21. https://doi.org/10.1038/nrneurol.2017.148
86. Shao Y, Sztainberg Y, Wang Q, Bajikar SS, Trostle AJ, Wan Y-W, Jafar-Nejad P, Rigo F, Liu Z, Tang J et al (2021) Antisense oligonucleotide therapy in a humanized mouse model of MECP2 duplication syndrome. Sci Transl Med 13:eaaz7785. https://doi.org/10.1126/scitranslmed.aaz7785
87. Lancaster MA, Corsini NS, Wolfinger S, Gustafson EH, Phillips AW, Burkard TR, Otani T, Livesey FJ, Knoblich JA (2017) Guided self-organization and cortical plate formation in human brain organoids. Nat Biotechnol 35:659–666. https://doi.org/10.1038/nbt.3906
88. Nickels SL, Modamio J, Mendes-Pinheiro B, Monzel AS, Betsou F, Schwamborn JC (2020) Reproducible generation of human midbrain organoids for in vitro modeling of Parkinson's disease. Stem Cell Res 46:101870. https://doi.org/10.1016/j.scr.2020.101870
89. Muguruma K, Nishiyama A, Kawakami H, Hashimoto K, Sasai Y (2015) Self-organization of polarized cerebellar tissue in 3D culture of human pluripotent stem cells. Cell Rep 10:537–550. https://doi.org/10.1016/j.celrep.2014.12.051
90. Onesto MM, Kim J-I, Pasca SP (2024) Assembloid models of cell-cell interaction to study tissue and disease biology. Cell Stem Cell 31:1563–1573. https://doi.org/10.1016/j.stem.2024.09.017
91. Kodera T, Takeuchi RF, Takahashi S, Suzuki K, Kassai H, Aiba A, Shiozawa S, Okano H, Osakada F (2023) Modeling the marmoset brain using embryonic stem cell-derived cerebral assembloids. Biochem Biophys Res Commun 657:119–127. https://doi.org/10.1016/j.bbrc.2023.03.019
92. Qian X, Song H, Ming G-L (2019) Brain organoids: advances, applications and challenges. Development 146:dev166074. https://doi.org/10.1242/dev.166074
93. Qian X, Jacob F, Song MM, Nguyen HN, Song H, Ming G-L (2018) Generation of human brain region-specific organoids using a miniaturized spinning bioreactor. Nat Protoc 13:565–580. https://doi.org/10.1038/nprot.2017.152
94. Shi Y, Sun L, Wang M, Liu J, Zhong S, Li R, Li P, Guo L, Fang A, Chen R et al (2020) Vascularized human cortical organoids (vOrganoids) model cortical development in vivo. PLoS Biol 18:e3000705. https://doi.org/10.1371/journal.pbio.3000705

Chapter 3
Genome Editing and Human-Induced Pluripotent Stem Cells Facilitate Parkinson's Disease Modeling

Catalina M. Polanco, Jose Inzunza, Ivan Nalvarte, and Jonathan Arias

Abstract In this chapter, we provide a comprehensive overview of the wide assortment of gene-editing tools currently available to the human stem cell field, aiming to serve as a guide for both the Parkinson's disease (PD) research community and the broader regenerative medicine field. This chapter highlights the contributions of gene editing to PD models, to illustrate their broader applicability. While extensive literature exists on PD using model organisms and non-human cell systems, our focus is uniquely placed on human embryonic stem cells (ESCs) and human-induced pluripotent stem (iPS) cells due to their direct relevance to human etiology.

We emphasize the critical role of gene editing in addressing challenges posed by genetic and phenotypic variability, demonstrating how these tools enable the creation of isogenic cell models to elucidate PD pathophysiology. Additionally, we explore the potential of advanced gene-editing solutions, including CRISPR-Cas9, base editors, prime editors, and epigenetic modulators to address diverse experimental questions with precision. We discuss how iPS cell models enhanced by programmable genetic and epigenetic modulators have been used to dissect underlying

C. M. Polanco
VU LSC - EMBL Partnership Institute for Gene Editing Technologies, Laboratory of Nuclease-Enabled Cell Therapies, Vilnius University, Vilnius, Lithuania

J. Inzunza
Department of Laboratory Medicine, Division of Clinical Microbiology, Karolinska Institutet Stem Cell Organoid (KISCO) facility, Karolinska Institutet, Huddinge, Sweden

I. Nalvarte
Department of Neurobiology, Care Sciences and Society, Division of Neurogeriatrics, Karolinska Institutet, Stockholm, Sweden

J. Arias (✉)
VU LSC - EMBL Partnership Institute for Gene Editing Technologies, Laboratory of Nuclease-Enabled Cell Therapies, Vilnius University, Vilnius, Lithuania

Department of Laboratory Medicine, Division of Clinical Microbiology, Karolinska Institutet Stem Cell Organoid (KISCO) facility, Karolinska Institutet, Huddinge, Sweden
e-mail: jonathan.arias@gmc.vu.lt

R. C. B. Wong and Y. Hayashi (eds.), *Advances in Using Stem Cells to Treat Neural Diseases*, Stem Cell Biology and Regenerative Medicine 77,
https://doi.org/10.1007/978-3-031-94101-6_3

mechanisms of PD, develop regenerative medicine prototypes, and address immune compatibility challenges in cell transplantation.

Keywords Programmable nucleases · Gene editing · Disease modeling · Human iPS cells · Prime editors · Base editors · CRISPR Cas · TALEN · ZFN · Regenerative medicine · Parkinson's disease

3.1 Introduction

There are multiple biobank collections of patient-derived human induced pluripotent stem (iPS) cells available to the regenerative medicine community [1]. Notable efforts include RIKEN BioResource Center (BRC) in Japan, which houses over 4,000 iPS cell samples; the WiCell research institute and the California Institute of Regenerative Medicine (CIRM) in collaboration with Fujifilm cellular dynamics in the US, each maintaining more than 1,500 iPS cells and embryonic stem cell (ESC) samples; and the European Biobank for induced pluripotent stem cells (EBiSC) in Europe, approximately 1,000 iPS cell samples. Additionally, the human-induced pluripotent stem cells initiative (HipSci) in the UK includes nearly 900 iPS cell clones, which are stored at the EBiSC and the European Collection of Authenticated Cell Cultures (ECACC).

Other significant contributors to the global iPS cell efforts include the Bioresource Collection and Research Center (BCRC) in Taiwan, the Korea National Stem Cell Biobank (KSCB) in South Korea, and the National Biobank of cell lines (BNCL) in Spain. One of the most comprehensive resources available to the iPS cell community is the human pluripotent stem cell registry (hPSCreg) [2], which currently hosts information on over 6,400 iPS cell clones, including genetically engineered variants.

Among these biobanks, repositories, and publications that have created human iPS cells, more than 730 cell lines encompassing familial and sporadic forms of PD are available (Table 3.1), as cataloged by the Integrated Collection of Stem Cell Banks (ICSCB-II). Furthermore, numerous initiatives compile cells from PD patients and related donors, providing a compatible source for the creation of iPS cell lines for disease modeling and therapeutic prototyping. Collectively, these biobanks represent an invaluable resource for the regenerative medicine and gene-editing communities.

Some of the first human iPS cell lines derived from PD donors were established in 2008 and 2009 for sporadic [3] and idiopathic conditions [4]. Pioneering studies using patient-derived iPS cells provided critical proof-of-concept evidence that strong phenotypic differences observed in well-defined hereditary forms of PD are faithfully retained in these cell models [5]. Examples include *SNCA* triplication [6], *SNCA* p.A53T [7–9], *PINK1* p.Q456X [10–12], *PINK1* p.V170G [10], *LRRK2* p.G2019S [11, 13–17], *LRRK2* p.R1441C [11], *PARK2* domains deletions [18–21],

Table 3.1 Biobank collections and repositories of human iPS cell lines, including those derived from PD patients, available to the regenerative medicine community

Name	Link	Country	Number of iPS cells (PD iPS cell samples)
RIKEN BRC	https://www.brc.riken.jp/en/	Japan	4,111 (34)
WiCell research institute[a]	https://www.wicell.org/	US	1,531 (34)
CIRM—Fujifilm cellular dynamics[c]	https://www.fujifilmcdi.com/cirm-ipsc-products/	US	1,554 (NA)
EBiSC	https://ebisc.org/	EU	963 (154)
HipSci	https://www.hipsci.org/	UK	840 (NA)
hPSCreg[b]	https://hpscreg.eu/	EU	6,488 (447)
BCRC	https://catalog.bcrc.firdi.org.tw/	Taiwan	112 (7)
KSCB	http://nih.go.kr/ncsr	Korea	155 (NA)
BNCL	https://www.isciii.es/	Spain	277 (18)

NA not available information
[a] Also containing ESC samples
[b] Registry
[c] As of July 16, 2025, CIRM is closed

PARK2 p.R42P and p.R275W [21], *PARK2* p.V324fsX110 [12], *GBA* double heterozygous p.L483P p.A495P (RecNcil) [22, 23], as well as *GBA* single heterozygous p.L444P [22] and p.N370S [22, 24].

In the hereditary forms of PD, phenotypic readouts are often pronounced when compared to healthy control lines from blood-related or unrelated donors. These readouts encompass a wide range of cellular hallmarks of PD [5] including differences in viability, mitochondria morphology and membrane potential, lysosomal pH, autophagic capacity, neuronal morphology, and electrophysiological properties. Complementary efforts have focused on generating healthy human iPS cell lines as standardized reference points [25].

Although comparisons between patients and other patients, as well as patients and healthy-controls, were crucial for developing phenotypic readout assays, it soon became evident that subtle or less prominent phenotypes were obscured by the "*genetic background noise.*" This variability, introduced by the multitude of SNPs unique to each genome, is particularly pronounced when comparing unrelated donor samples. While large cohorts of individuals allowed for statistical significance in some cases, comparisons within the same *genetic background* were persistently desirable.

The importance of *genetic background* standardization was highlighted by studies on monozygotic twins carrying the *GBA* mutation p.N370S, who showed clinically discordant PD [24]. *Genetic background noise*, as well as epigenetic differences, persisted despite age- and gender-matching of patient-derived human iPS cells. These challenges highlighted the critical need for isogenic cell models, and emphasized the enabling role of gene-editing tools in advancing our understanding of developmental biology, neuroscience, physiology, and pathology in PD.

3.2 Diversity of Gene-Editing Tools Implemented in PD Modeling

One of the first isogenic cell models for PD was developed in 2012 using a nuclease-free, helper-dependent adenoviral vector system to correct the *LRRK2* p.G2019S mutation [14]. However, due to the low efficiency of genome editing in human cell systems, the vast majority of subsequent studies have implemented at least one nuclease or nuclease derivative. These gene-editing technologies include zinc finger nucleases (ZFN), transcription activator-like effector nucleases (TALEN), clustered regularly interspaced short palindromic repeats (CRISPR) nucleases. Within the CRISPR-based systems, specific tools such as CRISPR-Cas base editors and prime editors enable precise nucleotide editing, while CRISPR interference (CRISPRi) and CRISPR activation (CRISPRa) are used for gene regulation. Additionally, CRISPR epigenetic modulation tools, including CRISPRoff and CRISPRon, allow for reversible modifications of gene expression. Transposases have also been employed for efficient gain-of-function. These advancements have significantly expanded the toolkit for genome editing, as discussed in the following sections.

3.2.1 ZFN in PD Modeling

Zinc Finger DNA recognition domains were first discovered in the transcription factor IIIA from *Xenopus laevis* [26], with distinctive repeating units linked to DNA recognition subsequently identified [27]. These Zinc Finger motifs were later adopted as synthetic programmable protein modules capable of functioning as transcriptional activators or repressors [28]. Structural studies provided insights into the protein-to-DNA–helix interaction interfaces [29, 30]. The chimeric fusion of Zinc Fingers with the C-terminal domain of FokI type II restriction endonuclease [31] led to the development of ZFN [32]. As native FokI requires homodimerization for catalytic activity, two complementary ZFNs must be directed to the target site to facilitate DNA cleavage. The first application of ZFNs in a human cell line successfully restored the expression of a GFP reporter [33], and the first in vivo application rescued the *yellow* mutant phenotype in *D. melanogaster* [34].

To address the cytotoxicity caused by FokI homodimerization and reduce the off-target activity, mutational screenings led to the identification of compulsory heterodimeric versions of ZFNs that retained high nuclease activity [35, 36]. The optimized and widely adopted versions include the (−)ELD and (+)KKR mutants [37].

The first isogenic pairs of PD-patient-derived human iPS cells were generated in 2011 [7], using both wild-type FokI and the compulsory heterodimeric ELD/KKR ZFN variants [37]. This study demonstrated the correction of a disease-causing single nucleotide polymorphism (SNP). It also introduced the disease-causing mutations *SNCA* p.A53T and *SNCA* p.E46K into human ESCs with a healthy genetic background, using homology-directed repair (HDR) with double-stranded DNA (dsDNA) templates and single-stranded oligo deoxynucleotides (ssODN). This study highlighted that successful gene-editing campaigns targeting human pluripotent stem cells in genes located within regions of reduced chromatin accessibility often require screening large cohorts, sometimes up to 480 clones. Additionally, it emphasized the high rate of attrition in correctly targeted clones due to disruptions in the second allele, emphasizing the importance of screening for the intended SNP composition.

The use of composite tools such as ZFNs benefits from a reduced probability of off-target effects, as both components must bind at the same location to exert activity. Subsequent studies leveraged this approach to develop isogenic collections for *GBA*-related PD models [22] using ZFN and dsDNA HDR templates. This study leveraged positive selection modules integrated within the donor templates, enabling the efficient identification of successfully targeted human iPS cell clones.

In addition to creating isogenic cell models differing in a particular SNP, frequently encoding clinically-relevant missense or nonsense mutations, studies have also examined knock-out isogenic models to elucidate loss-of-function phenotypes within the same genomic context. For example, *PARK2* knock-out human iPS cell lines were generated using ZFNs [21]. The knock-out isogenic lines enabled the evaluation of gene dosage effects in PD, revealing a proportional reduction in mitochondrial function in heterozygous and homozygous clones.

3.2.2 TALEN in PD Modeling

The repeat domain of transcription activator-like effector (TALE) proteins was first described in the plant pathogens *Xanthomonas campestris* [38] and *X. oryzae* [39]. TALEs act on target plant cells by upregulating the expression of genes that facilitate infection [40]. These proteins were characterized as having an N-terminus responsible for translocation, a central repeat domain encompassing up to 33 complete repeats, and a C-terminus containing nuclear localization signal and transcriptional activator domain. The TALE repeats were shown to bind DNA [41] and encoded target specificity [42, 43]. These discoveries enabled the use of TALE repeat motifs as programmable modules to guide functional domains to specific genomic targets in *Animalia*.

The most frequent repeat motif is 34 amino acids in length, comprising two alpha helices flanking a di-amino acid motif [44–46], as elucidated in representative crystal structures [46]. The di-amino acid motif, located at position 12–13, is hypervariable and confers one-repeat-one-bp specificity [45, 47]. On average, TALE proteins contain 17 complete repeats, although shorter repeats domains have been shown to be functional. The fusion of FokI nuclease domains with TALE gave rise to TALEN. DNA-binding and cleavage assays in human cells demonstrated that retention of TALE N-terminus (N-terminus Δ152) and C-terminal 68 amino acids was required for functional nuclease activity [48, 49] in TALEN. This model has been widely adopted for disease modeling in human systems.

TALEN and other gene-editing tools have enabled compound screening campaigns using genetically encoded knock-in reporters. For example, the *GBA1* gene, which encoded β-glucocerebrosidase (GCase), was fused with the HiBiT tag [50] to identify GCase-stabilizing compounds [51]. In this study, the HiBiT-*GBA1* expression module was knocked into the *CLYBL safe harbor locus* using TALEN, ensuring consistent and constitutive expression levels required for library screening [51]. Genetically encoded reporter knock-in lines facilitate compound screening and neurophysiological studies. *Safe harbor loci*, euchromatic regions unaffected by cargo insertions, are highly desirable for the stable expression of exogenous transgenes and genetic circuits. The *H11 locus* at chromosome 22 has been identified as an amenable site for such purposes, supporting the differentiation of midbrain neuronal lineages. This *locus* has been validated using TALEN in human ESC lines [52] and shown to support simultaneous expression of multiple transgenes, such as *LMX1A*, *FOXA2,* and *OTX2*.

Lineage-specific endogenous promoter monitoring is commonly employed during differentiation campaigns to purify derivative populations. For instance, reporter knock-in at the Tyrosine hydroxylase (*TH*) gene C-terminus, achieved using TALEN in human ESCs, has been shown to be a reliable method for purifying dopaminergic neurons with properties similar to primary brain samples [53].

Additionally, the recently developed isogenic stem cell collection to research PD (iSCORE-PD) leverages a healthy genetic background ESC to install PD-causing mutations, including the *LRRK2* p.G2019S mutation which was engineered using TALEN [54].

3.2.3 CRISPR-Cas Nucleases in PD Modeling

CRISPR systems were first characterized as an anti-phage defense mechanism in microorganisms [55]. These systems were shown to incorporate DNA sequences from extrachromosomal origin into microbial genomes, providing programmable defense to plasmids and phages [56–58]. CRISPR arrays consist of spacers, typically 30–60 bp in length, flanked by repeats of 35–40 bp. These spacers match to sequences in the target genome, known as protospacer, which are proximal to a protospacer adjacent motif (PAM).

The CRISPR-Cas system can be classified into two main categories: Class 1, which utilizes multi-protein effectors, and class 2, which employs a single-protein effector [59]. The most widely distributed and commonly used CRISPR-Cas tools in eukaryotic biology and disease modeling are Cas9 and Cas12a, belonging to the class 2 type II and V, respectively. The most widely used Cas9 protein variant is derived from *S. pyogenes* (*Sp*Cas9) and recognizes a PAM sequence of NGG. Similarly, the most commonly used Cas12a proteins, derived from

Acidaminococcus sp. and *Lachnospiraceae bacterium*, recognize a TTTV PAM sequence [60, 61].

Early studies on Class 2 type II systems revealed that a trans-activating CRISPR RNA (tracrRNA) associates with the spacer-repeat containing CRISPR RNA (crRNA) [62], and both components are necessary for catalytic nuclease activity [63]. This discovery lead to the development of simplified expression systems by fusing crRNA and tracrRNA into a chimeric single-guide RNA (sgRNA) [64, 65]. The ability of CRISPR-Cas9 to create knock-out human cells was validated in cell line models [66–68], enabling loss-of-function disease modeling.

Further advancements included the characterization of Cas9 nuclease's mechanism of action and kinetics [69–71], as well as its crystal structure [72], paving the way for the development of high-fidelity nuclease variants with reduced off-target activity [73].

CRISPR nucleases facilitate the dissection of genetic causation in PD through loss-of-function studies, enabling exploration of interaction between physiological components and PD-linked mutations. Additionally, they allow the correction of disease-causing SNPs via HDR for use in screening campaigns.

Endogenous promoter knock-in reporters, enabled by CRISPR-Cas9 and other nucleases, are ideal for improving the conditions of differentiation protocols. This is particularly relevant in PD, where differentiated population can be used to study cell replacement prototypes. For example, reporter knock-in via HDR at the endogenous *TH* promoter allows enrichment of dopaminergic neurons during differentiation [74, 75], as previously described for TALEN [53]. Screening of clones with large cargoes (usually in the kb range), as endogenous knock-in reporters, or SNP changes is accelerated by droplet PCR in PD and other disease modeling applications [76, 77]. C-terminus reporter fusions at the *TH* gene have also been utilized as dopaminergic neuron markers to evaluate mitochondrial ultrastructure in pools of iPS cell-derived neurons [78].

Cas9 nuclease-based knock-in at endogenous *loci* has advanced PD research. For instance, knock-in of intermediate C9orf72 repeats demonstrated increased *C9orf72* expression in iPS cell-derived neuronal progenitors and revealed widespread changes in vesicle trafficking and autophagy metabolism [79]. Additionally, stable Cas9 expression from euchromatic *loci* has provided a reliable system for simultaneous genetic and compound screening in the context of neurodegeneration. A Cas9 gene knock-in at the *GAPDH locus* enabled the identification of neuroprotective and ER-stress reducing compounds [80]. Genetically encoded reporters, such as the pH-sensitive rosella protein, have also been employed for compound screening and phenotype stratification in PD-patient derived cell models and isogenic collections [81, 82].

Cas9-mediated knock-out studies have further elucidated PD mechanisms. Knock-out of the *LRP1* gene, which encodes Prolow-density lipoprotein receptor-related protein 1, confirmed its involvement in α-Synuclein uptake [83]. Homozygous loss of *DNAJC6* expression has been linked to early-onset PD [84]. Cas9 knock-out in human ESC demonstrated that loss of *DNAJC6* results in lysosomal dysfunction, alterations in midbrain-like organoids electrophysiology, and increased S129 phosphorylation of α-Synuclein [85]. A commonly employed strategy during knock-out cell line development involves the sorting for electroporation events. This is enabled by a fluorescent protein reporter co-expressed from the dual expression system plasmids. Such strategy is particularly useful in cases where no donor template is available.

Knock-out studies targeting *PARK2* (encoding PARKIN), *PARK7* (DJ1), and *ATP13A2* (polyamine-transporting ATPase 13A2) on the same genetic background human ESC cell line revealed critical insights into their role in PD. Homozygous *PARK2* knock-out dopaminergic neurons exhibited increased levels of α-Synuclein and heightened cell death [75], while knock-out of *PARK7* and *ATP13A2* resulted in increased lysosomal proteins levels in neurons. Similarly, knock-out of *RAB8A* and *RAB10* in PD human iPS cell models caused a reduction in lysosome frequency and opposing effects on lysosome pH [86].

The role of *SNCA* gene dosage in PD has been extensively studied, and an increase in *SNCA* copy number is a well-documented cause of early-onset PD [87]. Recent work explored *SNCA* dosage effects on protein aggregation and Lewy body formation using sets of human iPS cells with *SNCA* multiplication [88]. CRISPR-Cas9 allowed a stepwise reduction of *SNCA* gene dosage, from four copies to null level, via single-nuclease targeting and convergent targeting of two exons, which induced large exon-intron-exon deletions [88]. Interestingly, some studies report that *SNCA* knock-out cells retain healthy neuronal phenotypes when evaluated in isogenic in vitro *models [*89] and in vivo, suggesting that α-Synuclein may be dispensable [90, 91]. Human ESCs modified with Cas9 to knock-out *SNCA* exon 2 using paired sgRNAs exhibited resistance to synucleinopathy [92], as evidenced by reduced S129-phosphorylated α-Synuclein in derived dopaminergic neurons [92]. Cas9-mediated knock-out of SNPs in the *SNCA* distal enhancer confirmed the roles of transcriptional factors EMX2 and NKX6-1 in regulating *SNCA* expression in neuronal lineages derived from human ESCs [93]. The distal enhancer was targeted using simultaneously two sgRNAs flanking the region.

Cas9 nuclease and HDR donor template-based strategies have been used to create isogenic collections for *LRRK2* p.G2019S, demonstrating increased phosphorylation of α-Synuclein in iPS cell-derived dopaminergic neurons [94], as well as fragmented mitochondria with decreased membrane potential [95]. Similarly, Cas9-based gene-corrected isogenic collections for tetrahydrobiopterin metabolic conditions, *PTS* (6-pyruvoyl tetrahydrobiopterin synthase) c.243G > A and *QDPR* (dihydropteridine reductase) c.176C > A, have shown restored *TH* expression and dopamine levels in iPS cell-derived dopaminergic neurons [96].

Since obtaining the desired genotype composition through HDR often requires the screening of large clonal sets, bi-allelic editing assisted by fluorescent reporters accelerates iPS cell line development [97–99]. Bi-allelic editing has been successfully employed to create isogenic collections for *PINK1* p.Q456X, enabling the identification of small molecules capable of normalizing autophagy and mitophagy levels [82]. Donor template-based editing can also be used to evaluate the relevance of SNPs in non-coding regions, such as the *SNCA* distal enhancer. This strategy identified SNPs associated with EMX2 and NKX6-1 binding at the *SNCA cis* regulatory regions, linking them to GWAS datasets [93].

Recent gene-editing campaigns have focused on creating collections of PD-linked isogenic mutants for standardized phenotype comparison using a variety of CRISPR nuclease-based editing approaches [54]. The iSCORE-PD utilized Cas9-based HDR with ssODN templates to create mutants such as *PINK1* p.Q129X, *SYNJ1* p.R258Q/FS, *LRRK2* p.G2019S, *VPS13C* p.A444P, *ATP13A2* p.T367R/FS, *VPS13C* p.W395C, *FBXO7* p.R498X/FS, *GBA1* IVS2 + 1 G > A, and *DNAJC6* c.801 A > G + FS. For the generation of heterozygous genotype clones, competing donor templates are frequently employed. In the case of heterozygous *GBA1* IVS2 + 1 G > A mutants, competing ssODN templates facilitated the generation of correctly edited clones.

The use of ssODN templates and standard *Sp*Cas9 has also validated the roles of neurodevelopmental genes in PD. For example, *LIN28* heterozygous and homozygous p.R192G mutations were validated as loss-of-function mutations in ESC-derived dopaminergic neurons [100]. Since some PD conditions arise from exon-skipping mutations, convergent Cas9 targeting on both sides of an exon can induce large exon-encompassing deletions. This approach has been demonstrated with exon-skipping mutants *PARK2* Ex3del and *PARK7* Ex1-5del [54].

Although HDR-based gene editing is generally considered inefficient for therapeutic application in vivo [101], human iPS cell models offer robust clonal selection ex vivo, overcoming editing efficiency challenges. When robust clonal selection and screening are feasible, this strategy has proven to be reliable, as exemplified in the aforementioned studies.

3.2.4 CRISPR-Cas Base Editors (BE) in PD Modeling

CRISPR-Cas9 nuclease activity depends on its RuvC and HNH domains, which contain well-characterized catalytic amino acids. This understanding facilitated the repurposing of the effector protein to a nickase (Cas9(D10A) or Cas9(H840A)) or an inactive RNA-programmable DNA-binding domain, also called dead Cas9 (dCas9) [64], achieved by introducing D10A and H840A mutations. These variants have been linked to base-modifying proteins to expand their functionality. A pioneering BE for C*-G to U*-A (T*-A) conversion (CBE), known as BE3 [101], was developed by fusing *R. norvegicus* APOBEC1 cytidine deaminase, Cas9(D10A) nickase, and uracil glycosylase inhibitor (UGI) from *B. subtilis* bacteriophage PBS1. UGI prevents base excision repair (BER) by inhibiting uracil N-glycosylase (UNG) [102], thereby preserving editing efficiency. The Cas9(D10A) nickase in CBE nicks the non-edited strand, recruiting the DNA repair machinery and promoting retention of the installed mutation. CBEs have an editing window of approximately 5 bp, located distal to PAM [101].

Further improvements to the linker length and the addition of a second UGI domain at the C-terminus reduced the indel frequency and gave rise to BE4 architecture [103]. Cas9 with varying PAM specificities, such as *S. pyogenes* (NGG) and *S. aureus* (NNGRRT) [104] are effective with BE3 and BE4 architectures [103, 105].

Protein evolution of *E. coli* tRNA-ARG adenine deaminase (TadA) to acquire DNA specificity enabled the creation of adenine BEs (ABE) for A*-T to I*-C (G*-C) conversions [106]. ABE7.10, one of the most widely used ABEs, features an editing window of approximately 5 bp distal to the PAM. Its architecture includes a wild-type TadA fused to a recombinant effector TadA* and a Cas9(D10A) nickase. This configuration compensates for the native TadA homodimer structure [107]. The recombinant TadA* subunit is primarily responsible for ABE's catalytic activity and contains multiple mutations (W23R, H36L, P48A, R51L, L84F, A106V, D108N, H123Y, S146C, D147Y, R152P, E155V, I156F, K157N) that enhance its efficiency and broaden its editing window.

ABE7.10 was further optimized to expand its editing window, encompassing up to 10 bp of the protospacer region distal to the PAM, leading to the development of the ABE8 family [108]. One notable variant, ABE8.20-d, incorporates additional mutations (I76Y, V82S, Y123H, Y147R, Q154R) relative to ABE7.10. The substitution V106W, which is known to mitigate mRNA modification levels [109], was also incorporated into ABE8.20-d. While editing activity is similar between dimer and protomer architectures, dimer-based variants such as ABE8.20-d are preferred for in vitro disease modeling due to the occurrence of intermolecular dimerization in protomers [110, 111]. TadA-8e ABE (ABE8e) is an alternative to ABE8.20-d and includes modifications (A109S, T111R, D119N, H122N, Y147D, F149Y, T166I and D167N) on the TadA-7.10 architecture [112]. Cryo-EM studies of ABE8e have elucidated the interaction mechanism between TadA domains and the R-loop structure [111].

To address the relatively large size of APOBEC1 in CBE systems, alternative CBEs using smaller deoxycytidine deaminases have been developed. Phage-assisted protein evolution campaign created CBEs from recombinant TadA8e. The resulting family of TadA-CBE is capable of targeting deoxycytidine instead of deoxyadenosine [113]. These developments also resulted in TadA dual base editors (DE) [113]. The most widely used CBE in this family, TadA-CBEd, features mutations (E27K, V28A, M61I, H96N and V106W [109]) on the TadA-8e variant within a BE4max architecture [114] (TadA* - Cas9(D10A) - UGI - UGI).

Base editing events are subject to bystander effects, where all the bases matching the editor's specificity within the functional window may be converted [101, 115]. Despite this limitation, BEs are well-suited for the precise corrections required in disease modeling, especially since many genetic diseases are caused by SNP changes [116].

In the context of PD, base editors allow installing disease-causing mutations or gene-correcting changes without introducing double-strand breaks. Convergent editing with *Sp*Cas9 HDR and ABE facilitated the correction of *LRRK2* p.G2019S in PD patient-derived iPS cells [117]. This study demonstrated that LRRK2 mutants exhibited decreased neurite length compared to isogenic control dopaminergic neurons, consistent with observations in patient-derived neurons [118]. This editing approach utilized a relaxed PAM *Sp*Cas9 derivative, referred to as *Sp*Cas9-NG [119], within an ABE-BE4max architecture [114, 120].

3.2.5 CRISPR-Cas Prime Editors (PE) in PD Modeling

Beyond the transitions (C*-G to T*-A and A*-T to G*-C) enabled by BE systems, certain pathogenic mutations require transversions, non-stochastic insertions and or deletions for accurate disease modeling or therapeutic applications. Pioneering protein engineering studies that optimized the fusion of a nickase with reverse transcriptase (RT) gave rise to PE, enabling the installation of virtually any small mutation into a target genome [121].

The PE2 system consists of the nickase *Sp*Cas9(H840A) fused to a Moloney-Murine Leukemia virus (M-MLV) RT containing specific mutations (D200N, T306K, W313F, T330P, L603P) that enhance its activity. PE2 uses a 3′-elongated prime-editing guide RNA (pegRNA) that incorporates the conventional sgRNA structure, followed by the desired mutation, and a homology stretch known as the primer binding site (PBS). This homology stretch allows the 3′ flap from the protospacer to serve as a primer for the RT domain, facilitating the installation of the desired mutation into the genome.

To increase the efficiency of PE2, an additional nickase is co-expressed to nick the non-edited strand, a strategy known as PE3. The co-expression of PE2 or PE3 with a dominant negative DNA mismatch repair protein MLH1 (MLH1dn), is referred to as PE4 and PE5, respectively [122]. These strategies circumvent the suppression by DNA mismatch repair (MMR) machinery on PE. Further improvements, including the addition of an NLS linker, codon optimization of the RT domain, and point mutations in the nickase Cas9(H840A + R221K, N394K) resulted in the PEmax system, which demonstrated enhanced editing efficiency [122]. Remarkably, removing the RNaseH domain from RT in PEmax preserves editing efficiency in vivo [123] and enables the generation of more compact PE proteins compatible with AAV delivery systems.

Efforts to reduce the molecular weight of PE systems focused on optimizing the RT domains. Phage-assisted protein evolution identified active mutants from the *E. coli* Ec48 retron RT [124], and *S. pombe* Tf1 retrotransposon RT [125, 126]. These compact PEs, referred to as the PE6 family, have smaller sizes that allow for in vivo modifications using AAV delivery systems [126].

Besides editing efficiency, split protein variants are another trend in PE-based gene editing. Split PEs enable the packaging of these gene editing tools into delivery systems such as AAV (4.7kbp cargo capacity) [127]. Untethered PE variants have been developed using intein-flanked *Sp*Cas9 and RT domains [128], or independent RT domains [128, 129], as well as the orthologs *Sa*Cas9, *Cj*Cas9, and *Lb*Cas12a [128, 130, 131]. These designs allow efficient packaging into AAVs.

To enhance the recruitment of RT domains to editing complexes, trans-recruitment strategies using MS2 bacteriophage coat proteins (MCP) fusions have been developed. Indeed, pegRNAs linked to MS2-aptamers efficiently recruit MCP-fused RT domains, creating an architecture known as prime editing template RNA (petRNA) [129].

The PE gene modification approach is increasingly being adopted by the PD modeling community. Isogenic ESC lines created using the PE3 editing strategy include *SNCA* p.A53T, *SNCA* p.A30P, and *LRRK2* p.G2019S [54]. Similarly, the stable expression of PE2 from a safe harbor *locus* enables efficient editing of human iPS cells and to install PD-linked mutations, including *SNCA* p.A30P, *SNCA* p.A53T, and *LRRK2* p.G2019S [132]. Editing efficiencies observed with stable PE2 expression in human iPS cell lines are consistent with those achieved using alternative delivery methods such as RNPs or mRNA.

3.2.6 CRISPRi, CRISPRa, CRISPRoff, and CRISPRon in PD Modeling

Besides the permanent genetic modifications introduced by nucleases, BE and PE systems, there are instances when reversible changes in the gene expression or the epigenetic landscape are preferred for disease modeling. These approaches, which were previously available in other formats as zinc finger and TALE [28, 133], can generally be categorized into interference (including repression), activation, and epigenetic modification.

The inactive variants of Cas9, such as dCas9(D10A, N863A) or dCas9(D10A, H840A), can be programmed to interfere with gene expression through transcriptional collision, an approach commonly referred to as CRISPR-Cas interference (CRISPRi) [134]. For stronger inhibition, dCas9 can be fused to the KRAB (Krüppel-associated box) domain of ZNF10 or ZIM3, which suppresses transcription at the target location [135–138]. The repressive effect exerted by KRAB is stronger than that of other domains, such as chromo shadow (CS) domain of chromobox protein homolog 5 or the WRPW domain of transcriptional factor HES-1.

Complementary to CRISPRi, CRISPR-Cas activators (CRISPRa) can upregulate gene expression by fusing dCas9(D10A, H840A) to activation domains such as VP64, p65 [135, 139], or by using tripartite fusion activators like VP64-p65-Rta (VPR) [140, 141]. Activation can also be achieved through the reversible interaction of an MS2-aptamer in the sgRNA with MCP-fusion proteins [139]. This method leverages the outward orientation of sgRNA tetraloop and the stem-loop 2 [72]. Commonly used MCP-fusion transactivation domains include the NF-κB transactivating subunit p65, and the heat-shock factor 1 [142].

Epigenetic modulation at target *loci* can be achieved using dCas9 fused with DNA methyltransferases or methylcytosine dioxygenases. CRISPR-based programmable epigenome editors for methylation are termed CRISPRoff [143]. A commonly used CRISPRoff architecture includes dCas9 fused to the KRAB domain of ZNF10, DNA-methyltransferase 3A (DNMT3A), and DNA-methyltransferase 3 L

(DNMT3L) [143, 144]. Conversely, programmable demethylation editors, often referred to as CRISPRon [143], use dCas9 fused to the TET1 catalytic domain, scFv-guided TET1 domains [145], or MCP-TET1 fusion proteins [144, 146]. These fusion proteins can be recruited to the target region via sgRNA hybrids containing MS2-aptamers, as described for CRISPRa. CRISPRon has been successfully employed to activate a variety of genes in human cell systems [143].

CRISPRi has successfully been used to inhibit the expression of *SNCA* in triplication cell models [147]. This approach utilized the dCas9 ortholog *Sa*Cas9 fused to methyl CpG binding protein 2 (MeCp2) and KRAB [148]. Similarly, CRISPRoff employing *Sp*Cas9 fused to DNMT3A has been used to repress *SNCA* in iPS cell-derived dopaminergic neurons [149]. Hypermethylation of *SNCA* intron 1 using CRISPRoff has been demonstrated to effectively downregulate *SNCA* expression.

3.2.7 Transposon Systems in PD Modeling

The piggyBac and tagAlong transposon systems were initially identified as mobile genetic elements transferred from *Trichoplusia ni* cell lines to baculovirus in vitro [150, 151]. The piggyBac transposase (PBase) protein and its mechanism of nuclear targeting were characterized in the early 2000s [152], and its capacity to edit human cells was validated in vitro [153]. The structure of PBase, along with its insertion and excision mechanisms, has been extensively characterized [154]. Hyperactive PBase variants with enhanced activity were developed and validated through the creation of transgenic ESC and iPS cells [155]. To address the potential detrimental effects of reinsertion following excision, PBase variants lacking the capacity to reintegrate were engineered. This PBase variant is referred as excision-only PBase [156].

PiggyBac transposons have been utilized to create rapid inclusionopathy iPS cell models for studying α-Synuclein phenotypes. Transposon ITR-flanked cargo containing wild-type or mutant *SNCA* variants can be integrated and induced using a TetON system, followed by differentiation into NGN2-positive layer II/III cortical glutamatergic neurons and SOX9-positive astrocytes. This system enabled the identification of neuroprotective p62-positive inclusions as well as neurotoxic lipid-rich inclusions [157].

3.3 Cell Replacement Therapies for the Treatment of PD and Synergies with Gene Editing

The differentiation of dopaminergic neurons from human ESCs and iPS cells represents one of the most promising avenues for regenerative medicine-based cell replacement therapies today [158, 159]. A completed phase 1 clinical trial

(NCT04802733) on ESC-based cell therapy for advanced PD [160], and an active clinical trial (JMA-IIA00384, UMIN000033564) using iPS-cell-derived dopaminergic progenitors for PD [161], exemplify these ambitious and encouraging efforts. These developments are grounded in a deep understanding of nigrostriatal graft integration, informed by pioneering in vivo findings from 1979 [162, 163] and extensive clinical research involving fetal mesencephalic tissue transplantation into PD patients [164, 165].

Studies have shown that grafts survive and innervate the striatum upon transplantation [165], providing clinical benefits to PD patients [164]. However, a double-blind controlled trial of bilateral fetal-nigral transplantation demonstrated significant effects only in patients with mild disease at baseline [166]. Long-term analyses of embryonic nigral transplants, conducted 11–16 years post-transplantation [167–169], revealed that PD pathology can affect the graft, evidenced by the presence of Lewy body inclusions in grafted cells [167]. These findings were confirmed using thioflavin-S staining and electron microscopy [169], supporting the hypothesis of host-to-graft disease propagation [168]. This phenomenon could be circumvented by using *LRP1* [83] or *SNCA* [92] knock-out cell sources, and supported by gene editing tools. Remarkably, these studies also demonstrate that transplanted cells can survive and remain functional for up to 16 years in the host tissue. The allogenic nature of transplanted cells raises the possibility that immune matching could further enhance these outcomes.

The results of fetal-nigral transplantation have been recapitulated in rat models using human iPS cell-derived dopaminergic neurons [170], providing additional encouragement to the regenerative medicine field within PD. The functional integration of iPS cell-derived dopaminergic neurons has also been validated in vivo using the non-human primate *M. fascicularis* PD model [171]. Differentiation protocols to generate highly pure dopaminergic progenitor populations from human iPS cells are well-established [161]. These differentiated populations have been confirmed to be free of genetic abnormalities and residual undifferentiated cells, enabled by robust purification procedures based on the floor plate marker CORIN [172].

HLA immune matching has been shown to be beneficial for iPS cell-derived dopaminergic neuron transplantation in *M. fascicularis* in vivo models [173] and for solid tissue transplantation in general, including cardiac sheet transplantation supplemented with immune suppressants [174]. Accessing HLA-homozygous human iPS cell cohorts would enable the generation of dopaminergic neurons with broad population coverage. However, since NK cells are responsive to HLA-C presence and diversity [175], HLA-C is expected to add complexity to the immune matching process in populations with high HLA-C allele diversity. Gene editing of histocompatibility genes offers a potential solution to improve immune compatibility for cell replacement therapies, providing an alternative to unmodified cell sources.

Recent applications of gene editing for immune modulation include the knock-out of *B2M* [176–178]; combined knock-out of *B2M* and knock-in of *HLA-E*;

knock-out of *HLA-A*, *HLA-B,* and *CIITA* [179]; knock-out of HLA-A and HLA-B [180]; combined knock-out of *HLA-A*, *HLA-B*, *HLA-C,* and *CIITA* with concurrent expression of *PD-L1*, *HLA-G,* and *CD47*; knock-out of *B2M*, *TAP1*, *CD74*, *CIITA*, *MICA,* and *MICB* [181, 182]; and transgenic expression of *CD47* [178]. These interventions enable allogenic human iPS cells to be recognized as HLA-matched cells and are commonly referred to as hypo-immunogenic, *"universal"* pluripotent stem cells, or *"superdonor"* pluripotent stem cells. Genome-engineered human iPS cells are thus positioned to support the advancement of cell replacement therapies for PD and other regenerative medicine applications.

3.4 Conclusions

In this chapter, we explored the diverse gene-editing tools implemented for disease modeling in the context of PD, primarily utilizing human iPS cells and their derivatives. From the wide array of tools available, including ZFN, TALEN, CRISPR-Cas12, ABEs, CBEs, and PEs, the vast majority of editing interventions in the PD field are carried out using CRISPR-*Sp*Cas9 nucleases. This predominance likely stems from the simplicity and accessibility of its molecular biology, such as straightforward sgRNA cloning, ease of design, and the predictability of editing outcomes.

While the relatively delayed adoption of frontier gene-editing tools in PD research may be attributed to the rapid pace of tool development, outpacing their integration into existing workflows, this trend is gradually shifting. Importantly, advancements in gene editing have consistently demonstrated that engineered human iPS cells can be produced free from off-target or undesirable mutations, as validated across various editing platforms [183]. This reliability highlights the utility of gene-editing technologies in creating precise and reproducible models for PD research.

We anticipate that the greatest synergies between gene editing and regenerative medicine, particularly in midbrain neuroscience, will arise from advances in immune compatibility modulation of human iPS cell systems. As described in this chapter, immune compatibility engineering holds the potential to overcome barriers to allogenic cell transplantation and to accelerate regenerative medicine applications through *off-the-shelf* modalities. These emerging approaches promise to not only refine our understanding of PD pathophysiology but also to catalyze the development of transformative prototypes and therapeutic interventions.

Acknowledgements This work was conducted as part of the execution of project "Mission-driven Implementation of Science and Innovation Program" 02-002-P-0001, funded by the Economic Revitalization and Resilience Enhancement Plan "New Generation Lithuania" supporting the Arias Lab.

References

1. Chen Y, Sakurai K, Maeda S et al (2021) Integrated collection of stem cell bank data, a data portal for standardized stem cell information. Stem Cell Reports 16:997–1005
2. Kurtz A, Mah N, Chen Y et al (2022) Human pluripotent stem cell registry: operations, role and current directions. Cell Prolif 55:e13238
3. Park I-H, Arora N, Huo H et al (2008) Disease-specific induced pluripotent stem cells. Cell 134:877–886
4. Soldner F, Hockemeyer D, Beard C et al (2009) Parkinson's disease patient-derived induced pluripotent stem cells free of viral reprogramming factors. Cell 136:964–977
5. Antony PMA, Diederich NJ, Krüger R et al (2013) The hallmarks of Parkinson's disease. FEBS J 280:5981–5993
6. Devine MJ, Ryten M, Vodicka P et al (2011) Parkinson's disease induced pluripotent stem cells with triplication of the α-synuclein locus. Nat Commun 2:440
7. Soldner F, Laganière J, Cheng AW et al (2011) Generation of isogenic pluripotent stem cells differing exclusively at two early onset Parkinson point mutations. Cell 146:318–331
8. Ryan SD, Dolatabadi N, Chan SF et al (2013) Isogenic human iPSC Parkinson's model shows nitrosative stress-induced dysfunction in MEF2-PGC1α transcription. Cell 155:1351–1364
9. Dettmer U, Newman AJ, Soldner F et al (2015) Parkinson-causing α-synuclein missense mutations shift native tetramers to monomers as a mechanism for disease initiation. Nat Commun 6:7314
10. Seibler P, Graziotto J, Jeong H et al (2011) Mitochondrial Parkin recruitment is impaired in neurons derived from mutant PINK1 induced pluripotent stem cells. J Neurosci 31:5970–5976
11. Cooper O, Seo H, Andrabi S et al (2012) Pharmacological rescue of mitochondrial deficits in iPSC-derived neural cells from patients with familial Parkinson's disease. Sci Transl Med 4:141ra90. https://doi.org/10.1126/scitranslmed.3003985
12. Miller JD, Ganat YM, Kishinevsky S et al (2013) Human iPSC-based modeling of late-onset disease via Progerin-induced aging. Cell Stem Cell 13:691–705
13. Nguyen HN, Byers B, Cord B et al (2011) LRRK2 mutant iPSC-derived DA neurons demonstrate increased susceptibility to oxidative stress. Cell Stem Cell 8:267–280
14. Liu G-H, Qu J, Suzuki K et al (2012) Progressive degeneration of human neural stem cells caused by pathogenic LRRK2. Nature 491:603–607
15. Reinhardt P, Schmid B, Burbulla LF et al (2013) Genetic correction of a LRRK2 mutation in human iPSCs links parkinsonian neurodegeneration to ERK-dependent changes in gene expression. Cell Stem Cell 12:354–367
16. Sánchez-Danés A, Richaud-Patin Y, Carballo-Carbajal I et al (2012) Disease-specific phenotypes in dopamine neurons from human iPS-based models of genetic and sporadic Parkinson's disease. EMBO Mol Med 4:380–395
17. Fernández-Santiago R, Carballo-Carbajal I, Castellano G et al (2015) Aberrant epigenome in iPSC-derived dopaminergic neurons from Parkinson's disease patients. EMBO Mol Med 7:1529–1546
18. Imaizumi Y, Okada Y, Akamatsu W et al (2012) Mitochondrial dysfunction associated with increased oxidative stress and α-synuclein accumulation in PARK2 iPSC-derived neurons and postmortem brain tissue. Mol Brain 5:35
19. Jiang H, Ren Y, Yuen EY et al (2012) Parkin controls dopamine utilization in human midbrain dopaminergic neurons derived from induced pluripotent stem cells. Nat Commun 3:668
20. Ren Y, Jiang H, Hu Z et al (2015) Parkin mutations reduce the complexity of neuronal processes in iPSC-derived human neurons. Stem Cells 33:68–78
21. Shaltouki A, Sivapatham R, Pei Y et al (2015) Mitochondrial alterations by PARKIN in dopaminergic neurons using PARK2 patient-specific and PARK2 knockout isogenic iPSC lines. Stem Cell Reports. 4:847–859

22. Schöndorf DC, Aureli M, McAllister FE et al (2014) iPSC-derived neurons from GBA1-associated Parkinson's disease patients show autophagic defects and impaired calcium homeostasis. Nat Commun 5:4028
23. Zampieri S, Cattarossi S, Bembi B et al (2017) GBA analysis in next-generation era. J Mol Diagn 19:733–741
24. Woodard CM, Campos BA, Kuo S-H et al (2014) iPSC-derived dopamine neurons reveal differences between monozygotic twins discordant for Parkinson's disease. Cell Rep 9:1173–1182
25. Pantazis CB, Yang A, Lara E et al (2022) A reference human induced pluripotent stem cell line for large-scale collaborative studies. Cell Stem Cell 29:1685–1702.e22
26. Ginsberg AM, King BO, Roeder RG (1984) Xenopus 5S gene transcription factor, TFIIIA: characterization of a cDNA clone and measurement of RNA levels throughout development. Cell 39:479–489
27. Miller J, McLachlan AD, Klug A (1985) Repetitive zinc-binding domains in the protein transcription factor IIIA from Xenopus oocytes. EMBO J 4:1609–1614
28. Choo Y, Sánchez-García I, Klug A (1994) In vivo repression by a site-specific DNA-binding protein designed against an oncogenic sequence. Nature 372:642–645
29. Fairall L, Schwabe JWR, Chapman L et al (1993) The crystal structure of a two zinc-finger peptide reveals an extension to the rules for zinc-finger/DNA recognition. Nature 366:483–487
30. Pavletich NP, Pabo CO (1991) Zinc finger-DNA recognition: crystal structure of a Zif268-DNA complex at 2.1 Å. Science 252:809–817
31. Li L, Wu LP, Chandrasegaran S (1992) Functional domains in Fok I restriction endonuclease. Proc Natl Acad Sci USA 89:4275–4279
32. Kim YG, Cha J, Chandrasegaran S (1996) Hybrid restriction enzymes: zinc finger fusions to Fok I cleavage domain. Proc Natl Acad Sci USA 93:1156–1160
33. Porteus MH, Baltimore D (2003) Chimeric nucleases stimulate gene targeting in human cells. Science 300:763–763
34. Bibikova M, Beumer K, Trautman JK et al (2003) Enhancing gene targeting with designed zinc finger nucleases. Science 300:764–764
35. Miller JC, Holmes MC, Wang J et al (2007) An improved zinc-finger nuclease architecture for highly specific genome editing. Nat Biotechnol 25:778–785
36. Szczepek M, Brondani V, Büchel J et al (2007) Structure-based redesign of the dimerization interface reduces the toxicity of zinc-finger nucleases. Nat Biotechnol 25:786–793
37. Doyon Y, Vo TD, Mendel MC et al (2011) Enhancing zinc-finger-nuclease activity with improved obligate heterodimeric architectures. Nat Methods 8:74–79
38. Bonas U, Stall RE, Staskawicz B (1989) Genetic and structural characterization of the avirulence gene avrBs3 from Xanthomonas campestris pv. vesicatoria. Mol Gen Genet 218:127–136
39. Hopkins CM (1992) Identification of a family of Avirulence genes from *Xanthomonas oryzae* pv. *oryzae*. MPMI 5:451
40. Boch J, Bonas U (2010) *Xanthomonas* AvrBs3 family-type III effectors: discovery and function. Annu Rev Phytopathol 48:419–436
41. Yang B, Zhu W, Johnson LB et al (2000) The virulence factor AvrXa7 of *Xanthomonas oryzae* pv. *oryzae* is a type III secretion pathway-dependent nuclear-localized double-stranded DNA-binding protein. Proc Natl Acad Sci USA 97:9807–9812
42. Kay S, Hahn S, Marois E et al (2007) A bacterial effector acts as a plant transcription factor and induces a cell size regulator. Science 318:648–651
43. Römer P, Hahn S, Jordan T et al (2007) Plant pathogen recognition mediated by promoter activation of the pepper *Bs3* resistance gene. Science 318:645–648
44. Mak AN-S, Bradley P, Cernadas RA et al (2012) The crystal structure of TAL effector PthXo1 bound to its DNA target. Science 335:716–719
45. Deng D, Yan C, Wu J et al (2014) Revisiting the TALE repeat. Protein Cell 5:297–306

46. Gao H, Wu X, Chai J et al (2012) Crystal structure of a TALE protein reveals an extended N-terminal DNA binding region. Cell Res 22:1716–1720
47. Boch J, Scholze H, Schornack S et al (2009) Breaking the code of DNA binding specificity of TAL-type III effectors. Science 326:1509–1512
48. Miller JC, Tan S, Qiao G et al (2011) A TALE nuclease architecture for efficient genome editing. Nat Biotechnol 29:143–148
49. Christian M, Cermak T, Doyle EL et al (2010) Targeting DNA double-Strand breaks with TAL effector nucleases. Genetics 186:757–761
50. Schwinn MK, Machleidt T, Zimmerman K et al (2018) CRISPR-mediated tagging of endogenous proteins with a luminescent peptide. ACS Chem Biol 13:467–474
51. Williams D, Glasstetter LM, Jong TT et al (2024) High-throughput screening for small-molecule stabilizers of misfolded glucocerebrosidase in Gaucher disease and Parkinson's disease. Proc Natl Acad Sci USA 121:e2406009121
52. Zhu F, Gamboa M, Farruggio AP et al (2014) DICE, an efficient system for iterative genomic editing in human pluripotent stem cells. Nucleic Acids Res 42:e34–e34
53. Xia N, Fang F, Zhang P et al (2017) A knockin reporter allows purification and characterization of mDA neurons from heterogeneous populations. Cell Rep 18:2533–2546
54. Busquets O, Li H, Syed KM et al (2025) iSCORE-PD: an isogenic stem cell collection to research Parkinson's disease. Neuroscience. https://doi.org/10.1101/2024.02.12.579917
55. Patterson AG, Yevstigneyeva MS, Fineran PC (2017) Regulation of CRISPR–Cas adaptive immune systems. Curr Opin Microbiol 37:1–7
56. Bolotin A, Quinquis B, Sorokin A et al (2005) Clustered regularly interspaced short palindrome repeats (CRISPRs) have spacers of extrachromosomal origin. Microbiology 151:2551–2561
57. Makarova KS, Grishin NV, Shabalina SA et al (2006) A putative RNA-interference-based immune system in prokaryotes: computational analysis of the predicted enzymatic machinery, functional analogies with eukaryotic RNAi, and hypothetical mechanisms of action. Biol Direct 1:7
58. Barrangou R, Fremaux C, Deveau H et al (2007) CRISPR provides acquired resistance against viruses in prokaryotes. Science 315:1709–1712
59. Makarova KS, Wolf YI, Iranzo J et al (2020) Evolutionary classification of CRISPR–Cas systems: a burst of class 2 and derived variants. Nat Rev Microbiol 18:67–83
60. Zetsche B, Gootenberg JS, Abudayyeh OO et al (2015) Cpf1 is a single RNA-guided endonuclease of a class 2 CRISPR-Cas system. Cell 163:759–771
61. Zetsche B, Heidenreich M, Mohanraju P et al (2017) Multiplex gene editing by CRISPR–Cpf1 using a single crRNA array. Nat Biotechnol 35:31–34
62. Deltcheva E, Chylinski K, Sharma CM et al (2011) CRISPR RNA maturation by trans-encoded small RNA and host factor RNase III. Nature 471:602–607
63. Gasiunas G, Barrangou R, Horvath P et al (2012) Cas9–crRNA ribonucleoprotein complex mediates specific DNA cleavage for adaptive immunity in bacteria. Proc Natl Acad Sci USA 109:E2579–E2586. https://doi.org/10.1073/pnas.1208507109
64. Jinek M, Chylinski K, Fonfara I et al (2012) A programmable dual-RNA–guided DNA endonuclease in adaptive bacterial immunity. Science 337:816–821
65. Knott GJ, Doudna JA (2018) CRISPR-Cas guides the future of genetic engineering. Science 361:866–869
66. Mali P, Yang L, Esvelt KM et al (2013) RNA-guided human genome engineering via Cas9. Science 339:823–826
67. Jinek M, East A, Cheng A et al (2013) RNA-programmed genome editing in human cells. elife 2:e00471
68. Cong L, Ran FA, Cox D et al (2013) Multiplex genome engineering using CRISPR/Cas systems. Science 339:819–823
69. Sternberg SH, Redding S, Jinek M et al (2014) DNA interrogation by the CRISPR RNA-guided endonuclease Cas9. Nature 507:62–67

70. Szczelkun MD, Tikhomirova MS, Sinkunas T et al (2014) Direct observation of R-loop formation by single RNA-guided Cas9 and Cascade effector complexes. Proc Natl Acad Sci USA 111:9798–9803
71. Sternberg SH, LaFrance B, Kaplan M et al (2015) Conformational control of DNA target cleavage by CRISPR–Cas9. Nature 527:110–113
72. Nishimasu H, Ran FA, Hsu PD et al (2014) Crystal structure of Cas9 in complex with guide RNA and target DNA. Cell 156:935–949
73. Kleinstiver BP, Pattanayak V, Prew MS et al (2016) High-fidelity CRISPR–Cas9 nucleases with no detectable genome-wide off-target effects. Nature 529:490–495
74. Calatayud C, Carola G, Fernández-Carasa I et al (2019) CRISPR/Cas9-mediated generation of a tyrosine hydroxylase reporter iPSC line for live imaging and isolation of dopaminergic neurons. Sci Rep 9:6811
75. Ahfeldt T, Ordureau A, Bell C et al (2020) Pathogenic pathways in early-onset autosomal recessive Parkinson's disease discovered using isogenic human dopaminergic neurons. Stem Cell Reports 14:75–90
76. Überbacher C, Obergasteiger J, Volta M et al (2019) Application of CRISPR/Cas9 editing and digital droplet PCR in human iPSCs to generate novel knock-in reporter lines to visualize dopaminergic neurons. Stem Cell Res 41:101656
77. Miyaoka Y, Chan AH, Judge LM et al (2014) Isolation of single-base genome-edited human iPS cells without antibiotic selection. Nat Methods 11:291–293
78. Yokota M, Kakuta S, Shiga T et al (2021) Establishment of an in vitro model for analyzing mitochondrial ultrastructure in PRKN-mutated patient iPSC-derived dopaminergic neurons. Mol Brain 14:58
79. Cali CP, Patino M, Tai YK et al (2019) C9orf72 intermediate repeats are associated with corticobasal degeneration, increased C9orf72 expression and disruption of autophagy. Acta Neuropathol 138:795–811
80. Pavlou S, Foskolou S, Patikas N et al (2023) CRISPR-Cas9 genetic screen leads to the discovery of L-Moses, a KAT2B inhibitor that attenuates Tunicamycin-mediated neuronal cell death. Sci Rep 13:3934
81. Arias-Fuenzalida J, Jarazo J, Walter J et al (2019) Automated high-throughput high-content autophagy and mitophagy analysis platform. Sci Rep 9:9455
82. Jarazo J, Barmpa K, Modamio J et al (2022) Parkinson's disease phenotypes in patient neuronal cultures and brain organoids improved by 2-Hydroxypropyl-β-Cyclodextrin treatment. Mov Disord 37:80–94
83. Chen K, Martens YA, Meneses A et al (2022) LRP1 is a neuronal receptor for α-synuclein uptake and spread. Mol Neurodegener 17:57
84. Olgiati S, Quadri M, Fang M et al (2016) DNAJC 6 Mutations Associated with Early- Onset Parkinson's disease. Ann Neurol 79:244–256
85. Wulansari N, Darsono WHW, Woo H-J et al (2021) Neurodevelopmental defects and neurodegenerative phenotypes in human brain organoids carrying Parkinson's disease-linked *DNAJC6* mutations. Sci Adv 7:eabb1540
86. Mamais A, Sanyal A, Fajfer A et al (2024) The LRRK2 kinase substrates RAB8a and RAB10 contribute complementary but distinct disease-relevant phenotypes in human neurons. Stem Cell Reports 19:163–173
87. Wurster I, Quadalti C, Rossi M et al (2022) Linking the phenotype of SNCA triplication with PET-MRI imaging pattern and alpha-synuclein CSF seeding. NPJ Parkinsons Dis 8:117
88. Iannielli A, Luoni M, Giannelli SG et al (2022) Modeling native and seeded Synuclein aggregation and related cellular dysfunctions in dopaminergic neurons derived by a new set of isogenic iPSC lines with SNCA multiplications. Cell Death Dis 13:881
89. Inoue S, Nishimura K, Gima S et al (2023) CRISPR-Cas9-edited *SNCA* knockout human induced pluripotent stem cell-derived dopaminergic neurons and their vulnerability to neurotoxicity. Biol Pharm Bull 46:517–522

90. Dauer W, Kholodilov N, Vila M et al (2002) Resistance of α-synuclein null mice to the parkinsonian neurotoxin MPTP. Proc Natl Acad Sci USA 99:14524–14529
91. Klivenyi P, Siwek D, Gardian G et al (2006) Mice lacking alpha-synuclein are resistant to mitochondrial toxins. Neurobiol Dis 21:541–548
92. Chen Y, Dolt KS, Kriek M et al (2019) Engineering synucleinopathy-resistant human dopaminergic neurons by CRISPR-mediated deletion of the SNCA gene. Eur J Neurosci 49:510–524
93. Soldner F, Stelzer Y, Shivalila CS et al (2016) Parkinson-associated risk variant in distal enhancer of α-synuclein modulates target gene expression. Nature 533:95–99
94. Qing X, Walter J, Jarazo J et al (2017) CRISPR/Cas9 and piggyBac-mediated footprint-free LRRK2-G2019S knock-in reveals neuronal complexity phenotypes and α-Synuclein modulation in dopaminergic neurons. Stem Cell Res 24:44–50
95. Walter J, Bolognin S, Antony PMA et al (2019) Neural stem cells of Parkinson's disease patients exhibit aberrant mitochondrial morphology and functionality. Stem Cell Reports 12:878–889
96. Ishikawa T, Imamura K, Kondo T et al (2016) Genetic and pharmacological correction of aberrant dopamine synthesis using patient iPSCs with BH4 metabolism disorders. Hum Mol Genet 25:5188–5197
97. Arias-Fuenzalida J, Jarazo J, Qing X et al (2017) FACS-assisted CRISPR-Cas9 genome editing facilitates Parkinson's disease modeling. Stem Cell Reports 9:1423–1431
98. Gomez-Giro G, Arias-Fuenzalida J, Jarazo J et al (2019) Synapse alterations precede neuronal damage and storage pathology in a human cerebral organoid model of CLN3-juvenile neuronal ceroid lipofuscinosis. Acta Neuropathol Commun 7:222
99. Barbuti P, Antony P, Santos B et al (2020) Using high-content screening to generate single-cell gene-corrected patient-derived iPS clones reveals excess alpha-Synuclein with familial Parkinson's disease point mutation A30P. Cells 9:2065
100. Chang M, Oh B, Choi J et al (2019) LIN 28A loss of function is associated with Parkinson's disease pathogenesis. EMBO J 38:e101196
101. Komor AC, Kim YB, Packer MS et al (2016) Programmable editing of a target base in genomic DNA without double-stranded DNA cleavage. Nature 533:420–424
102. Pearl LH (2000) Structure and function in the uracil-DNA glycosylase superfamily. Mutation Research/DNA Repair 460:165–181
103. Komor AC, Zhao KT, Packer MS et al (2017) Improved base excision repair inhibition and bacteriophage Mu Gam protein yields C:G-to-T:A base editors with higher efficiency and product purity. Sci Adv 3:eaao4774
104. Ran FA, Cong L, Yan WX et al (2015) In vivo genome editing using Staphylococcus aureus Cas9. Nature 520:186–191
105. Kim YB, Komor AC, Levy JM et al (2017) Increasing the genome-targeting scope and precision of base editing with engineered Cas9-cytidine deaminase fusions. Nat Biotechnol 35:371–376
106. Gaudelli NM, Komor AC, Rees HA et al (2017) Programmable base editing of a•T to G•C in genomic DNA without DNA cleavage. Nature 551:464–471
107. Kim J, Malashkevich V, Roday S et al (2006) Structural and kinetic characterization of *Escherichia coli* TadA, the wobble-specific tRNA deaminase. Biochemistry 45:6407–6416
108. Gaudelli NM, Lam DK, Rees HA et al (2020) Directed evolution of adenine base editors with increased activity and therapeutic application. Nat Biotechnol 38:892–900
109. Rees HA, Wilson C, Doman JL et al (2019) Analysis and minimization of cellular RNA editing by DNA adenine base editors. Sci Adv 5:eaax5717
110. Hua K, Tao X, Liang W et al (2020) Simplified adenine base editors improve adenine base editing efficiency in rice. Plant Biotechnol J 18:770–778
111. Lapinaite A, Knott GJ, Palumbo CM et al (2020) DNA capture by a CRISPR-Cas9–guided adenine base editor. Science 369:566–571
112. Richter MF, Zhao KT, Eton E et al (2020) Phage-assisted evolution of an adenine base editor with improved Cas domain compatibility and activity. Nat Biotechnol 38:883–891

113. Neugebauer ME, Hsu A, Arbab M et al (2023) Evolution of an adenine base editor into a small, efficient cytosine base editor with low off-target activity. Nat Biotechnol 41:673–685
114. Koblan LW, Doman JL, Wilson C et al (2018) Improving cytidine and adenine base editors by expression optimization and ancestral reconstruction. Nat Biotechnol 36:843–846
115. Wang Q, Yang J, Zhong Z et al (2021) A general theoretical framework to design base editors with reduced bystander effects. Nat Commun 12:6529
116. Landrum MJ, Lee JM, Benson M et al (2016) ClinVar: public archive of interpretations of clinically relevant variants. Nucleic Acids Res 44:D862–D868
117. Chang K-H, Huang C-Y, Ou-Yang C-H et al (2021) In vitro genome editing rescues parkinsonism phenotypes in induced pluripotent stem cells-derived dopaminergic neurons carrying LRRK2 p.G2019S mutation. Stem Cell Res Ther 12:508
118. Korecka JA, Talbot S, Osborn TM et al (2019) Neurite collapse and altered ER Ca2+ control in human Parkinson disease patient iPSC-derived neurons with LRRK2 G2019S mutation. Stem Cell Reports 12:29–41
119. Nishimasu H, Shi X, Ishiguro S et al (2018) Engineered CRISPR-Cas9 nuclease with expanded targeting space. Science 361:1259–1262
120. Huang S, Liao Z, Li X et al (2019) Developing ABEmax-NG with precise targeting and expanded editing scope to model pathogenic splice site mutations in vivo. iScience 15:640–648
121. Anzalone AV, Randolph PB, Davis JR et al (2019) Search-and-replace genome editing without double-strand breaks or donor DNA. Nature 576:149–157
122. Chen PJ, Hussmann JA, Yan J et al (2021) Enhanced prime editing systems by manipulating cellular determinants of editing outcomes. Cell 184:5635–5652.e29
123. Davis JR, Banskota S, Levy JM et al (2024) Efficient prime editing in mouse brain, liver and heart with dual AAVs. Nat Biotechnol 42:253–264
124. Millman A, Bernheim A, Stokar-Avihail A et al (2020) Bacterial retrons function in antiphage defense. Cell 183:1551–1561.e12
125. Kirshenboim N, Hayouka Z, Friedler A et al (2007) Expression and characterization of a novel reverse transcriptase of the LTR retrotransposon Tf1. Virology 366:263–276
126. Doman JL, Pandey S, Neugebauer ME et al (2023) Phage-assisted evolution and protein engineering yield compact, efficient prime editors. Cell 186:3983–4002.e26
127. Li C, Samulski RJ (2020) Engineering adeno-associated virus vectors for gene therapy. Nat Rev Genet 21:255–272
128. Grünewald J, Miller BR, Szalay RN et al (2023) Engineered CRISPR prime editors with compact, untethered reverse transcriptases. Nat Biotechnol 41:337–343
129. Liu B, Dong X, Cheng H et al (2022) A split prime editor with untethered reverse transcriptase and circular RNA template. Nat Biotechnol 40:1388–1393
130. Liang R, He Z, Zhao KT et al (2024) Prime editing using CRISPR-Cas12a and circular RNAs in human cells. Nat Biotechnol 42:1867–1875
131. Lan T, Chen H, Tang C et al (2023) Mini-PE, a prime editor with compact Cas9 and truncated reverse transcriptase. Mol Ther Nucleic Acids 33:890–897
132. Li H, Busquets O, Verma Y et al (2022) Highly efficient generation of isogenic pluripotent stem cell models using prime editing. elife 11:e79208
133. Maeder ML, Angstman JF, Richardson ME et al (2013) Targeted DNA demethylation and activation of endogenous genes using programmable TALE-TET1 fusion proteins. Nat Biotechnol 31:1137–1142
134. Qi LS, Larson MH, Gilbert LA et al (2013) Repurposing CRISPR as an RNA-guided platform for sequence-specific control of gene expression. Cell 152:1173–1183
135. Gilbert LA, Larson MH, Morsut L et al (2013) CRISPR-mediated modular RNA-guided regulation of transcription in eukaryotes. Cell 154:442–451
136. Replogle JM, Bonnar JL, Pogson AN et al (2022) Maximizing CRISPRi efficacy and accessibility with dual-sgRNA libraries and optimal effectors. elife 11:e81856

137. Alerasool N, Segal D, Lee H et al (2020) An efficient KRAB domain for CRISPRi applications in human cells. Nat Methods 17:1093–1096
138. Mandegar MA, Huebsch N, Frolov EB et al (2016) CRISPR interference efficiently induces specific and reversible gene silencing in human iPSCs. Cell Stem Cell 18:541–553
139. Konermann S, Brigham MD, Trevino AE et al (2015) Genome-scale transcriptional activation by an engineered CRISPR-Cas9 complex. Nature 517:583–588
140. Chavez A, Scheiman J, Vora S et al (2015) Highly efficient Cas9-mediated transcriptional programming. Nat Methods 12:326–328
141. Nguyen DP, Miyaoka Y, Gilbert LA et al (2016) Ligand-binding domains of nuclear receptors facilitate tight control of split CRISPR activity. Nat Commun 7:12009
142. Marinho HS, Real C, Cyrne L et al (2014) Hydrogen peroxide sensing, signaling and regulation of transcription factors. Redox Biol 2:535–562
143. Nuñez JK, Chen J, Pommier GC et al (2021) Genome-wide programmable transcriptional memory by CRISPR-based epigenome editing. Cell 184:2503–2519.e17
144. Liu XS, Wu H, Ji X et al (2016) Editing DNA methylation in the mammalian genome. Cell 167:233–247.e17
145. Morita S, Noguchi H, Horii T et al (2016) Targeted DNA demethylation in vivo using dCas9–peptide repeat and scFv–TET1 catalytic domain fusions. Nat Biotechnol 34:1060–1065
146. Xu X, Tao Y, Gao X et al (2016) A CRISPR-based approach for targeted DNA demethylation. Cell Discov 2:16009
147. Sun Z, Kantor B, Chiba-Falek O (2024) Neuronal-type-specific epigenome editing to decrease SNCA expression: implications for precision medicine in synucleinopathies. Mol Ther Nucleic Acids 35:102084
148. Sastre D, Zafar F, Torres CAM et al (2023) Inactive S. Aureus Cas9 downregulates alpha-synuclein and reduces mtDNA damage and oxidative stress levels in human stem cell model of Parkinson's disease. Sci Rep 13:17796
149. Kantor B, Tagliafierro L, Gu J et al (2018) Downregulation of SNCA expression by targeted editing of DNA methylation: a potential strategy for precision therapy in PD. Mol Ther 26:2638–2649
150. Potter JN, Faulkner P, MacKinnon EA (1976) Strain selection during serial passage of Trichoplusia in nuclear polyhedrosis virus. J Virol 18:1040–1050
151. Cary LC, Goebel M, Corsaro BG et al (1989) Transposon mutagenesis of baculoviruses: analysis of Trichoplusia ni transposon IFP2 insertions within the FP-locus of nuclear polyhedrosis viruses. Virology 172:156–169
152. Keith JH, Fraser TS, Fraser MJ (2008) Analysis of the piggyBac transposase reveals a functional nuclear targeting signal in the 94 c-terminal residues. BMC Mol Biol 9:72
153. Wilson MH, Coates CJ, George AL (2007) PiggyBac transposon-mediated gene transfer in human cells. Mol Ther 15:139–145
154. Chen Q, Luo W, Veach RA et al (2020) Structural basis of seamless excision and specific targeting by piggyBac transposase. Nat Commun 11:3446
155. Yusa K, Zhou L, Li MA et al (2011) A hyperactive *piggyBac* transposase for mammalian applications. Proc Natl Acad Sci USA 108:1531–1536
156. Li X, Burnight ER, Cooney AL et al (2013) piggyBac transposase tools for genome engineering. Proc Natl Acad Sci USA 110:E2279–E2287. https://doi.org/10.1073/pnas.1305987110
157. Lam I, Ndayisaba A, Lewis AJ et al (2024) Rapid iPSC inclusionopathy models shed light on formation, consequence, and molecular subtype of α-synuclein inclusions. Neuron 112:2886–2909.e16
158. Takahashi J (2020) iPS cell-based therapy for Parkinson's disease: a Kyoto trial. Regen Ther 13:18–22
159. Barker RA, Parmar M, Studer L et al (2017) Human trials of stem cell-derived dopamine neurons for Parkinson's disease: Dawn of a new era. Cell Stem Cell 21:569–573

160. Piao J, Zabierowski S, Dubose BN et al (2021) Preclinical efficacy and safety of a human embryonic stem cell-derived midbrain dopamine progenitor product, MSK-DA01. Cell Stem Cell 28:217–229.e7
161. Doi D, Magotani H, Kikuchi T et al (2020) Pre-clinical study of induced pluripotent stem cell-derived dopaminergic progenitor cells for Parkinson's disease. Nat Commun 11:3369
162. Björklund A, Stenevi U (1979) Reconstruction of the nigrostriatal dopamine pathway by intracerebral nigral transplants. Brain Res 177:555–560
163. Perlow MJ, Freed WJ, Hoffer BJ et al (1979) Brain grafts reduce motor abnormalities produced by destruction of Nigrostriatal dopamine system. Science 204:643–647
164. Kordower JH, Freeman TB, Chen E et al (1998) Fetal nigral grafts survive and mediate clinical benefit in a patient with Parkinson's disease. Mov Disord 13:383–393
165. Kordower JH, Freeman TB, Snow BJ et al (1995) Neuropathological evidence of graft survival and striatal Reinnervation after the transplantation of Fetal mesencephalic tissue in a patient with Parkinson's disease. N Engl J Med 332:1118–1124
166. Olanow CW, Goetz CG, Kordower JH et al (2003) A double-blind controlled trial of bilateral fetal nigral transplantation in Parkinson's disease. Ann Neurol 54:403–414
167. Kordower JH, Chu Y, Hauser RA et al (2008) Lewy body–like pathology in long-term embryonic nigral transplants in Parkinson's disease. Nat Med 14:504–506
168. Li J-Y, Englund E, Holton JL et al (2008) Lewy bodies in grafted neurons in subjects with Parkinson's disease suggest host-to-graft disease propagation. Nat Med 14:501–503
169. Li J, Englund E, Widner H et al (2010) Characterization of Lewy body pathology in 12- and 16-year-old intrastriatal mesencephalic grafts surviving in a patient with Parkinson's disease. Mov Disord 25:1091–1096
170. Cai J, Yang M, Poremsky E et al (2010) Dopaminergic neurons derived from human induced pluripotent stem cells survive and integrate into 6-OHDA-Lesioned rats. Stem Cells Dev 19:1017–1023
171. Kikuchi T, Morizane A, Doi D et al (2017) Human iPS cell-derived dopaminergic neurons function in a primate Parkinson's disease model. Nature 548:592–596
172. Doi D, Samata B, Katsukawa M et al (2014) Isolation of human induced pluripotent stem cell-derived dopaminergic progenitors by cell sorting for successful transplantation. Stem Cell Reports 2:337–350
173. Morizane A, Kikuchi T, Hayashi T et al (2017) MHC matching improves engraftment of iPSC-derived neurons in non-human primates. Nat Commun 8:385
174. Kawamura T, Miyagawa S, Fukushima S et al (2016) Cardiomyocytes Derived from MHC-homozygous induced pluripotent stem cells exhibit reduced allogeneic immunogenicity in MHC-matched non-human primates. Stem Cell Reports 6:312–320
175. Ichise H, Nagano S, Maeda T et al (2017) NK cell alloreactivity against KIR-ligand-mismatched HLA-Haploidentical tissue derived from HLA haplotype-homozygous iPSCs. Stem Cell Reports 9:853–867
176. Gaykema LH, Van Nieuwland RY, Lievers E et al (2024) T-cell mediated immune rejection of Beta-2-microglobulin knockout induced pluripotent stem cell-derived kidney organoids. Stem Cells Transl Med 13:69–82
177. Gornalusse GG, Hirata RK, Funk SE et al (2017) HLA-E-expressing pluripotent stem cells escape allogeneic responses and lysis by NK cells. Nat Biotechnol 35:765–772
178. Deuse T, Hu X, Gravina A et al (2019) Hypoimmunogenic derivatives of induced pluripotent stem cells evade immune rejection in fully immunocompetent allogeneic recipients. Nat Biotechnol 37:252–258
179. Kitano Y, Nishimura S, Kato TM et al (2022) Generation of hypoimmunogenic induced pluripotent stem cells by CRISPR-Cas9 system and detailed evaluation for clinical application. Mol Ther Methods Clin Dev 26:15–25
180. Xu H, Wang B, Ono M et al (2019) Targeted disruption of HLA genes via CRISPR-Cas9 generates iPSCs with enhanced immune compatibility. Cell Stem Cell 24:566–578.e7

181. Pizzato HA, Alonso-Guallart P, Woods J et al (2024) Engineering human pluripotent stem cell lines to evade xenogeneic transplantation barriers. Stem Cell Reports 19:299–313
182. Han X, Wang M, Duan S et al (2019) Generation of hypoimmunogenic human pluripotent stem cells. Proc Natl Acad Sci USA 116:10441–10446
183. Veres A, Gosis BS, Ding Q et al (2014) Low incidence of off-target mutations in individual CRISPR-Cas9 and TALEN targeted human stem cell clones detected by whole-genome sequencing. Cell Stem Cell 15:27–30

Chapter 4
Orchestrating Stem Cell-Bioengineered Material Interactions: Treatment and Modeling of Neurodegenerative Diseases

Javad Harati and Peng-Yuan Wang

Abstract Endogenous stem cells in the brain have a limited capacity to replace lost or damaged cells in neurodegenerative diseases such as Alzheimer's and Parkinson's. Therapeutic interventions involving cell therapy, which employ lab-cultured stem cells and their derivatives, face several challenges, including inflammatory microenvironments, poor cell survival during and after transplantation, and inadequate functional integration. Additionally, traditional drug development pathways, which rely on conventional disease models, often fail to accurately mimic the complex pathophysiological conditions of the human brain in vivo. To overcome these challenges, advanced biomaterials such as nanoparticles and hydrogels along with biofabrication technologies like 3D bioprinting have emerged as powerful tools for precisely regulating stem cell fate and behavior. Innovations in biomaterials focus on enhancing key cellular interactions with the microenvironment, including adhesion to the extracellular matrix (ECM), cell–cell communication, and signaling through soluble factors. Furthermore, the versatility of biomaterials enables the development of solutions, such as hydrogels, which serve as 3D scaffolds, providing mechanical and structural support. These materials can be engineered to interact with cells, protect transplanted cells, and help mitigate inflammatory microenvironments. In disease modeling, bioengineering approaches, such as microfluidic platforms, integrate cells or organoids with physiological stimuli, including dynamic mechanical forces, gradient-dependent signaling, and spatiotemporal control of biochemical cues, to more accurately mimic native tissue microenvironments. These systems enable high-fidelity replication of neurodegenerative hallmarks, including protein aggregation and neuronal network dysfunction, offering a more predictive platform for drug discovery. This chapter examines how advancements in bioengineering ranging from custom biomaterials to biofabricated systems are advancing

J. Harati · P.-Y. Wang (✉)
Oujiang Laboratory, Key Laboratory of Alzheimer's Disease of Zhejiang Province, Institute of Aging, Wenzhou Medical University, Wenzhou, Zhejiang, People's Republic of China
e-mail: py.wang@ojlab.ac.cn

R. C. B. Wong and Y. Hayashi (eds.), *Advances in Using Stem Cells to Treat Neural Diseases*, Stem Cell Biology and Regenerative Medicine 77,
https://doi.org/10.1007/978-3-031-94101-6_4

the translation of stem cell-based therapies and drug development into clinical applications for neurodegenerative disorders.

Keywords ECM · Bioengineered materials · Hydrogel · 3D bioprinting · Cell reprogramming · Biomaterials · Mechanotransduction · Neurodegenerative diseases · Disease modeling · Microfluidic

4.1 Bioengineered Materials Are Rewiring the Brain's Capacity to Heal Through Stem Cell Modulation

The human brain, a labyrinth of 86 billion neurons orchestrating every thought, memory, and movement, is a paradox: unparalleled in complexity yet tragically fragile. When injured, its capacity to heal is severely limited, leaving lasting impacts on our lives. Because of this limitation, despite decades of research on neurodegenerative diseases such as Alzheimer's disease (AD) and Parkinson's disease (PD), we still lack curative treatments. Current therapies, including those developed through drug discovery, primarily focus on alleviating symptoms and delaying disease progression. Similarly, in conditions such as stroke, treatments may reduce acute damage but often fall short of addressing long-term deficits. Now, a revolution is brewing at the intersection of biology and engineering. Bioengineered materials provide a versatile platform equipped with multiple toolkits, offering innovative solutions to current challenges in neurodegenerative disease therapy. These tools don't just treat disease; they aim to rebuild what has been lost [1]. Gliadel, a carmustine-loaded wafer for brain tumors, stands alone as the only biomaterial-based therapy approved for the treatment of a brain-related disease. Yet this is just the tip of the iceberg. Imagine materials that not only deliver drugs but also nudge stem cells to regenerate neurons, reawaken dormant repair pathways, or recreate disease in a dish to accelerate the discovery of new drugs.

4.1.1 Why Adult Brain Stem Cells Fail to Regenerate Brain Damage

While stem cells make quick work of regenerating skin, liver, and bone marrow, the brain's evolution is the greatest masterpiece that barely gets a second chance. Why does the organ that gives us thought, memory, and consciousness lack the ability to heal itself? It all comes down to a trade-off: evolution favored intelligence over regeneration. For mammals, this choice comes at a cost, leaving the brain vulnerable to injury and disease, with scarce stem cells, a harsh microenvironment, and the relentless toll of aging standing in the way of recovery (Table 4.1).

Table 4.1 Factors limiting neural regeneration in mammals

Cause	Effects
Energy optimization and trade-offs	Higher animals have evolved energy-efficient brains that support complex cognition but limit regenerative capacity, whereas zebrafish exhibit extensive brain regeneration despite lower cognitive complexity [2]
Developmental shift	Neurogenesis declines during late embryonic and early postnatal stages as the brain shifts focus from growth to energy-efficient synaptic maturation and connectivity
Inhibitory microenvironment	In mammals, injury induces oxidative stress, glial scarring, and sustained inflammation, with cytotoxic factors such as reactive oxygen species and TGF-β impeding regeneration, unlike the rapid resolution of inflammation seen in axolotls
Age-related decline	As in other tissues, neurogenesis in the brain decreases significantly with age, further limiting adult regenerative potential

Additionally, neurogenesis is confined to a specific region of the brain. In the adult mammalian brain, neural stem cells (NSCs) reside in discrete neurogenic niches: the subventricular zone (SVZ) lining the lateral ventricles and the subgranular zone (SGZ) of the hippocampus. These niches maintain a delicate balance between plasticity and stability. Flooding the brain with new neurons risks chaos, akin to tossing unlabeled wires into a supercomputer. Evolution opted for strict quality control, but this leaves damaged regions stranded without repair crews.

The SGZ supports hippocampal-dependent memory and spatial learning by controlling the integration of new neurons into existing circuits. Similarly, the SVZ supplies interneurons to the olfactory bulb, enabling adaptive odor processing. Both niches feature unique ECM components (e.g., laminins, tenascins) that regulate the quiescence, activation, and differentiation of NSCs. In contrast, putative neurogenic regions, such as the hypothalamus and neocortex, lack such specialized microenvironments, which explain their limited regenerative activity [3].

4.1.2 Application of Stem Cells and Cellular Reprogramming in Brain Regeneration

If the adult brain's stem cells are too scarce, too dormant, or trapped in hostile terrain (Sect. 4.1.1), how can we circumvent the constraints of evolution? The answer lies in three bold strategies: awakening dormant NSCs, manufacturing new neurons in vitro, or rewriting the brain's cellular identity entirely. Each approach faces unique hurdles, and bioengineered materials may hold the key to their success:

- Activation of Endogenous NSCs: NSCs can be modulated through pharmacological agents and microenvironmental cues, including electrical stimulation. Bioengineered hydrogels loaded with growth factors (e.g., BDNF) or conductive scaffolds can provide localized, sustained activation while shielding NSCs from inhibitory signals.

- Pluripotent Stem Cells-derived Neurons: Induced pluripotent stem cells (iPSCs) and embryonic stem cells (ESCs) from both allogeneic and autologous sources can be differentiated into various neuronal subtypes for therapeutic transplantation. However, survival rates decline in the brain's hostile microenvironment. Biomaterial scaffolds that mimic the brain's ECM, with stiffness-matched matrices and neuroprotective cargo, could serve as "survival pods" to enhance graft integration and survival.
- Neural Direct Reprogramming: This approach involves converting non-neuronal cells, such as fibroblasts or astrocytes, directly into neurons [4, 5]. Nanoparticles can enhance the in vivo delivery of genetic factors required for this process, while synthetic niches improve control over neural fate.

Mesenchymal stem cells (MSCs), which were once overlooked in conversations about brain repair, are now gaining attention for their potential in this field. However, their classification as multipotent stem cells raises concerns about their ability to differentiate into neurons. The neuron-like properties observed in these cells may stem from stress responses rather than true neuronal differentiation [6]. The use of MSCs may occur indirectly through bioengineered scaffolds that are functionalized with chemotactic signals to attract MSCs to areas of injury. Once at these locations, MSCs can secrete neuroprotective factors that help maintain neural homeostasis [7].

Moreover, chronic diseases don't just kill neurons; they corrupt glial cells due to inflammation, turning allies into adversaries. Bioengineered materials could recalibrate this imbalance, restoring synaptic support and taming inflammation while new neurons integrate.

4.1.3 Standing on the Shoulders of Bioengineered Materials to Navigate Stem Cell Fate

Every cell, including stem cells and neurons, relies on a finely tuned interplay of biophysical and biochemical cues from its microenvironment to maintain functionality, particularly when removed from its native context. Bioengineered materials offer a flexible platform for simulating and modifying external signals under specific pathophysiological conditions, providing tools to enhance biological systems and improve therapeutic outcomes. These signals encompass not only cell–ECM interactions and direct cell–cell communications but also the action of soluble factors, including growth factors, cytokines, and chemokines. Thus, the strategic use of bioengineered materials offers significant potential for enhancing neuronal culture systems, preserving stemness, guiding differentiation, and regulating neuronal function and maturation (Fig. 4.1).

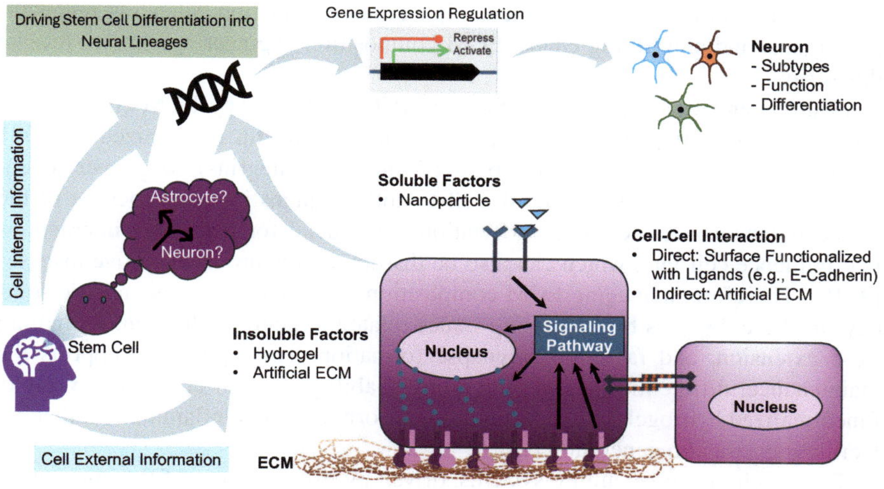

Fig. 4.1 The decision-making pathway of stem cells differentiating into neurons highlights key points where bioengineered materials can provide support and solutions. This includes the influence of soluble factors, insoluble factors, and cell–cell interactions in guiding stem cell fate

4.1.3.1 Cell–ECM Engineering

The brain's ECM is uniquely soft compared to other tissues due to its relatively low content of fibrous proteins such as collagen, fibronectin, and vitronectin. Instead, it is predominantly composed of glycosaminoglycans (GAGs, e.g., hyaluronan), proteoglycans (e.g., neurocan, brevican, versican, and aggrecan), and glycoproteins (e.g., tenascin-R). In addition, ECM composition varies across brain regions, particularly within NSC niches, such as the SVZ, where specialized structures known as "fractones" are enriched in components like collagen IV, laminin, nidogen, and heparan sulfate proteoglycans, characterized by complex, branching, fractal-like patterns [8].

By engineering the key attributes of ECMs, such as composition, structure, and mechanical properties, stem cell behavior, including the maintenance of stemness and the determination of differentiation fate, can be precisely regulated. This approach enables the investigation of specific biological processes and disease mechanisms, potentially leading to the development of improved therapeutic strategies.

4.1.3.1.1 ECM Composition Engineering

Assume the brain's ECM is a dynamic language, a molecular dialect spoken through laminin, fibronectin, and collagen. Neurons and stem cells "listen" to this language via integrin receptors, decoding instructions to survive, grow, or regenerate. But in neurodegenerative diseases like AD, this dialog turns toxic. Misfolded proteins, such as amyloid-beta (Aβ) oligomers, hijack the conversation, silencing regenerative signals and amplifying chaos. ECM composition engineering aims to rewrite this script.

Neurons and stem cells (i.e., NSCs and PSCs) rely on key ECM components, such as laminin (recognized by α6β1 integrin), fibronectin (recognized by α5β1 integrin), and collagen (recognized by α1β1 and α2β1 integrins), for proper attachment, growth, and function. Additionally, subtle changes in the properties of the extracellular matrix, such as glycosylation, are crucial for mediating interactions that influence cell fate, which can also be disrupted in neurodegenerative diseases [9, 10]. Therefore, managing ECM composition could serve as a therapeutic strategy for these diseases by addressing issues related to neuron adhesion, promoting axon extension, and facilitating synapse formation. This approach supports the maintenance, differentiation, and long-term viability of stem cells. For example, functionalized hydrogels can reduce glial scar formation and inflammation, thereby facilitating axonal regeneration [11].

The development of culture systems, driven by the discovery of the importance of ECM, led to the use of poly-L-lysine coatings for neurons and fibroblast feeder layers for pluripotent stem cells (PSCs). Subsequent methods, including bulk coatings with specific proteins (e.g., vitronectin and collagen), complex ECM mixtures (e.g., Matrigel), or decellularized ECM (dECM), have improved both stem cell and neural culture systems, thereby enhancing differentiation and network formation. Furthermore, highlighting the significance of ECM compositional changes during disease development and progression emphasizes the therapeutic potential of ECM compositional engineering [12–15]. Engineering strategies can target incorporating specific ECM components, such as GAGs [16, 17] or integrin-targeting motifs, such as KVAV, IKVAV, and HYD1 (derived from laminin), for brain tissue engineering applications. These components play a key role in guiding stem cell fate decisions, promoting the differentiation of neural progenitors into neurons while reducing the formation of astrocytes [18–20]. Alternatively, dECM leverages the inherent biochemical and structural diversity of native tissues to replicate complex, tissue-specific niches required for the maintenance of NSCs [8]. However, challenges such as batch-to-batch variability, immunogenicity, and reduced reproducibility relative to synthetic scaffolds must be addressed to facilitate clinical translation.

The composition of scaffolds can be specifically designed to address particular microenvironmental challenges. In the case of AD, Aβ oligomers function like molecular landmines, disrupting synapses and erasing memories. KLVFF-infused scaffolds serve as mine-clearance teams, binding to and neutralizing Aβ, thereby protecting neurons from hyperexcitability while also supporting neural stem cells [21–24].

4.1.3.1.2 ECM Mechanical Engineering

The biomechanical properties of the microenvironment, particularly stiffness, are closely linked to stem cell fate decisions, including whether to maintain stemness or differentiate, as well as decisions regarding neuron lineage commitment [25, 26].

Alterations in the mechanical properties of the brain have been shown to impact its regenerative capacity during development, aging, and disease [27, 28]. Accordingly, modulating this feature of the brain microenvironment may enhance or restore regenerative potential by activating NSCs [29]. Although direct in vivo evidence for modulating stiffness via bioengineered materials is limited, in vitro studies robustly demonstrate that ECM mechanics significantly affect NSC behavior and differentiation efficiency. Moreover, mechanical changes in the microenvironment can influence processes such as axon growth, synapse formation, and the maturation of differentiated neurons. Examples indicate that relative soft conditions that can mimic brain mechanical fracture promote neural fate during differentiation and reprogramming [30–32], while these softer conditions also have the potential to support neural maturation [33, 34].

4.1.3.1.3 ECM Topography and Structure Engineering

Injury disrupts the brain's architectural "roadmap," leaving neurons directionless. Bioengineered topographies restore this guidance system, offering physical cues to navigate hostile terrain. In their natural microenvironment, neurons and NSCs are constantly exposed to subtle changes, such as those in the microglial surface, which can potentially trigger distinct signaling pathways in neurons [35]. Several strategies, including surface coating, imprinting, soft lithography, particle self-assembly, and 3D printing, can be employed to engineer the roughness and topography of cell culture surfaces at the micro- and nanoscales [36–39].

These physical cues encompass various structural, mechanical, and compositional features of the microenvironment (Sects. 4.1.3.1.1 and 4.1.3.1.2), which influence cell shape, cytoskeletal tension, and adhesion. These changes activate specific signaling cascades, leading to differential gene expression and ultimately shaping cellular behavior and fate [38, 40, 41]. Moreover, compared to a flat surface, spatial patterning with roughness and topography in the microenvironment provides contact guidance cues that can direct cellular alignment and offer additional signals to steer stem cell fate toward a neuronal lineage [42]. In support of this, various modifications to cell culture substrate topography, including elongation-promoting cues like grooved substrates or aligned fibers, have been shown to enhance neural differentiation and function effectively [43–46].

4.1.3.1.4 ECM Electrical Conductivity

Neurons primarily communicate through electrical signals, known as action potentials, and chemical signaling via the release of neurotransmitters at synapses. Action potentials arise from the movement of ions, particularly sodium (Na^+) and potassium (K^+), through voltage-gated ion channels in the neuronal membrane. These signals enable the transmission of information along neurons and across synapses. Neuropathological conditions can disrupt these electrical processes, impairing neural function.

While the brain's ECM has inherently low or negligible electrical conductivity, it indirectly influences conductivity by acting as a scaffold for ion channels and regulating extracellular ion distribution. For example, sulfated GAGs such as heparan sulfate (HS), chondroitin sulfate (CS), and dermatan sulfate (DS) support mineralization and structure in bone. In the brain, they contribute to biochemical and biophysical functions, including presenting growth factors and ions to neurons and stem cells. Consequently, brain conductivity primarily depends on the presence of ions and electrolytes in the extracellular fluid [47].

Engineered ECMs with electrical conductivity can target stem cell functions through two primary mechanisms: intrinsic material conductivity and external stimulation via electrical currents. Differentiation of neural stem cells can be promoted without electrical stimulation by providing conductivity in a neural tissue range, which impacts calcium signaling and epigenetics [48]. Like a maestro conducting an orchestra, controlled calcium influx (via a control on ECM conductivity) orchestrates the differentiation of stem cells into functional neurons while silencing glial fate programs [48–51]. To achieve these effects, a diverse array of conductive bioengineered materials has been developed. These include substrates based on metals, silicon, and carbon-based compounds, as well as electrically conductive polymers and naturally derived materials, such as modified silk proteins and chitosan polysaccharides, which offer both electrical conductivity and biocompatibility [50, 52, 53]. While direct evidence on the impact of electrical conductivity manipulation on stem cell function under pathophysiological conditions remains unclear, conductive biomaterials, such as graphene-based constructs, may offer benefits in neurodegenerative conditions, including the prevention of α-synucleinopathy in PD and the promotion of stem cell differentiation [54, 55].

Electrical stimulation, which is already approved as a treatment for diseases like PD through implants, would work by activating stem cells [56, 57]. Alternative electrical stimulation approaches utilizing piezoelectric materials, such as PVDF, ZnO, and $BaTiO_3$ offer a promising non-invasive strategy for promoting neurogenesis. For instance, piezoelectric materials can be systemically introduced to the brain; subsequent ultrasound stimulation of specific brain regions generates localized electrical signals, thereby activating stem cell functions [58–61].

4.1.3.2 Cell–Cell Interactions

Injury and disease fragment the brain's cellular "conversations," leaving neurons and stem cells isolated and impairing their coordinated, collective behavior. Bioengineered materials restore these dialogs, mimicking the dynamic chemical, mechanical, and electrical signals that orchestrate collective behavior, turning chaotic microenvironments into synchronized hubs of regeneration. Neurons participate in a wide array of cell–cell interactions, including neuron–neuron, neuron–glia, neuron–endothelial, and neuron–astrocyte. Each of these interactions represents a potential target for modulation through bioengineering approaches. Table 4.2 summarizes the key cellular components involved in neural cell–cell interactions,

Table 4.2 Key cellular components involved in cell–cell interactions in the brain and their role in cell fate decisions following stimulation by signals from bioengineered materials. Feedback responses of receiver cells are not included in the schematic presentation

Key contributors	Functions	Bioengineering strategy example
Cadherins 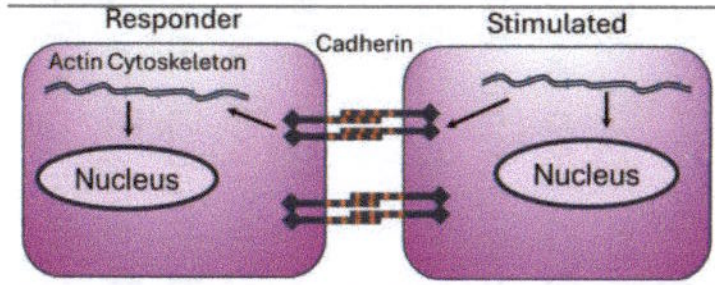	Primarily, N-cadherin plays a crucial role in neural development, synaptic organization, and maintaining the structural integrity of the nervous system. By mediating homophilic interactions between adjacent neural progenitor cells, N-cadherin facilitates the formation of cohesive cell clusters that synchronize differentiation signals and coordinate the transition of progenitors into mature neurons or glial cells [62, 63]	**HAV peptide-functionalized hydrogels/scaffolds** to mimic N-cadherin-mediated adhesion
Gap junctions 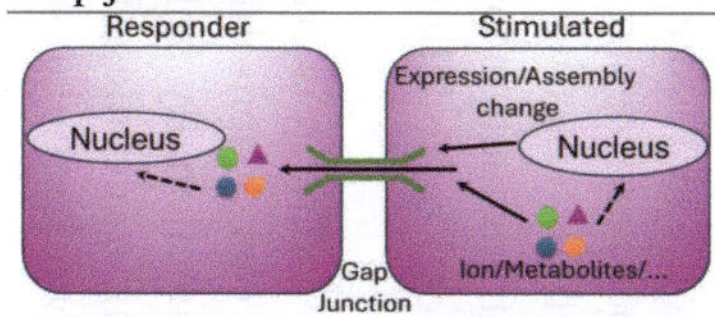	Connexins are the fundamental subunits of gap junction channels, facilitating direct electrical and metabolic coupling that underpins synchronized neural activity and helps coordinate the transition from a proliferative state to differentiation, thereby promoting collective and synchronized maturation of neural cells. This coupling supports astrocyte-neuron interactions and regulates NSC fate, thereby facilitating proper neuronal integration [64, 65]	**Connexin-mimetic nanoparticles** (e.g., C×43-functionalized scaffolds) are used to restore intercellular communication
Tight junctions, desmosomes, and ephrins 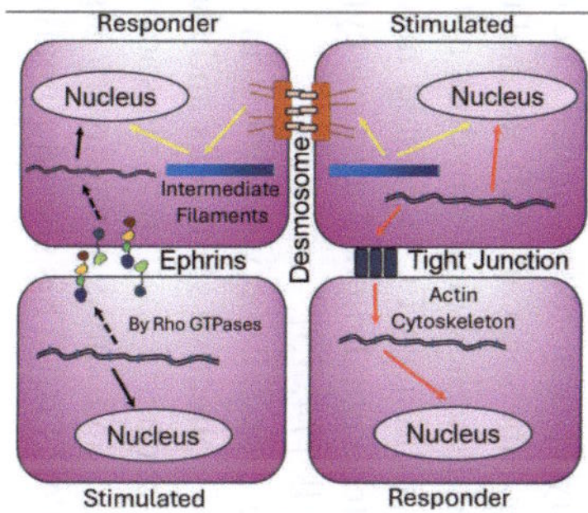	**Tight junctions** maintain apical-basal polarity in neural progenitors, a prerequisite for asymmetric cell division and fate determination. While **desmosomes** are less prominent in neural tissues, they play a crucial role in maintaining the physical cohesion of progenitor layers. Together, these junctions preserve the integrity of the NSC niche, indirectly supporting collective differentiation. **Eph receptors** and their membrane-bound ephrin ligands mediate juxtacrine signaling that regulates cell positioning, boundary formation, and migration in NSC niches. These interactions determine whether stem cells remain proliferative or begin differentiating and guide their collective migration to form organized neural circuits	Ephrin-functionalized scaffolds to guide NSC migration away from inhibitory scars, and stiffness-tunable hydrogels to regulate tight junction formation indirectly

emphasizing their ability to transduce biophysical signals either directly through cytoskeletal linkage between adjacent cells or indirectly, by engaging signaling molecules or modulating signaling pathways in neighboring cells.

Bioengineered material strategies targeting cell–cell interactions primarily involve two approaches. The first utilizes synthetic peptides, such as histidine-alanine-valine (HAV), derived from the extracellular domain of N-cadherin. These peptides emulate N-cadherin's adhesive domain and are commonly immobilized on bioengineered material surfaces, including hydrogels and scaffolds, to promote neural differentiation [66]. The second approach focuses on modifications to the ECM, particularly its mechanical and structural properties (discussed in previous sections), which affect cell–cell communication through interactions with adhesion molecules, such as cadherins. For instance, alterations in the surface roughness and stiffness of bioengineered materials have been shown to regulate spheroid formation, highlighting the significance of biophysical cues in shaping cell–cell interactions [67].

4.1.3.2.1 Cadherins

Available evidence indicates that N-cadherin functions by mechanically coupling cells through their actin cytoskeletons, promoting collective synchronization. This supports both the maintenance of undifferentiated stem cell states and the promotion of differentiation and maturation into neural cells under appropriate conditions [68–72]. Consequently, bioengineered materials that leverage N-cadherin can effectively regulate stem cell function and neural differentiation.

In one scenario, biomaterials engineered by incorporating N-cadherin have been shown to support the formation of synaptic connections with synchronized activity [71]. Similarly, substrates with engineered N-cadherin enhance NSC differentiation and regulate stem cell fate. For example, Fc-tagged recombinant N-cadherins (N-Cad-Fc), combined with gelatin-based hydrogels, enable cortical organoid cultures to achieve tunable cell fate, diversity, maturation, and commitment [72, 73].

In a different scenario, the mechanical properties and structure of the microenvironment can be engineered to modulate cell–cell interactions mediated by N-cadherin, thus influencing collective cellular behavior. This mechanism is particularly critical for stem cell fate decisions, as modulation of N-cadherin expression or function can shift the balance between neuronal and glial differentiation [63]. Although N-cadherin does not directly interact with ECM components, it indirectly influences ECM-related mechanotransduction by competing with integrin-mediated signaling and altering cytoskeletal remodeling through integrin-cadherin crosstalk [74, 75].

The crosstalk between integrin and N-cadherin is crucial in regulating the functions of collective stem cells and the nervous system. Integrins (cell–ECM sensors) and N-cadherin (cell–cell sensors) are functionally linked through the actin cytoskeleton and share downstream signaling molecules, including RhoA, FAK, and YAP/TAZ [76]. This shared signaling network regulates receptor spatial distribution and intracellular force dynamics, which are crucial for coordinating cell migration and tissue development. By balancing ECM-driven mechanotransduction with cell–cell interactions, this dual signaling mechanism enables the synchronized differentiation of neurons and the formation of organized tissue patterns.

4.1.3.2.2 Gap Junctions, Tight Junctions, Desmosomes, and Ephrins in Bioengineered Microenvironments

Gap junctions, composed of connexin proteins (e.g., C×36 and C×43), enable direct intercellular communication by facilitating the exchange of ions, metabolites, and small signaling molecules with molecular weights of less than 1 kDa. These junctions are crucial for synchronizing the activity of neural progenitors and coordinating the differentiation of neurons or glia. Mechanistically, β-catenin, a transcriptional co-activator in Wnt signaling, can be regulated through Gap Junction-mediated Ca^{2+} flux, which drives the expression of genes essential for neuronal fate [77, 78].

While the bioengineering of modulation of tight junctions (e.g., claudin/occludin-based barriers), desmosomes (e.g., cadherin-mediated adhesion), and ephrins (e.g., bidirectional signaling in tissue patterning) remains understudied, systematic studies are necessary to elucidate these interactions and utilize them for brain tissue engineering.

4.1.3.3 Soluble Factors

Soluble factors are the brain's molecular orchestra conductors directing stem cells to grow, migrate, or transform. However, in the chaos of neurodegenerative disease, this symphony falls silent. Bioengineered materials, such as nanoparticles, when employed as particulate carriers, enable the delivery and controlled release of bioactive components, thereby influencing the neural fate of stem cells and promoting brain regeneration [79–81].

Beyond their delivery capabilities, nano- and microparticles play a crucial role in modulating reactive oxygen species (ROS), which serve as essential signaling molecules that govern stem cell proliferation, differentiation, and survival. For instance, titanium dioxide (TiO_2) nanoparticles generate ROS upon light exposure, thereby activating critical pathways such as PI3K/Akt and MAPK/ERK to promote neuronal differentiation [82]. Similarly, as another example, silver nanoparticles modulate ROS production by regulating the activity of ERK and AKT kinases, thereby enhancing neuronal differentiation further [83]. Moreover, nanoparticles can induce remodeling of the ECM by releasing enzymes or growth factors that alter its stiffness and composition. These ECM modifications, in turn, impact mechanotransduction pathways, such as the YAP/TAZ signaling pathway, which are central to determining stem cell fate [84]. Consequently, bioengineered material-based nanoparticles offer a versatile platform for controlling and modulating brain function and regeneration.

Additionally, nanoparticles facilitate the efficient delivery of bioactive molecules across the blood–brain barrier (BBB). Through precise surface functionalization, these particles can be engineered to target-specific brain regions, such as the hippocampus or SVZ, where they modulate NSC activity. For instance, the delivery of microRNA-222 via chitosan nanoparticles and retinoic acid-loaded polymeric nanoparticles has demonstrated enhanced neural differentiation as proof of concept [85, 86]. Moreover, the integration of Cu_2-Se nanoparticles (CS NPs) with second near-infrared (NIR-II) laser irradiation further underscores the versatility of nanoparticle applications. Under NIR-II stimulation, CS NPs activate voltage-gated calcium channels (VGCCs) on NSCs, thereby initiating the Ca^{2+}/CaMK/CREB/c-Fos signaling pathway, which promotes neural differentiation and survival [87].

Despite these promising applications, the potential neurotoxicity of specific nanoparticles, such as silver nanoparticles, remains a concern. These particles can cross the BBB and accumulate in the brain, potentially inducing oxidative stress, neuroinflammation, and neuronal cell death. In contrast, natural polymer-based nanoparticles (e.g., those derived from chitosan) generally present a lower toxicity risk; however, variations in their size and surface charge may affect their reliability and reproducibility in experimental settings.

4.1.4 Targeting Microglia for Modulating Stem Cell Fate in the Brain

Microglia are the brain's double-edged swords: guardians turned destroyers in AD. Bioengineered nanoparticles offer a truce, rewiring these cells from inflammatory arsonists to regenerative firefighters. In pathological conditions, such as AD, microglia become chronically activated due to the persistent accumulation of Aβ. This chronic activation shifts them into a pro-inflammatory state (M1 phenotype), reducing their phagocytic ability and impairing their capacity to clear Aβ. Prolonged inflammation further inhibits the proliferation and differentiation of NSCs, leading to reduced neurogenesis. For example, pro-inflammatory cytokines could suppress the production and function of the brain-derived neurotrophic factor (BDNF), which is essential for supporting neurogenesis and neuronal survival [88, 89].

Modulating the inflammatory milieu of the brain using bioengineered materials presents a promising strategy for delivering anti-inflammatory agents to damaged neural tissue. For example, nanoparticles can be functionalized with therapeutic compounds such as resveratrol, curcumin, quercetin, or minocycline, or designed to exhibit biomimetic properties. These approaches have demonstrated anti-inflammatory effects across various models, notably by significantly reducing the secretion of pro-inflammatory cytokines from macrophages activated by Aβ fibrils. Additionally, nanoparticles can function as nanozymes, mimicking the enzymatic activity of catalase, superoxide dismutase, and peroxidase. These nanozymes play a role in scavenging ROS and alleviating oxidative stress, thereby mitigating both inflammatory and oxidative damage [90]. For instance, gold nanoclusters (AuNCs) functionalized with dihydrolipoic acid (DHLA) have been shown to promote M2 polarization in microglial cells, characterized by reduced reactive oxygen species (ROS) levels, diminished nuclear factor-κB (NF-κB) signaling, and enhanced autophagy. These combined effects contribute to neuroprotection and facilitate neuronal regeneration [91].

4.2 Hydrogels in Stem Cell and Neuron Transplantation for Brain Regeneration

4.2.1 Challenges and Hydrogel Solutions in Stem Cells-Derived Neuron Therapy

Transplanting neurons into the brain is akin to replanting a forest after a wildfire, but how can we protect these fragile seedlings in a hostile, scarred landscape? Hydrogels, the brain's synthetic soil, offer a lifeline, anchoring neurons, shielding them from storms, and nurturing their growth. The advent of iPSCs, coupled with the development of robust neural differentiation protocols, has provided a viable and ethically sound source for generating neurons. Despite these advances, the clinical translation of iPSC-derived neurons faces significant challenges spanning in vitro cellular manipulation, delivery processes, and final neural integration into the brain (Fig. 4.2). Key challenges related to cell transplantation into the brain include:

- **Neuron Sensitivity During Delivery:** Neurons/neuronal progenitors are highly vulnerable to damage during transplantation due to mechanical and shear stresses. Maintaining their viability and functionality post-transplantation remains a critical hurdle.
- **Complexity of Differentiation Protocols:** Differentiating iPSCs into functional neurons involves intricate and often variable protocols, resulting in a heterogeneous population of neurons. Optimization is essential for generating homogeneous, therapeutically relevant neuronal populations.
- **Multifaceted Pathophysiology of Brain Disorders:** The complex interplay of heterogeneous pathological mechanisms, including microenvironmental alterations, poses significant challenges to developing standardized therapeutic strategies that employ iPSC-derived neurons.
- **Cell Washout and Cavity Formation**: When neurons or stem cells are transplanted into the brain, two significant issues can arise from the dynamics of local fluid flow: first, cells may be displaced by interstitial or cerebrospinal fluid (CSF) forces, and second, if many cells are lost from the intended transplant site, a cavity may form in the brain. Embedding cells in hydrogels or other scaffolding materials can help "anchor" them at the target site. Such scaffolds mimic the native ECM, provide physical support, and may reduce the risk of cavity formation by maintaining a more continuous tissue structure.

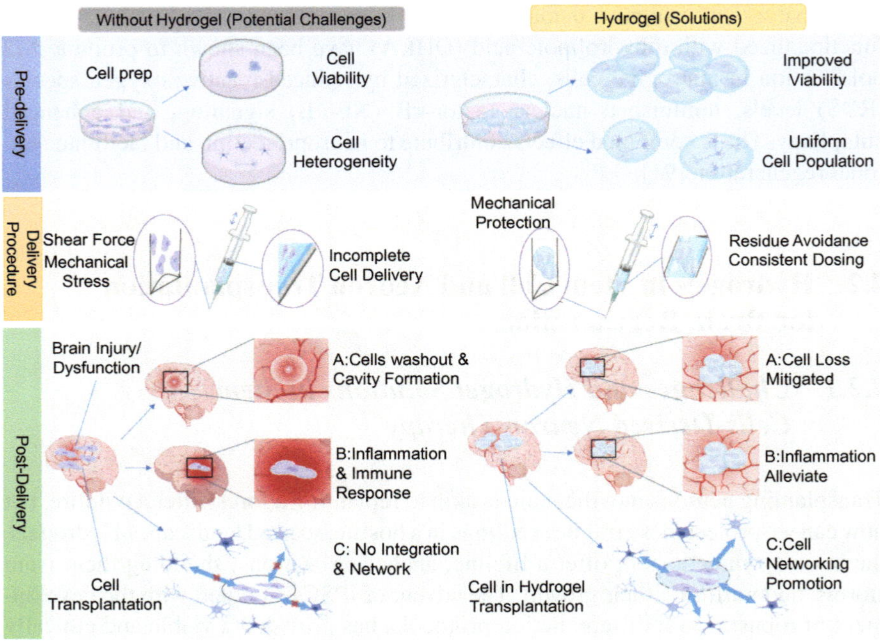

Fig. 4.2 Hydrogels support brain regeneration at various stages of cell transplantation, including before, during, and after cell delivery

Hydrogels are the Swiss Army knives of brain repair: they shield neurons during delivery, sculpt stem cells into precise lineages, and even rewire the brain's hostile terrain. Specifically, hydrogels can:

- **Enhance In Vitro Manipulation:** By providing a biomimetic microenvironment, hydrogels support stem cell differentiation and expansion. For example, peptide-functionalized hydrogels (e.g., those containing laminin-mimetic peptides) enhance neuronal maturation by replicating ECM cues.
- **Enhance Cell Delivery:** Hydrogels serve as protective carriers that shield transplanted neurons from shear forces during injections and mitigate oxidative stress following transplantation. Thermoresponsive hydrogels, such as poly(N-isopropylacrylamide), facilitate minimally invasive delivery by transitioning from a liquid to a gel state at body temperature.
- **Promote Functional Integration:** Hydrogels engineered with neurotrophic factors (e.g., BDNF) or immunomodulatory molecules (e.g., CX3CL1) can reduce immune rejection and foster synaptic connectivity between transplanted and host neurons.

4.2.2 Microenvironmental Considerations in AD and PD

Dysfunctional microenvironments have a significant impact on the outcomes of cell therapy and must be carefully considered in solutions that utilize a biomaterial approach. Transplanting NSCs into the AD-affected brain is like sending firefighters into a blaze without protective gear: Aβ plaques smother integration, stiffened ECM strangles growth, and inflammation scorches survival. Hydrogel-based strategies don't just deliver cells; they arm them for battle. By neutralizing toxins and reprogramming the hostile terrain, these materials transform transplantation from a gamble into a targeted rescue mission [92–94].

The versatility of bioengineering allows for disease-specific strategies. In AD, a self-perpetuating cascade, including Aβ plaque accumulation, tau tangles, and persistent maladaptive inflammation, drives neurodegeneration. In PD, the progressive loss of dopaminergic neurons in the substantia nigra leads to both motor and non-motor dysfunctions. While both diseases involve microenvironmental dysfunction, the challenge is particularly severe in AD and must be addressed alongside factors like immune rejection.

The distinct pathological landscapes can be targeted using specific bioengineering approaches for treatment. For instance, in AD, one strategy involves the clearance of amyloid-beta (Aβ). Various delivery systems, such as hydrogels and polymeric nanoparticles, can be engineered to deliver Aβ-binding peptides (e.g., KLVFF), which function as molecular "sponges" to sequester plaques. This approach may also enhance the functionality of neural stem cells (NSCs) [95–99]. In PD, bioengineering strategies can incorporate glial cell line-derived neurotrophic factor (GDNF) into hydrogels to enhance the survival and function of dopaminergic neurons [95, 100, 101].

4.2.3 Enhanced Hydrogel Functionality for Brain Neural Regeneration

4.2.3.1 Targeting Hypoxia

To address this, hydrogels can be engineered for controlled, sustained oxygen release to support cell survival and differentiation [102, 103]. For example, integrating myoglobin (Mb) into peptide-based hydrogels mimics natural oxygen transport: Mb binds oxygen in high-oxygen environments during implantation and releases it under hypoxic conditions [104]. Alternative strategies include embedding metal-organic frameworks (MOFs), which deliver exogenous oxygen or generate it endogenously [105], and perfluorocarbon (PFC) emulsions, known for their high oxygen-carrying capacity [106]. These approaches synergize with hydrogels to alleviate hypoxia-related challenges.

4.2.3.2 Combating Oxidative Stress

Mitochondrial dysfunction in conditions like ischemia and AD drives excessive ROS production, exacerbated by hypoxia. Neurons, being post-mitotic, are highly susceptible to ROS damage. Hydrogels can counteract this by encapsulating antioxidants broadly defined as ROS scavengers, molecules that prevent ROS formation, repair oxidative damage, or enhance endogenous defenses (e.g., vitamins). For instance, polyphenols and gallic acid are directly incorporated into hydrogels or tethered to polymer chains for localized, sustained release [107, 108].

4.2.3.3 Engineering Hydrogel Composition, Structure, and Degradation

Hydrogels are not the one-size-fits-all solution. Their composition must evolve with the brain's needs:

Degradation Dynamics: Degradation kinetics balance the self-renewal and differentiation of NSCs. Rapidly degradable formulations, such as gelatin-methacryloyl (GelMA), create temporary microenvironments that help preserve the stemness of neural stem cells during proliferation. Conversely, slow-degrading hydrogels (e.g., hyaluronic acid (HA)-based, modified to control their degradation rates) provide lasting structural support to guide neuronal maturation [109, 110].

Bioactive Functionalization: Incorporating neurotrophic factors, such as BDNF and GDNF, into hydrogel mimics endogenous signaling to enhance NSC survival and differentiation. These cues can guide differentiation, for example, suppress glial overgrowth while promoting functional neuronal lineage specification [111].

Structural Engineering: Advanced fabrication techniques, such as 3D bioprinting, enable spatial control over hydrogel architecture, allowing for the embedding of gradients of growth factors or ECM components to replicate region-specific brain niches (e.g., cortical vs. hippocampal microenvironments).

The convergence of tunable degradation, bioactive signaling, and microstructural engineering allows hydrogels to function as dynamic, four-dimensional (4D) scaffolds. These systems not only provide physical support but also evolve in sync with tissue regeneration, releasing neurotrophic factors during critical phases or gradually ceding space to nascent neural networks. Such designs align with developmental principles, where transient cues and mechanical feedback orchestrate tissue morphogenesis.

4.3 Bioengineered Materials in Brain Disease Modeling

4.3.1 Significance of In Vitro Modeling in Neurodegenerative Research

Neurodegenerative diseases, such as AD, PD, and Huntington's disease (HD), are characterized by multifactorial pathologies that rely on reductionist experimental approaches for drug development. For instance, although several AD therapies, including Memantine (2003), Aducanumab (2021), and Lecanemab (2022), have received regulatory approval, the overall success rate in translating novel treatments remains low. This gap highlights the limitations of traditional in vitro models, which often fail to recapitulate the dynamic cell–cell interactions, region-specific spatial heterogeneity, and metabolic demands inherent in the evolving physiology of the human brain microenvironment.

4.3.1.1 Limitations of Conventional In Vitro and Animal Models (Reductionist Models)

Traditional two-dimensional (2D) cell culture systems, such as cells grown in a monolayer on plastic dishes, offer a simple and accessible method for in vitro studies. However, they fail to recapitulate the complex three-dimensional (3D) architecture of brain tissue, limiting their ability to accurately mimic both physiological and pathological conditions. In particular, these models lack realistic cell–cell and cell–ECM interactions, leading to altered gene expression profiles and unrealistic pathological features, such as β-amyloid peptide aggregation and diffusion. Animal models, although widely used, pose ethical concerns and introduce species-specific differences, such as the distinct handling of amyloid-β and divergent immune responses in rodents, compromising their translational relevance. Collectively, these limitations contribute to the frequent failure of preclinical findings to translate into effective clinical outcomes.

4.3.1.2 Inadequate Recapitulation of the Dynamic Cellular Microenvironment

Neurodegenerative diseases are characterized by ongoing changes in the cellular microenvironment, including alterations in ECM composition, tissue stiffness, and inflammatory signaling. For example, in AD, the accumulation of amyloid-β plaques is accompanied by ECM stiffening. This change can alter stem cell differentiation, favoring the formation of astrocytes over neurons and affecting responses to therapeutic agents [26, 112–114].

Traditional in vitro models often fail to capture these intricate biochemical and biophysical dynamics, resulting in experimental outcomes that do not fully reflect the clinical scenario. By contrast, integrative platforms that combine, for example, stem cells, tunable hydrogels, and microfluidic technologies now enable precise control over dynamic microenvironmental cues. These advanced systems faithfully recreate disease-specific microenvironments, such as those involving amyloid-β and matrix stiffness dynamics in AD, thereby narrowing the gap between preclinical research and clinical applications [1]. The question is no longer whether bioengineered models outperform traditional ones but rather how soon they can revolutionize drug development. When therapies are tested in systems that replicate the brain's mechanical, biochemical, and temporal complexity, will the near 100% failure rate finally drop?

4.3.2 Components of the Disease Modeling Platform

Modern disease modeling platforms integrate five pillars to bridge the translational gap between drug discovery and clinical application (Table 4.3). Stem cells are the central element, significantly influenced by the other components. For example, 3D bioprinting improves the organization of cells and other components, including stem cells, with increased precision. This enhanced organization affects the development and behavior of these cells when exposed to environments that closely mimic those of the human body, rather than simply mixing them in bulk.

Table 4.3 Components of disease modeling in a modern platform

Key contributors	Functions	Innovation bioengineering example
Cellular components	Stem cells provide the essential foundation by differentiating into various neural lineages, including neurons and glial cells	Development of brain organoids and assembloids that recapitulate tissue-level architecture and functional heterogeneity
Hydrogels/scaffolds	Modulation of microenvironmental cues can mimic natural and diseased conditions	Use of functionalized and smart hydrogels (e.g., amyloid-β-functionalized hydrogels) that induce disease-specific neuronal stress
Microfluidic devices	Simulation of fluid shear stress, nutrient gradients, and cell–cell interactions	Development of integrated systems such as BBB models and gut–axis–brain chips (i.e., organs-on-chips)
Advanced bioengineering	Construction of anatomically and physiologically relevant tissue-like structures with precise genetic control	3D bioprinting techniques and CRISPR-based gene editing
Computational and analytical tools	Analyzing multi-omics datasets to predict disease mechanisms	Machine learning approaches for biomarker identification [115]

4.3.3 Bioengineered Materials-Assisted Brain Organoids

Brain organoids are three-dimensional, self-organizing structures derived from pluripotent stem cell aggregates embedded in ECM-rich matrices. They provide exceptional platforms for disease modeling, drug discovery, and cell therapy by partially recapitulating the human brain's structural and functional architecture. However, to meet the needs for disease modeling, persistent challenges such as inherent heterogeneity (due to stochastic differentiation), lack of vascularization, and incomplete cellular maturation continue to restrict their translational potential. Innovations in bioengineering materials, including tunable ECM-mimetic scaffolds, vascularization strategies, and dynamic culture systems, remain crucial for bridging these gaps and unlocking the full potential of brain organoids in research and clinical applications (Fig. 4.3).

To address the limitations of brain organoids, advanced bioengineering now allows for precise control over micro- and nanoscale environmental cues. For instance, while hydrogels are a fundamental component in organoid generation, traditional bulk biomaterial scaffolds, such as Matrigel, silk fibroin, and collagen hydrogels often do not effectively mimic in vivo conditions. This is primarily due to their lack of spatiotemporal complexity, which is essential for replicating the brain's ECM. To improve this, engineering strategies are being developed to modify hydrogels by incorporating micro- and nanoscale features such as aligned fibers,

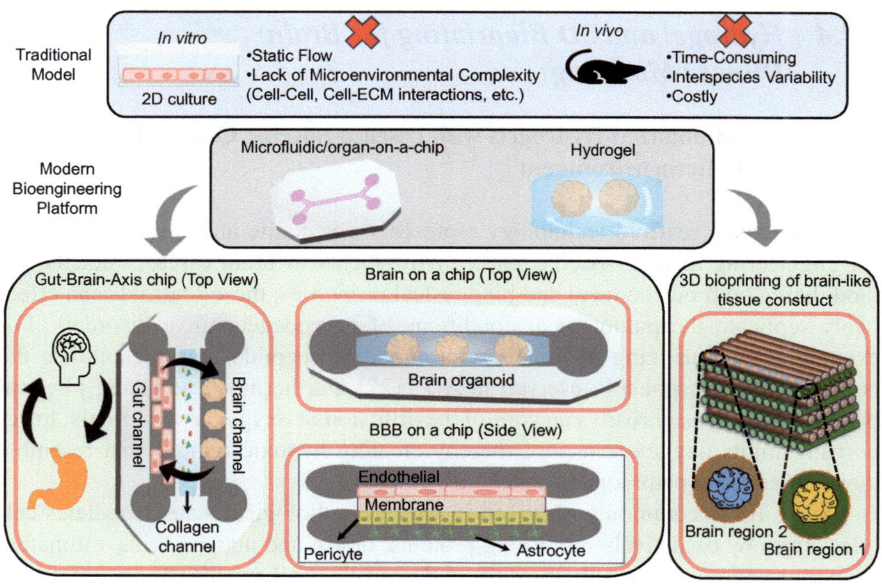

Fig. 4.3 Bioengineering Approaches for Disease Modeling. New bioengineering approaches and tools, such as microfluidic devices, hydrogels, and 3D bioprinting, address the drawbacks of traditional disease modeling systems

patterned topographies, and biochemical gradients. These modifications mimic region-specific ECM properties observed in both healthy and diseased states, albeit in different ways [116, 117]. For instance, aligned fibrous networks can mimic the brain's anisotropic tissue architecture, guiding directional neurite outgrowth, while spatially patterned stiffness or adhesive motifs may replicate cortical layering. By transforming passive scaffolds into dynamic, instructive microenvironments, this engineering approach enables temporally controlled maturation and organization of brain organoids into layered structures that closely resemble the cerebral cortex [118].

The combination of spatiotemporal engineering (e.g., smart hydrogels), advanced 3D bioprinting, and microfluidic technologies has led to remarkable spatial precision in creating complex tissue constructs [119–121]. By allowing the controlled deposition of cell-laden bioinks and spatial biochemical distribution, these techniques enable the creation of region-specific domains, for instance, replicating distinct hippocampal or cortical territories [122, 123]. The resulting increase in reproducibility and structural complexity renders the organoids more reflective of in vivo brain architecture. Additionally, integrating biomimetic vascular networks and engineered microstructures such as microcapsules and fiber network constructs enhances the distribution of nutrients and oxygen within the organoid. This not only supports long-term cellular viability but also facilitates further maturation of neural tissue [124–126].

4.3.4 Hydrogel and 3D Bioprinting for Brain Disease Modeling

4.3.4.1 Functionalized Hydrogels with Disease-Specific Compartment and Microenvironment

Hydrogel-based scaffold technology represents a versatile and essential platform for engineering disease-specific microenvironments in brain disease models. By modulating both biochemical and biophysical properties, these scaffolds can effectively replicate the pathological conditions of neurodegenerative disorders. For example, functionalizing hydrogels with amyloid-β peptides can recapitulate the neuronal stress responses observed in AD [127]. Furthermore, adjusting physical parameters such as porosity can control the diffusion of oxygen and nutrients; lower porosity limits oxygen transport, thereby creating hypoxic regions that resemble those observed in pathological states.

Excess ROS, commonly observed in various pathological states, mediates cellular injury by oxidatively modifying proteins, lipids, and nucleic acids, ultimately triggering cell death through apoptotic and necrotic mechanisms. By incorporating elements that mimic oxidative stress (e.g., Catechol-ZnO) [128], hydrogel scaffolds provide a controlled microenvironment that accurately replicates the pathological features of diseases. An additional level of complexity can be achieved through the application of advanced 3D bioprinting techniques, which enable the fabrication of these hydrogels into intricate, biomimetic architectures that closely replicate the native ECM of the brain, along with its pathological nuances. Such constructs not only promote neural cell growth and precise spatial organization but also facilitate the accurate recreation of disease-relevant microenvironments, thereby enhancing the robustness of in vitro modeling and therapeutic screening [129–131].

4.3.4.2 Stimuli-Responsive Hydrogels: Targeting the Dynamic Microenvironmental of the Brain During Disease Development

Stimuli-responsive hydrogels constitute a sophisticated platform for elucidating cellular responses to dynamic biochemical and mechanical cues within the microenvironment during disease progression. These hydrogels are designed to respond to a variety of external stimuli, including temperature, pH, light, electric and magnetic fields, pressure, and specific chemical triggers. Upon activation, they exhibit precise and reversible changes in their mechanical properties (e.g., stiffness modulation or degradation) and biochemical characteristics (e.g., controlled release of bioactive molecules), allowing them to mimic the dynamic nature of physiological and pathological conditions [132]. For instance, azobenzene-containing hydrogels can undergo reversible transitions between states when exposed to UV and visible light, allowing for dynamic control over scaffold stiffness. This capability is particularly relevant for modeling neurodegenerative diseases such as AD and PD, where mechanical properties of the ECM play a critical role during disease progression [133, 134]. Additionally, stimuli-responsive hydrogels can be designed to release morphogens (e.g., Sonic Hedgehog [SHH] or Wnt proteins) in a controlled manner, guiding region-specific organoid development in a spatiotemporal manner to enhance the model accuracy [135, 136].

4.3.4.3 3D Bioprinting for Brain Disease Modeling

The brain's architecture comprises specialized cells organized into distinct laminar structures, highlighting the utility of 3D bioprinting as a powerful tool for modeling brain anatomy and pathology. This technique enables tissue organization at micro- to nanoscale resolutions, facilitating the arrangement of neurons, astrocytes, oligodendrocytes, and microglia into structures that closely mimic native brain tissue [137, 138].

In brain disease modeling, 3D bioprinting provides a robust platform for replicating pathologies associated with disrupted tissue architecture and cellular organization. This technology can effectively address challenges such as impaired neuronal layering, abnormal glial distribution, and defective axonal pathfinding. For example, in neurodegenerative disorders like AD, pathological hallmarks extend beyond amyloid plaques and neurofibrillary tangles to include synaptic loss and reactive gliosis. By precisely controlling the composition and spatial organization of the bioprinted constructs, these models can faithfully recapitulate key disease features, including protein aggregation, synaptic deficits, and inflammatory responses [139–141].

4.3.5 Microfluidic devices for Brain Disease Modeling

Reductionist and static models, including both traditional cell cultures and organoid systems often fail to faithfully recapitulate the complexity of brain, thereby limiting the accuracy of disease modeling and negatively impact drug development. Microfluidic platforms are poised to provide solutions through improved control over microenvironmental signals and enabling advanced co-culture systems, including dynamic cell–cell interactions. Operating at the microscale within channels and chambers ranging from tens to hundreds of micrometers, these systems allow for precise regulation of fluid flow, shear stress, and the spatiotemporal distribution of signaling molecules, oxygen, and nutrients. Although microfluidic devices are not inherently categorized as bioengineered materials, their integration with hydrogels, nanoparticles, fibers, and sheets significantly enhances their functionality. This synergy increases the physiological relevance and dynamic nature of in vitro models, enabling more accurate simulation of the microenvironments present in brain tissue and associated pathologies [142].

4.3.5.1 Brain-on-a-Chip (BoC)

Brain-on-a-chip systems merge physiologically relevant microfluidic environments with the co-culture of diverse brain cell types, recreating the human brain's complex microenvironment. These platforms replicate critical tissue features, such as dynamic cell–cell interactions, tissue–tissue interfaces, and mechanical cues like shear stress, to closely approximate the brain's in vivo conditions.

In 2D microfluidic configuration, neurons or glial cells are cultured on flat surfaces within devices featuring multiple interconnected chambers or channels. This design enables the separation of different cell types and the controlled introduction of biochemical factors, allowing for the study of the impact of these signals on cell–cell interactions. This setup allows for in-depth investigations into neuronal activity, synaptic transmission, and electrophysiological properties. Additionally, these devices enable the precise introduction of disease-related factors with high precision, such as amyloid-beta peptides, tau, and alpha-synuclein. This enables the development of robust models to investigate the mechanisms underlying neurodegenerative diseases [126, 143–146].

In a 3D microfluidic setup, two different scenarios can occur. The first scenario involves embedding cells within a hydrogel matrix inside a microfluidic device, which has the advantage of overlap, previously discussed in the hydrogel section. The second scenario refers to incorporating pre-formed organoids into a microfluidic platform to create specific BoC models. This includes models of the entire brain, as well as particular regions such as the cerebral cortex, hippocampus, hypothalamus, and choroid plexus. By combining organoids with microfluidic systems (whether in mono- or co-culture), we can enhance their functionality and improve the reliability of the resulting models [123, 146, 147]. To address the common challenge of vascularization in brain organoids, microfluidic systems are used to emulate the brain's vascular network [148, 149]. Utilizing advanced microfabrication techniques, microfluidic systems provide a precisely controlled platform for regulating biophysical signals, such as fluid flow and shear stress, that are crucial for the development of the vascular system. In these devices, channels are engineered to replicate the dimensions of natural microvasculature and are seeded with endothelial cells (ECs). Under carefully regulated flow conditions, ECs adhere to the channel walls, align, proliferate, and form a continuous endothelium. Notably, the controlled biophysical environment not only supports the development of these engineered vascular channels but also stimulates EC invasion into adjacent brain organoids. This integration facilitates the formation of functional vascular networks within the organoid, thereby enhancing nutrient and oxygen delivery, waste removal, and overall tissue maturation, and ultimately creating a more physiologically relevant in vitro model of neurovascular interactions [150].

The BoC models can be categorized into axonal growth models, neuron–glia interaction models, and BBB models, each facilitating the study of various brain pathological conditions. Furthermore, the range of models extends beyond these examples. In addition to these models, a diverse range of chip designs exists. For instance, multiorgan-on-a-chip platforms (such as the neuromuscular junction chip) and glymphatics-on-a-chip systems, where human astrocytes and blood vessels are co-cultured in 3D gel-embedded parallel microchannels, offer further insights into the glymphatic system's role in disease modeling [151].

4.3.5.1.1 Axonal Chip

Axonal dysfunction is a hallmark of various neurological disorders, with diffuse axonal injury (DAI) being a prominent example. DAI, often resulting from traumatic brain injuries, primarily affects white matter tracts, leading to a spectrum of neurological impairments, ranging from minor cognitive deficits to severe coma states [152]. Chips designed to study axonal defects typically incorporate compartments that physically separate axons from neural somas via microchannels. This configuration allows for controlled investigations of axonal behavior and responses to various stimuli. For example, when integrated with electrical stimulation, these devices provide valuable insights into the mechanisms governing axonal growth, directional guidance, and degeneration processes that are critical in the progression of neurodegenerative diseases [152–154]. Furthermore, the integration of bioengineered materials allows for the precise patterning of microenvironmental topographies, such as nanoscale cues, which more accurately replicate the complexities of a dysfunctional brain microenvironment [155, 156].

4.3.5.1.2 Neuron–Glia Chip

Neuron–glia chips represent a powerful platform for investigating the intricate interplay between neurons and glial cells. These interactions are crucial to both normal brain function and the progression of neurological disorders. These platforms are mainly instrumental in elucidating mechanisms underlying neuroinflammation, neuronal metabolism, and synaptic regulation. In the context of AD, for example, microglia become activated in response to amyloid-beta plaques, leading to the release of pro-inflammatory cytokines and reactive oxygen species. This inflammatory cascade can precipitate synapse loss and neuronal death, thereby accelerating disease progression [157].

Developed chip models have demonstrated the feasibility of studying the metabolic interactions between neurons and glial cells [158] as well as examining synaptic activity [159] through the co-culture systems. To enhance the relevance of in vitro platforms for studying neurodegenerative diseases, such as AD, 3D co-culture models have been developed that incorporate human neurons, astrocytes, and microglia. This integrated approach enables a comprehensive analysis of key AD-related features, including amyloid-beta aggregation, tau phosphorylation, and neuroinflammatory responses, thereby providing valuable insights into the underlying mechanisms driving the disease [160].

4.3.5.1.3 Blood–Brain Barrier (BBB) Chip

BBB chips are designed to mimic the complex interactions between cerebral microvessels and glial cells. It addresses two major challenges: reproducing the selective permeability of the BBB and overcoming the limitations of traditional drug development models. Unlike conventional transwell co-culture assays that often simplify cell-to-cell interactions and lack dynamic fluid forces, the BBB chip utilizes carefully controlled shear stresses ranging from 10 to 20 dyn/cm^2 under normal physiological conditions and up to 40 dyn/cm^2 under elevated stress. This innovative design features a three-dimensional multicellular architecture that more accurately reflects the in vivo microenvironment [161, 162].

The typical device consists of a semi-porous polymer membrane, such as PET, combined with a hydrogel-based extracellular matrix, which may include collagen, fibrin, or Matrigel. This design creates distinct yet interactive chambers. On one side of the membrane, brain microvascular endothelial cells are cultured, while astrocytes and pericytes are situated in the opposing compartment. This arrangement promotes robust bidirectional signaling, reflecting the natural organization of the neurovascular unit. Additionally, microfluidic flow within the system activates endothelial mechanosensors. This action influences junctional complexes by modulating the expression of claudin-5 in response to various shear conditions, ultimately enhancing barrier integrity—a feature that static systems often fail to achieve [162]. Furthermore, shear stress induced by chip systems within physiologically relevant ranges significantly impacts endocytosis and transcytosis; for instance, it may enhance the formation of vesicles responsible for transporting molecules across cellular barriers [163].

4.3.5.2 Multi Organ-on-a-Chip

In an ideal scenario, microfluidic systems could integrate diverse organoids, cells, and tissues to replicate the interconnected nature of human organs, facilitating the development of comprehensive and physiologically relevant disease models. This approach is crucial for understanding organ–organ interactions, which play a pivotal role in the progression of complex diseases and the advancement of drug development [164]. Currently, practical configurations in microphysiological platforms primarily focus on modeling binary interactions between the brain and other organ systems, such as immune cells, the liver, the gut, and muscle tissue. These models provide valuable insights into the intricate communication networks between the brain and peripheral organs, thereby deepening our understanding of disease mechanisms and guiding the development of potential therapeutic strategies.

4.3.5.2.1 Gut–Brain Axis-on-a-Chip

Accumulating evidences suggest a significant link between gut microbiome alterations (dysbiosis) and the pathogenesis of AD through mechanisms involving intestinal permeability, BBB disruption, oxidative stress, inflammation, and immune dysregulation [165–168]. The gut–brain axis (GBA), a bidirectional communication pathway connecting the gastrointestinal system and the brain, encompasses multiple pathways, including neural (brain), immune, endocrine, and metabolic systems. By replicating these interactions, GBA-on-a-chip models offer a controlled platform for studying the physiological and pathological mechanisms underlying neurodegenerative disorders, inflammatory bowel diseases (IBD), and psychiatric conditions [166, 169, 170].

State-of-the-art microfluidic platforms now offer innovative ways to model this axis in vitro. In these systems, a modular chip is designed with distinct yet interconnected compartments that mimic the gut and brain environments. One compartment is lined with intestinal epithelial cells to replicate the gut's barrier and secretory functions, while adjacent sections house neural cells or brain tissue constructs, creating a controlled interface between these systems. A finely tuned perfusion system simulates dynamic fluid flow, creating chemical gradients and shear forces that closely resemble in vivo conditions. This arrangement facilitates the assessment of microbial byproducts and exosome transport across the gut barrier, providing insight into their potential roles in modulating neural function and contributing to disease pathology [171, 172].

4.3.5.2.2 Neuromuscular Junction (NMJ) Chip

These chips are designed to study interactions between neurons and muscles, particularly for modeling neuromuscular junction disorders such as amyotrophic lateral sclerosis (ALS). ALS is a progressive neurodegenerative disease characterized by dysfunction and degeneration of motor neurons (MNs) in the brain and spinal cord, ultimately leading to muscle atrophy and paralysis. NMJ chips aim to replicate the physiological environment of the neuromuscular junction to investigate disease mechanisms and test therapies. A typical NMJ chip platform features three interconnected compartments to mimic in vivo conditions:

1. **Motor neuron compartment**: Optimized for neuronal survival and maturation, this chamber contains the cell bodies of motor neurons.
2. **Muscle compartment**: Cultured under conditions that promote muscle cell differentiation and contractile function.
3. **Axon guidance microchannels**: A narrow network connecting the two compartments, allowing axons from motor neurons to extend directionally toward muscle cells.

In this setup, functional neuromuscular junctions can form, enabling bidirectional signaling between neurons and muscles. This system can also be integrated with assembloids, such as cortico-motor assembloids, to connect neuronal circuits with skeletal muscle function. Furthermore, advanced engineering strategies can enhance this system. For instance, incorporating optical or electrical stimulation enables precise control of neuromuscular activity, which improves disease modeling and drug screening capabilities [173, 174].

4.4 Beyond Neurodegeneration: Integration of Bioengineered Materials with Stem Cells in Other Brain Disorders

Bioengineering research extends beyond neurodegenerative diseases, enhancing stem cell therapies and modeling for various neurological disorders. While many approaches overlap, such as improving neuronal delivery for cell therapy and replicating inflammatory environments during modeling, each pathological condition presents unique features. For instance, traumatic brain injury (TBI) involves specific axonal damage, necessitating tailored bioengineering strategies. Similarly, hemorrhagic stroke, resulting from cerebral blood vessel rupture, requires distinct approaches. Table 4.4 provides an illustrative example of how bioengineered strategy, integrated with stem cell functionality, is applied in both treatment and modeling across these disorders.

Table 4.4 Bioengineered examples for brain diseases: treatment and modeling

Brain disease	Clinical/therapeutic example	Modeling example
1. Traumatic brain injury (TBI) Axonal injury and vascular disruption	**Antioxidant hydrogels** locally scavenge excess ROS at injury sites, mitigating oxidative stress and supporting tissue repair [107]	**Axon-on-a-chip** platforms simulate the mechanical forces of TBI using biomaterials, enabling real-time tracking of axonal damage [152]
2. Stroke Obstruction and rupture of cerebral blood vessels	**Hyaluronic acid (HA)** hydrogels enhance the survival and differentiation of iPSC-derived neural progenitors in stroke cores [175]	**3D-bioprinted neurovascular units (NVUs)** mimic BBB breakdown **Microfluidic oxygen/glucose-deprived devices** model ischemic reperfusion injury [176]
3. Brain tumor	**Tumoricidal neural stem cells** (tNSCs) in combination with hydrogel to treat post-operative glioblastoma [177]	The **glioblastoma-on-a-chip** replicates patient-specific resistance to concurrent chemoradiation and temozolomide, enabling the identification of drug combinations that achieve enhanced tumor cell killing [178]
4. Genetic disorders	**Lipid nanoparticles** are utilized for delivering gene-editing tools to neural stem and progenitor cells in the brain [179].	By integrating specific cells with defined genetic dysfunctions (e.g., stem cell-derived neurons or endothelial cells) into a chip, it is possible to model genetic disorders, such as those associated with Huntington's disease [180]
5. Brain infections	Bioengineered materials, especially **nanotechnology for drug delivery,** can help combat brain infections, which can impact neurogenesis [181, 182]	**NVU-on-a-chip** integrates fungal/bacterial agents to study BBB infection dynamics [183]
6. Brain–computer interfaces (BCI)	**Organoid-BCI systems** integrate human brain organoids with recording/stimulation devices [184]	The **shell and E-flower shape-based MEA** help in the assessment of organoid (disease model) electrical activity [185, 186]
7. Neurodegenerative diseases	**3D hiPSC-derived neural microtissues** (e.g., dopaminergic neurons) for cell replacement [187]	**BBB leakage models** induce amyloid-β/tau pathology and synaptic loss [188]

Acknowledgement PYW thanks the support from the Ministry of Science and Technology of China (2022YFA1105101), the Zhejiang Provincial Natural Science Foundation of China (LZ23C070004), and the Wenzhou High-Level Innovation Team Program (2024R2001). We thank Huiyuan Chen and Cheryl Yeap for their invaluable assistance in drawing the figures.

Conflict of Interest **The authors declare that they have no conflict of interest.**

References

1. Harati J, Wang P (2024) Leveraging integrative technologies to translate stem cell and cell reprogramming potential for neurodegenerative diseases. Eur Cell Mater 48:151–155. https://doi.org/10.22203/eCM.v048a09
2. Magistretti PJ, Allaman I (2015) A cellular perspective on brain energy metabolism and functional imaging. Neuron 86(4):883–901. https://doi.org/10.1016/j.neuron.2015.03.035
3. Llorente V, Velarde P, Desco M, Gómez-Gaviro MV (2022) Current understanding of the neural stem cell niches. Cells 11(19):3002. https://doi.org/10.3390/cells11193002
4. Harati J, Liu K, Shahsavarani H, Du P, Galluzzi M, Deng K, Mei J, Chen H-Y, Bonakdar S, Aflatoonian B, Hou G, Zhu Y, Pan H, Wong RCB, Shokrgozar MA, Song W, Wang P-Y (2023) Defined physicochemical cues steering direct neuronal reprogramming on colloidal self-assembled patterns (CSAPs). ACS Nano 17(2):1054–1067. https://doi.org/10.1021/acsnano.2c07473
5. Wang P, Song W (2023) Direct neuronal reprogramming for central nervous system regeneration. Brain-X 1(3). https://doi.org/10.1002/brx2.36
6. Scuteri A, Miloso M, Foudah D, Orciani M, Cavaletti G, Tredici G (2011) Mesenchymal stem cells neuronal differentiation ability: a real perspective for nervous system repair? Curr Stem Cell Res Ther 6(2):82–92. https://doi.org/10.2174/157488811795495486
7. Ma T, Wu J, Mu J, Gao J (2022) Biomaterials reinforced MSCs transplantation for spinal cord injury repair. Asian J Pharm Sci 17(1):4–19. https://doi.org/10.1016/j.ajps.2021.03.003
8. Hoshiba T, Sugano Y, Yokoyama N (2018) Murine neural stem cell (NSC) line, MEB5-derived decellularized matrix as an in vitro extracellular matrix model in NSC niche. Chem Lett 47(12):2016:6397820. https://doi.org/10.1246/cl.180788
9. Yale AR, Kim E, Gutierrez B, Hanamoto JN, Lav NS, Nourse JL, Salvatus M, Hunt RF, Monuki ES, Flanagan LA (2023) Regulation of neural stem cell differentiation and brain development by MGAT5-mediated N-glycosylation. Stem Cell Reports 18(6):1340–1354. https://doi.org/10.1016/j.stemcr.2023.04.007
10. Zhang Q, Ma C, Chin L-S, Li L (2020) Integrative glycoproteomics reveals protein N-glycosylation aberrations and glycoproteomic network alterations in Alzheimer's disease. Sci Adv 6(40):eabc5802. https://doi.org/10.1126/sciadv.abc5802
11. Liu Q, Peng S, Tang Q, Li C, Chen J, Pang P, Liu W, Zhou X, Cai X, Lin H, Ji X, Ji Z (2024) A N-cadherin nano-antagonist hydrogel enhances recovery from spinal cord injury by impeding glial scarring. Adv Funct Mater 34(39). https://doi.org/10.1002/adfm.202404563
12. Hußler W, Höhn L, Stolz C, Vielhaber S, Garz C, Schmitt FC, Gundelfinger ED, Schreiber S, Seidenbecher CI (2022) Brevican and neurocan cleavage products in the cerebrospinal fluid—differential occurrence in ALS, epilepsy and small vessel disease. Front Cell Neurosci 16:838432. https://doi.org/10.3389/fncel.2022.838432
13. Fontanil T, Mohamedi Y, Moncada-Pazos A, Cobo T, Vega JA, Cobo JL, García-Suárez O, Cobo J, Obaya ÁJ, Cal S (2019) Neurocan is a new substrate for the ADAMTS12 metalloprotease: potential implications in neuropathies. Cell Physiol Biochem 52(5):1003–1016. https://doi.org/10.33594/000000069
14. Liang Y-R, Zhang X-H (2024) Inhibition of GluN2B-containing NMDA receptors in early life combined with social stress in adulthood leads to alterations in prefrontal PNNs in mice. Acta Physiologica Sinica 76(1):1–11. https://doi.org/10.13294/j.aps.2023.0056
15. Morawski M, Pavlica S, Seeger G, Grosche J, Kouznetsova E, Schliebs R, Brückner G, Arendt T (2010) Perineuronal nets are largely unaffected in alzheimer model Tg2576 mice. Neurobiol Aging 31(7):1254–1256. https://doi.org/10.1016/j.neurobiolaging.2008.07.023
16. Garrudo FFF, Mikael PE, Xia K, Silva JC, Ouyang Y, Chapman CA, Hoffman PR, Yu Y, Han X, Rodrigues CAV, Ferreira FC, Linhardt RJ (2021) The effect of electrospun scaffolds on the glycosaminoglycan profile of differentiating neural stem cells. Biochimie 182:61–72. https://doi.org/10.1016/j.biochi.2021.01.001

17. Li F, Ducker M, Sun B, Szele FG, Czernuszka JT (2020) Interpenetrating polymer networks of collagen, hyaluronic acid, and chondroitin sulfate as scaffolds for brain tissue engineering. Acta Biomater 112:122–135. https://doi.org/10.1016/j.actbio.2020.05.042
18. Silva GA, Czeisler C, Niece KL, Beniash E, Harrington DA, Kessler JA, Stupp SI (2004) Selective differentiation of neural progenitor cells by high-epitope density nanofibers. Science 303(5662):1352–1355. https://doi.org/10.1126/science.1093783
19. Silva J, Bento AR, Barros D, Laundos TL, Sousa SR, Quelhas P, Sousa MM, Pêgo AP, Amaral IF (2017) Fibrin functionalization with synthetic adhesive ligands interacting with A6β1 integrin receptor enhance neurite outgrowth of embryonic stem cell-derived neural stem/progenitors. Acta Biomater 59:243–256. https://doi.org/10.1016/j.actbio.2017.07.013
20. Farrukh A, Ortega F, Fan W, Marichal N, Paez JI, Berninger B, Campo AD, Salierno MJ (2017) Bifunctional hydrogels containing the laminin motif IKVAV promote neurogenesis. Stem Cell Reports 9(5):1432–1440. https://doi.org/10.1016/j.stemcr.2017.09.002
21. Dineley KT, Kayed R, Neugebauer V, Fu Y, Zhang W, Reese LC, Taglialatela G (2010) Amyloid-β oligomers impair fear conditioned memory in a calcineurin-dependent fashion in mice. J Neurosci Res 88(13):2923–2932. https://doi.org/10.1002/jnr.22445
22. Palop JJ, Mucke L (2010) Synaptic depression and aberrant excitatory network activity in Alzheimer's disease: two faces of the same coin? Neuromolecular Med 12(1):48–55. https://doi.org/10.1007/s12017-009-8097-7
23. Arai T, Sasaki D, Araya T, Sato T, Sohma Y, Kanai M (2014) A cyclic KLVFF-derived peptide aggregation inhibitor induces the formation of less-toxic off-pathway amyloid-β oligomers. ChemBioChem 15(17):2577–2583. https://doi.org/10.1002/cbic.201402430
24. Bernabeu-Zornoza A, Coronel R, Palmer C, López-Alonso V, Liste I (2021) Oligomeric and fibrillar species of Aβ42 diversely affect human neural stem cells. Int J Mol Sci 22(17):9537. https://doi.org/10.3390/ijms22179537
25. Wang PY, Tsai WB, Voelcker NH (2012) Screening of rat mesenchymal stem cell behaviour on polydimethylsiloxane stiffness gradients. Acta Biomater 8(2):519–530. https://doi.org/10.1016/j.actbio.2011.09.030
26. Keung AJ, De Juan-Pardo EM, Schaffer DV, Kumar S (2011) Rho GTPases mediate the mechanosensitive lineage commitment of neural stem cells. Stem Cells 29(11):1886–1897. https://doi.org/10.1002/stem.746
27. Murphy MC, Jones DT, Jack CR, Glaser KJ, Senjem ML, Manduca A, Felmlee JP, Carter RE, Ehman RL, Huston J (2016) Regional brain stiffness changes across the Alzheimer's disease spectrum. Neuroimage Clin 10:283–290. https://doi.org/10.1016/j.nicl.2015.12.007
28. Takamura T, Motosugi U, Sasaki Y, Kakegawa T, Sato K, Glaser KJ, Ehman RL, Onishi H (2020) Influence of age on global and regional brain stiffness in young and middle-aged adults. J Magn Reson Imaging 51(3):727–733. https://doi.org/10.1002/jmri.26881
29. Hu Y, Jia Y, Wang S, Ma Y, Huang G, Ding T, Feng D, Genin GM, Wei Z, Xu F (2023) An ECM-mimicking, injectable, viscoelastic hydrogel for treatment of brain lesions. Adv Healthc Mater 12(1):2201594. https://doi.org/10.1002/adhm.202201594
30. Iwashita M, Ohta H, Fujisawa T, Cho M, Ikeya M, Kidoaki S, Kosodo Y (2019) Brain-stiffness-mimicking tilapia collagen gel promotes the induction of dorsal cortical neurons from human pluripotent stem cells. Sci Rep 9(1):3068. https://doi.org/10.1038/s41598-018-38395-5
31. Keung AJ, Asuri P, Kumar S, Schaffer DV (2012) Soft microenvironments promote the early neurogenic differentiation but not self-renewal of human pluripotent stem cells. Integr Biol (Camb) 4(9):1049–1058. https://doi.org/10.1039/c2ib20083j
32. Xu Z, Li Y, Li P, Sun Y, Lv S, Wang Y, He X, Xu J, Xu Z, Li L, Li Y (2022) Soft substrates promote direct chemical reprogramming of fibroblasts into neurons. Acta Biomater 152:255–272. https://doi.org/10.1016/j.actbio.2022.08.049
33. Kreysing E, Gautier H, Humphrey R, Mooslehner K, Muresan L, Haarhoff D, Mukherjee S, Zhao X, Winkel A, Dimitracopoulos A, Pillai EK, Káradóttir RT, Franze K (2024) Substrate stiffness regulates neuronal maturation via Piezo1-mediated TTR activity. BioRxiv. https://doi.org/10.1101/2024.07.25.605128

34. Teixeira AI, Ilkhanizadeh S, Wigenius JA, Duckworth JK, Inganäs O, Hermanson O (2009) The promotion of neuronal maturation on soft substrates. Biomaterials 30(27):4567–4572. https://doi.org/10.1016/j.biomaterials.2009.05.013
35. Weissmüller G, Garcia-Abreu J, Mascarello Bisch P, Moura Neto V, Cavalcante LA (2000) Glial cells with differential neurite growth-modulating properties probed by atomic force microscopy. Neurosci Res 38(2):217–220. https://doi.org/10.1016/S0168-0102(00)00159-0
36. Wang P-Y, Thissen H, Kingshott P (2016) Modulation of human multipotent and pluripotent stem cells using surface nanotopographies and surface-immobilised bioactive signals: a review. Acta Biomater 45:31–59. https://doi.org/10.1016/j.actbio.2016.08.054
37. Zhang Z, Lv Y, Harati J, Song J, Du P, Ou P, Liang J, Wang H, Wang PY (2023) Submicron-grooved films modulate the directional alignment and biological function of Schwann cells. J Funct Biomater 14(5):238. https://doi.org/10.3390/jfb14050238
38. Harati J, Tao X, Shahsavarani H, Du P, Galluzzi M, Liu K, Zhang Z, Shaw P, Shokrgozar MA, Pan H, Wang PY (2022) Polydopamine-mediated protein adsorption alters the epigenetic status and differentiation of primary human adipose-derived stem cells (HASCs). Front Bioeng Biotechnol 10:934179. https://doi.org/10.3389/fbioe.2022.934179
39. Ding S, Kingshott P, Thissen H, Pera M, Wang PY (2017) Modulation of human mesenchymal and pluripotent stem cell behavior using biophysical and biochemical cues: a review. Biotechnol Bioeng 114(2):260–280. https://doi.org/10.1002/bit.26075
40. Lin J, Zeng J, Sun W, Liu K, Enkhbat M, Yi D, Harati J, Liu J, Kingshott P, Chen B, Ma F, Wang PY (2021) Colloidal self-assembled patterns maintain the pluripotency and promote the hemopoietic potential of human embryonic stem cells. Front Cell Dev Biol 9:771773. https://doi.org/10.3389/fcell.2021.771773
41. Deng K, Du P, Liu K, Tao X, Harati J, Jhang JW, Kim J, Wang PY (2021) Programming colloidal self-assembled patterns (CSAPs) into thermo-responsible hybrid surfaces for controlling human stem cells and macrophages. ACS Appl Mater Interfaces 13(16):18563–18580. https://doi.org/10.1021/acsami.1c02969
42. Lim SH, Liu XY, Song H, Yarema KJ, Mao HQ (2010) The effect of nanofiber-guided cell alignment on the preferential differentiation of neural stem cells. Biomaterials 31(34):9031–9039. https://doi.org/10.1016/j.biomaterials.2010.08.021
43. Ghazali ZS, Eskandari M, Bonakdar S, Renaud P, Mashinchian O, Shalileh S, Bonini F, Uckay I, Preynat-Seauve O, Braschler T (2021) Neural priming of adipose-derived stem cells by cell-imprinted substrates. Biofabrication 13(3). https://doi.org/10.1088/1758-5090/abc66f
44. Li Z, Wang W, Kratz K, Küchler J, Xu X, Zou J, Deng Z, Sun X, Gossen M, Ma N, Ma N, Lendlein A (2016) Influence of surface roughness on neural differentiation of human induced pluripotent stem cells. Clin Hemorheol Microcirc 64(3):355–366. https://doi.org/10.3233/CH-168121
45. Litowczenko J, Wychowaniec JK, Załęski K, Marczak Ł, Edwards-Gayle CJC, Tadyszak K, Maciejewska BM (2023) Micro/nano-patterns for enhancing differentiation of human neural stem cells and fabrication of nerve conduits via soft lithography and 3d printing. Biomater Adv 154:213653. https://doi.org/10.1016/j.bioadv.2023.213653
46. Yang K, Jung H, Lee H-R, Lee JS, Kim SR, Song KY, Cheong E, Bang J, Im SG, Cho S-W (2014) Multiscale, hierarchically patterned topography for directing human neural stem cells into functional neurons. ACS Nano 8(8):7809–7822. https://doi.org/10.1021/nn501182f
47. Nicholson C, Hrabětová S (2017) Brain extracellular space: the final frontier of neuroscience. Biophys J 113(10):2133–2142. https://doi.org/10.1016/j.bpj.2017.06.052
48. Li Y-M, Ji Y, Meng Y-X, Kim Y-J, Lee H, Kurian AG, Park J-H, Yoon J-Y, Knowles JC, Choi Y, Kim Y-S, Yoon B-E, Singh RK, Lee H-H, Kim H-W, Lee J-H (2024) Neural tissue-like, not supraphysiological, electrical conductivity stimulates neuronal lineage specification through calcium signaling and epigenetic modification. Adv Sci 11(35):2400586. https://doi.org/10.1002/advs.202400586
49. Koppes AN, Keating KW, McGregor AL, Koppes RA, Kearns KR, Ziemba AM, McKay CA, Zuidema JM, Rivet CJ, Gilbert RJ, Thompson DM (2016) Robust neurite extension following

exogenous electrical stimulation within single walled carbon nanotube-composite hydrogels. Acta Biomater 39:34–43. https://doi.org/10.1016/j.actbio.2016.05.014

50. Wang S, Guan S, Sun C, Liu H, Liu T, Ma X (2023) Electrical stimulation enhances the neuronal differentiation of neural stem cells in three-dimensional conductive scaffolds through the voltage-gated calcium ion channel. Brain Res 1798:148163. https://doi.org/10.1016/j.brainres.2022.148163
51. Wei H, Gu Y, Li A, Song P, Liu D, Sun F, Ma X, Qian X (2024) Conductive 3D Ti3C2Tx MXene-Matrigel hydrogels promote proliferation and neuronal differentiation of neural stem cells. Colloids Surf B Biointerfaces 233:113652. https://doi.org/10.1016/j.colsurfb.2023.113652
52. Li K, Li P, Fan Y (2019) The assembly of silk fibroin and graphene-based nanomaterials with enhanced mechanical/conductive properties and their biomedical applications. J Mater Chem B 7(44):6890–6913. https://doi.org/10.1039/c9tb01733j
53. Yang B, Wang P-B, Mu N, Ma K, Wang S, Yang C-Y, Huang Z-B, Lai Y, Feng H, Yin G-F, Chen T-N, Hu C-S (2021) Graphene oxide-composited chitosan scaffold contributes to functional recovery of injured spinal cord in rats. Neural Regen Res 16(9):1829–1835. https://doi.org/10.4103/1673-5374.306095
54. Fang Q, Zhang Y, Chen X, Li H, Cheng L, Zhu W, Zhang Z, Tang M, Liu W, Wang H, Wang T, Shen T, Chai R (2020) Three-dimensional graphene enhances neural stem cell proliferation through metabolic regulation. Front Bioeng Biotechnol:7. https://doi.org/10.3389/fbioe.2019.00436
55. Kim D, Yoo JM, Hwang H, Lee J, Lee SH, Yun SP, Park MJ, Lee MJ, Choi S, Kwon SH, Hong BH, Ko HS (2018) Graphene quantum dots prevent α-synucleinopathy in Parkinson's disease. Nat Nanotechnol 13(9):812–818. https://doi.org/10.1038/s41565-018-0179-y
56. Sefton E, Iwasa SN, Morrison T, Naguib HE, Popovic MR, Morshead CM (2020) Electric field application in vivo regulates neural precursor cell behavior in the adult mammalian forebrain. eNeuro 7(4):1–12. https://doi.org/10.1523/ENEURO.0273-20.2020
57. Yang H, Su Y, Sun Z, Ma B, Liu F, Kong Y, Sun C, Li B, Sang Y, Wang S, Geng Z, Liu H (2022) Gold nanostrip array-mediated wireless electrical stimulation for accelerating functional neuronal differentiation. Adv Sci 9(22):e2202376. https://doi.org/10.1002/advs.202202376
58. James EC, Tomaskovic-Crook E, Crook JM (2024) Engineering 3D scaffold-free nanoparticle-laden stem cell constructs for piezoelectric enhancement of human neural tissue formation and function. Adv Sci 11(40):e2310010. https://doi.org/10.1002/advs.202310010
59. Zhang Y, Jiang J, Xie X, Jiang L, Wu C, Sun J, Zheng T, Fan H, Wu J (2024) Cell-anchored lead-free piezoelectric KNN NPs resisting washout for low-intensity ultrasound driven neuromodulation. Adv Funct Mater 34(42). https://doi.org/10.1002/adfm.202406919
60. Bargero A, Battaglini M, Curiale T, Carmignani A, Montorsi M, Labardi M, Pucci C, Marino A, Ciofani G (2025) Ultrasound-responsive polymeric piezoelectric nanoparticles for remote activation and neuronal differentiation of human neural stem cells. Small Sci 5(2):2400354. https://doi.org/10.1002/smsc.202400354
61. Kim T, Kim HJ, Choi W, Lee YM, Pyo JH, Lee J, Kim J, Kim J, Kim J-H, Kim C, Kim C, Kim WJ (2023) Deep brain stimulation by blood–brain-barrier-crossing piezoelectric nanoparticles generating current and nitric oxide under focused ultrasound. Nat Biomed Eng 7(2):149–163. https://doi.org/10.1038/s41551-022-00965-4
62. Hirano S, Takeichi M (2012) Cadherins in brain morphogenesis and wiring. Physiol Rev 92(2):597–634. https://doi.org/10.1152/physrev.00014.2011
63. Yagita Y, Sakurai T, Tanaka H, Kitagawa K, Colman DR, Shan W (2009) N-cadherin mediates interaction between precursor cells in the subventricular zone and regulates further differentiation. J Neurosci Res 87(15):3331–3342. https://doi.org/10.1002/jnr.22044
64. Dossi E, Zonca L, Pivonkova H, Milior G, Moulard J, Vargova L, Chever O, Holcman D, Rouach N (2024) Astroglial gap junctions strengthen hippocampal network activity by sustaining afterhyperpolarization via KCNQ channels. Cell Rep 43(5):114158. https://doi.org/10.1016/j.celrep.2024.114158

65. Fróes MM, Correia AHP, Garcia-Abreu J, Spray DC, Campos De Carvalho AC, Neto VM (1999) Gap-junctional coupling between neurons and astrocytes in primary central nervous system cultures. Proc Natl Acad Sci U S A 96(13):7541–7546. https://doi.org/10.1073/pnas.96.13.7541
66. Cherry JF, Bennett NK, Schachner M, Moghe PV (2014) Engineered N-cadherin and L1 biomimetic substrates concertedly promote neuronal differentiation, neurite extension and neuroprotection of human neural stem cells. Acta Biomater 10(10):4113–4126. https://doi.org/10.1016/j.actbio.2014.06.001
67. Chou S-F, Lai J-Y, Cho C-H, Lee C-H (2016) Relationships between surface roughness/stiffness of chitosan coatings and fabrication of corneal keratocyte spheroids: effect of degree of deacetylation. Colloids Surf B Biointerfaces 142:105–113. https://doi.org/10.1016/j.colsurfb.2016.02.051
68. Long KR, Ffrench-Constant C (2014) Neural stem cell quiescence comes to an un-sticky end. Nat Cell Biol 16(7):625–627. https://doi.org/10.1038/ncb3003
69. Liu W, Xu B, Xue W, Yang B, Fan Y, Chen B, Xiao Z, Xue X, Sun Z, Shu M, Zhao Y, Dai J (2020) A functional scaffold to promote the migration and neuronal differentiation of neural stem/progenitor cells for spinal cord injury repair. Biomaterials 243:119941. https://doi.org/10.1016/j.biomaterials.2020.119941
70. Paerhati P, Ito A, Yoshioka K, Iwamoto K, Fujiwara S, Horie M, Kawabe Y, Kamihira M (2019) Neural differentiation of mouse induced pluripotent stem cells using cadherin gene-engineered PA6 feeder cells. J Biosci Bioeng 127(5):633–640. https://doi.org/10.1016/j.jbiosc.2018.10.009
71. O'Grady BJ, Balotin KM, Bosworth AM, McClatchey PM, Weinstein RM, Gupta M, Poole KS, Bellan LM, Lippmann ES (2020) Development of an N-cadherin biofunctionalized hydrogel to support the formation of synaptically connected neural networks. ACS Biomater Sci Eng 6(10):5811–5822. https://doi.org/10.1021/acsbiomaterials.0c00885
72. Cavanaugh M, Willits RK (2025) Mechanotransductive N-cadherin binding induces differentiation in human neural stem cells. Mechanobiol Med 3(1):100099. https://doi.org/10.1016/j.mbm.2024.100099
73. Baek J, Kumar S, Schaffer DV, Im SG (2022) N-cadherin adhesive ligation regulates mechanosensitive neural stem cell lineage commitment in 3D matrices. Biomater Sci 10(23):6768–6777. https://doi.org/10.1039/d2bm01349e
74. Mui KL, Bae YH, Gao L, Liu SL, Xu T, Radice GL, Chen CS, Assoian RK (2015) N-cadherin induction by ECM stiffness and FAK overrides the spreading requirement for proliferation of vascular smooth muscle cells. Cell Rep 10(9):1477–1486. https://doi.org/10.1016/j.celrep.2015.02.023
75. Barcelona-Estaje E, Oliva MAG, Cunniffe F, Rodrigo-Navarro A, Genever P, Dalby MJ, Roca-Cusachs P, Cantini M, Salmeron-Sanchez M (2024) N-cadherin crosstalk with integrin weakens the molecular clutch in response to surface viscosity. Nat Commun 15(1):8824. https://doi.org/10.1038/s41467-024-53107-6
76. Mui KL, Chen CS, Assoian RK (2016) The mechanical regulation of integrin-cadherin crosstalk organizes cells, signaling and forces. J Cell Sci 129(6):1093–1100. https://doi.org/10.1242/jcs.183699
77. Rinaldi F, Hartfield EM, Crompton LA, Badger JL, Glover CP, Kelly CM, Rosser AE, Uney JB, Caldwell MA (2014) Cross-regulation of connexin43 and β-catenin influences differentiation of human neural progenitor cells. Cell Death Dis 5(1):e1017. https://doi.org/10.1038/cddis.2013.546
78. Lemcke H, Nittel ML, Weiss DG, Kuznetsov SA (2013) Neuronal differentiation requires a biphasic modulation of gap junctional intercellular communication caused by dynamic changes of Connexin43 expression. Eur J Neurosci 38(2):2218–2228. https://doi.org/10.1111/ejn.12219
79. Ferguson EL, Naseer S, Powell LC, Hardwicke J, Young FI, Zhu B, Liu Q, Song B, Thomas DW (2018) Controlled release of dextrin-conjugated growth factors to support growth and

differentiation of neural stem cells. Stem Cell Res 33:69–78. https://doi.org/10.1016/j.scr.2018.10.008

80. Chung CY, Yang JT, Kuo YC (2013) Polybutylcyanoacrylate nanoparticle-mediated neurotrophin-3 gene delivery for differentiating IPS cells into neurons. Biomaterials 34(22):5562–5570. https://doi.org/10.1016/j.biomaterials.2013.04.006
81. Cho Y-W, Jee S, Suhito IR, Lee J-H, Park CG, Choi KM, Kim T-H (2022) Single metal-organic framework–embedded nanopit arrays: a new way to control neural stem cell differentiation. Sci Adv 8(16):eabj7736. https://doi.org/10.1126/sciadv.abj7736
82. Liu X, Ren X, Deng X, Huo Y, Xie J, Huang H, Jiao Z, Wu M, Liu Y, Wen T (2010) A protein interaction network for the analysis of the neuronal differentiation of neural stem cells in response to titanium dioxide nanoparticles. Biomaterials 31(11):3063–3070. https://doi.org/10.1016/j.biomaterials.2009.12.054
83. Dayem AA, Kim B, Gurunathan S, Choi HY, Yang G, Saha SK, Han D, Han J, Kim K, Kim JH, Cho SG (2014) Biologically synthesized silver nanoparticles induce neuronal differentiation of SH-SY5Y cells via modulation of reactive oxygen species, phosphatases, and kinase signaling pathways. Biotechnol J 9(7):934–943. https://doi.org/10.1002/biot.201300555
84. Zhou Z, Zhang Y, Zeng Y, Yang D, Mo J, Zheng Z, Zhang Y, Xiao P, Zhong X, Yan W (2024) Effects of nanomaterials on synthesis and degradation of the extracellular matrix. ACS Nano 18(11):7688–7710. https://doi.org/10.1021/acsnano.3c09954
85. Li Z, Meng Z, Zhao Z (2022) Silk fibroin nanofibrous scaffolds incorporated with MicroRNA-222 loaded chitosan nanoparticles for enhanced neuronal differentiation of neural stem cells. Carbohydr Polym 277:118791. https://doi.org/10.1016/j.carbpol.2021.118791
86. Santos T, Ferreira R, Maia J, Agasse F, Xapelli S, Cortes L, Bragancia J, Malva JO, Ferreira L, Bernardino L (2012) Polymeric nanoparticles to control the differentiation of neural stem cells in the subventricular zone of the brain. ACS Nano 6(12):10463–10474. https://doi.org/10.1021/nn304541h
87. Gao Y, Liu H, Zheng Q, Yang L, Cao G, Yuan J, Hu S, Li Z (2024) Modulating efficient differentiation of neural stem cells into neurons by using plasmonic nanoparticles and the NIR II irradiation to boost therapy of Parkinson's disease. Nano Today 57:102392. https://doi.org/10.1016/j.nantod.2024.102392
88. Chen JJ, Wang T, Di An C, Jiang CY, Zhao J, Li S (2016) Brain-derived neurotrophic factor: a mediator of inflammation-associated neurogenesis in Alzheimer's disease. Rev Neurosci 27(8):793–811. https://doi.org/10.1515/revneuro-2016-0017
89. Sung P-S, Lin P-Y, Liu C-H, Su H-C, Tsai K-J (2020) Neuroinflammation and neurogenesis in Alzheimer's disease and potential therapeutic approaches. Int J Mol Sci 21(3):701. https://doi.org/10.3390/ijms21030701
90. Cheng F, Kotha S, Fu M, Yang Q, Wang H, He W, Mao X (2024) Nanozyme enabled protective therapy for neurological diseases. Nano Today 54:102142. https://doi.org/10.1016/j.nantod.2023.102142
91. Xiao L, Wei F, Zhou Y, Anderson GJ, Frazer DM, Lim YC, Liu T, Xiao Y (2020) Dihydrolipoic acid-gold nanoclusters regulate microglial polarization and have the potential to alter neurogenesis. Nano Lett 20(1):478–495. https://doi.org/10.1021/acs.nanolett.9b04216
92. Zou P, Wu C, Liu TC-Y, Duan R, Yang L (2023) Oligodendrocyte progenitor cells in Alzheimer's disease: from physiology to pathology. Transl Neurodegener 12(1):52. https://doi.org/10.1186/s40035-023-00385-7
93. Cheng Q, Ma X, Liu J, Feng X, Liu Y, Wang Y, Ni W, Song M (2023) Pharmacological inhibition of the asparaginyl endopeptidase (AEP) in an Alzheimer's disease model improves the survival and efficacy of transplanted neural stem cells. Int J Mol Sci 24(9):7739. https://doi.org/10.3390/ijms24097739
94. Xin Y, Fang F, Yue Q, Luo Y, Tian S, Cheng L, Wang X, Yang X, Luo L, Meng F (2024) Microenvironment modulating nanogels by shiitake-derived lentinan and a reactive oxygen species scavenging conjugated polymer for the treatment of Alzheimer's disease. Nano Today 55:102178. https://doi.org/10.1016/j.nantod.2024.102178

95. Kumaraswamy P, Sethuraman S, Krishnan UM (2014) Development of a dual nanocarrier system as a potential stratagem against amyloid-induced toxicity. Exp Opin Drug Deliv 11(8):1131–1147. https://doi.org/10.1517/17425247.2014.912211
96. Chafekar SM, Malda H, Merkx M, Meijer EW, Viertl D, Lashuel HA, Baas F, Scheper W (2007) Branched KLVFF tetramers strongly potentiate inhibition of β-amyloid aggregation. ChemBioChem 8(15):1857–1864. https://doi.org/10.1002/cbic.200700338
97. Pederzoli F, Ruozi B, Duskey J, Hagmeyer S, Sauer AK, Grabrucker S, Coelho R, Oddone N, Ottonelli I, Daini E, Tosi G, Grabrucker AM (2019) Nanomedicine against Aβ aggregation by β–sheet breaker peptide delivery: in vitro evidence. Pharmaceutics 11(11):572. https://doi.org/10.3390/pharmaceutics11110572
98. Qu A, Huang F, Li A, Yang H, Zhou H, Long J, Shi L (2017) The synergistic effect between KLVFF and self-assembly chaperones on both disaggregation of beta-amyloid fibrils and reducing consequent toxicity. Chem Commun 53(7):1289–1292. https://doi.org/10.1039/C6CC07803F
99. Yau W-M, Tycko R (2018) Depletion of amyloid-β peptides from solution by sequestration within fibril-seeded hydrogels. Protein Sci 27(7):1218–1230. https://doi.org/10.1002/pro.3387
100. Ucar B, Humpel C (2019) Therapeutic efficacy of glial cell-derived neurotrophic factor loaded collagen scaffolds in ex vivo organotypic brain slice Parkinson's disease models. Brain Res Bull 149:86–95. https://doi.org/10.1016/j.brainresbull.2019.04.012
101. Moriarty N, Pandit A, Dowd E (2017) Encapsulation of primary dopaminergic neurons in a gdnf-loaded collagen hydrogel increases their survival, re-innervation and function after intrastriatal transplantation. Sci Rep 7(1):16033. https://doi.org/10.1038/s41598-017-15970-w
102. Zhang X, Le W (2010) Pathological role of hypoxia in Alzheimer's disease. Exp Neurol 223(2):299–303. https://doi.org/10.1016/j.expneurol.2009.07.033
103. Ortega JA, Sirois CL, Memi F, Glidden N, Zecevic N (2017) Oxygen levels regulate the development of human cortical radial glia cells. Cerebral Cortex 27(7):3736–3751. https://doi.org/10.1093/cercor/bhw194
104. Wang Y, Zoneff ER, Thomas JW, Hong N, Tan LL, McGillivray DJ, Perriman AW, Law KCL, Thompson LH, Moriarty N, Jackson CJ, Nisbet DR (2023) Hydrogel oxygen reservoirs increase functional integration of neural stem cell grafts by meeting metabolic demands. Nat Commun 14(1):457. https://doi.org/10.1038/s41467-023-36133-8
105. Zhou S, Jia F, Wei Y, Du J, Liu J, Dong W, Sui J, Xue W, Yang Y, Chen L, Meng X, Yu S (2024) Metal–organic framework nanocomposites in conquering hypoxia for tumor therapy. Adv Funct Mater 34(17). https://doi.org/10.1002/adfm.202314012
106. Abutarboush R, Mullah SH, Saha BK, Haque A, Walker PB, Aligbe C, Pappas G, Tran Ho LTV, Arnaud FG, Auker CR, Scultetus AH, Moon-Massat P (2018) Brain oxygenation with a non-vasoactive perfluorocarbon emulsion in a rat model of traumatic brain injury. Microcirculation 25(3):e12441. https://doi.org/10.1111/micc.12441
107. Qian F, Han Y, Han Z, Zhang D, Zhang L, Zhao G, Li S, Jin G, Yu R, Liu H (2021) In situ implantable, post-trauma microenvironment-responsive, ROS depletion hydrogels for the treatment of traumatic brain injury. Biomaterials 270:120675. https://doi.org/10.1016/j.biomaterials.2021.120675
108. Zhang D, Ren Y, He Y, Chang R, Guo S, Ma S, Guan F, Yao M (2022) In situ forming and biocompatible hyaluronic acid hydrogel with reactive oxygen species-scavenging activity to improve traumatic brain injury repair by suppressing oxidative stress and neuroinflammation. Mater Today Bio 15:100278. https://doi.org/10.1016/j.mtbio.2022.100278
109. Cruz EM, Machado LS, Zamproni LN, Bim LV, Ferreira PS, Pinto LA, Pessan LA, Backes EH, Porcionatto MA (2023) A gelatin methacrylate-based hydrogel as a potential bioink for 3D bioprinting and neuronal differentiation. Pharmaceutics 15(2):627. https://doi.org/10.3390/pharmaceutics15020627
110. Kivijärvi T, Goksøyr Ø, Yassin MA, Jain S, Yamada S, Morales-López A, Mustafa K, Finne-Wistrand A (2022) Hybrid material based on hyaluronan hydrogels and poly(L-Lactide-

Co-1,3-Trimethylene Carbonate) scaffolds toward a cell-instructive microenvironment with long-term in vivo degradability. Mater Today Bio 17:100483. https://doi.org/10.1016/j.mtbio.2022.100483

111. Cook DJ, Nguyen C, Chun HN, Llorente IL, Chiu AS, Machnicki M, Zarembinski TI, Carmichael ST (2017) Hydrogel-delivered brain-derived neurotrophic factor promotes tissue repair and recovery after stroke. J Cereb Blood Flow Metab 37(3):1030–1045. https://doi.org/10.1177/0271678X16649964
112. Qiao E, Fulmore CA, Schaffer DV, Kumar S (2024) Substrate stress relaxation regulates neural stem cell fate commitment. Proc Natl Acad Sci U S A 121(28):e2317711121. https://doi.org/10.1073/pnas.2317711121
113. Yan Y, Bejoy J, Marzano M, Li Y (2019) The use of pluripotent stem cell-derived organoids to study extracellular matrix development during neural degeneration. Cells 8(3):242. https://doi.org/10.3390/cells8030242
114. Itokazu Y, Yu RK (2014) Amyloid β-Peptide 1–42 modulates the proliferation of mouse neural stem cells: upregulation of fucosyltransferase IX and notch signaling. Mol Neurobiol 50(1):186–196. https://doi.org/10.1007/s12035-014-8634-8
115. Peng B, Yao X, Risacher SL, Saykin AJ, Shen L, Ning X (2020) Cognitive biomarker prioritization in Alzheimer's disease using brain morphometric data. BMC Med Inform Decis Mak 20(1):319. https://doi.org/10.1186/s12911-020-01339-z
116. Morawski M, Brückner G, Jäger C, Seeger G, Matthews RT, Arendt T (2012) Involvement of perineuronal and perisynaptic extracellular matrix in Alzheimer's disease neuropathology. Brain Pathology 22(4):547–561. https://doi.org/10.1111/j.1750-3639.2011.00557.x
117. Downs M, Zaia J, Sethi MK (2023) Mass spectrometry methods for analysis of extracellular matrix components in neurological diseases. Mass Spectrometry Rev 42(5):1848–1875. https://doi.org/10.1002/mas.21792
118. Tang-Schomer MD, White JD, Tien LW, Schmitt LI, Valentin TM, Graziano DJ, Hopkins AM, Omenetto FG, Haydon PG, Kaplan DL (2014) Bioengineered functional brain-like cortical tissue. Proc Natl Acad Sci U S A 111(38):13811–13816. https://doi.org/10.1073/pnas.1324214111
119. Soliman BG, Chin IL, Li Y, Ishii M, Ho MH, Doan VK, Cox TR, Wang PY, Lindberg GCJ, Zhang YS, Choi YS, Lim KS (2024) Droplet-based microfluidics for engineering shape-controlled hydrogels with stiffness gradient. Biofabrication 16(4). https://doi.org/10.1088/1758-5090/ad6d8e
120. Lee S-H, Li Z, Zhang EY, Kim DH, Huang Z, Heo Y, Lee SJ, Kang H-W, Burdick JA, Mauck RL, Mauck RL, Heo SC (2025) Precision repair of zone-specific meniscal injuries using a tunable extracellular matrix-based hydrogel system. Bioact Mater 48:400–413. https://doi.org/10.1016/j.bioactmat.2025.02.013
121. Neumann M, di Marco G, Iudin D, Viola M, van Nostrum CF, van Ravensteijn BGP, Vermonden T (2023) Stimuli-responsive hydrogels: the dynamic smart biomaterials of tomorrow. Macromolecules 56(21):8377–8392. https://doi.org/10.1021/acs.macromol.3c00967
122. Roth JG, Brunel LG, Huang MS, Liu Y, Cai B, Sinha S, Yang F, Paşca SP, Shin S, Heilshorn SC (2023) Spatially controlled construction of assembloids using bioprinting. Nat Commun 14(1):4346. https://doi.org/10.1038/s41467-023-40006-5
123. Zhu Y, Zhang X, Sun L, Wang Y, Zhao Y (2023) Engineering human brain assembloids by microfluidics. Adv Mater 35(14):e2210083. https://doi.org/10.1002/adma.202210083
124. Xu L, Ding H, Wu S, Xiong N, Hong Y, Zhu W, Chen X, Han X, Tao M, Wang Y, Wang D, Xu M, Huo D, Gu Z, Liu Y (2024) Artificial meshed vessel-induced dimensional breaking growth of human brain organoids and multiregional assembloids. ACS Nano 18(38):26201–26214. https://doi.org/10.1021/acsnano.4c07844
125. Charles S, Jackson-Holmes E, Sun G, Zhou Y, Siciliano B, Niu W, Han H, Nikitina A, Kemp ML, Wen Z, Wen Z, Lu H (2025) Non-invasive quality control of organoid cultures using mesofluidic CSTR bioreactors and high-content imaging. Adv Mater Technol 10(3):2400473. https://doi.org/10.1002/admt.202400473

126. Abdulla A, Chen S, Chen Z, Wang Y, Yan H, Chen R, Ahmad KZ, Liu K, Yan C, He J, Jiang L, Ding X (2023) Three-dimensional microfluidics with dynamic fluidic perturbation promotes viability and uniformity of human cerebral organoids. Biosens Bioelectron 240:115635. https://doi.org/10.1016/j.bios.2023.115635
127. Mathes TG, Monirizad M, Ermis M, de Barros NR, Rodriguez M, Kraatz H-B, Jucaud V, Khademhosseini A, Falcone N (2024) Effects of amyloid-β-mimicking peptide hydrogel matrix on neuronal progenitor cell phenotype. Acta Biomater 183:89–100. https://doi.org/10.1016/j.actbio.2024.05.020
128. Hwang C, Choi MH, Kim HE, Jeong SH, Park JU (2022) Reactive oxygen species-generating hydrogel platform for enhanced antibacterial therapy. NPG Asia Mater 14(1):72. https://doi.org/10.1038/s41427-022-00420-5
129. Haki M, Shafaei N, Moeini M (2025) In situ gelling silk fibroin/ECM hydrogel with sustained oxygen release for neural tissue engineering applications. J Biomed Mater Res A 113(1):e37837. https://doi.org/10.1002/jbm.a.37837
130. Zambuto SG, Serrano JF, Vilbert AC, Lu Y, Harley BAC, Pedron S (2020) Response of neuroglia to hypoxia-induced oxidative stress using enzymatically crosslinked hydrogels. MRS Commun 10(1):83–90. https://doi.org/10.1557/mrc.2019.159
131. Olufunmilayo EO, Gerke-Duncan MB, Holsinger RMD (2023) Oxidative stress and antioxidants in neurodegenerative disorders. Antioxidants 12(2):517. https://doi.org/10.3390/antiox12020517
132. Homma K, Chang AC, Yamamoto S, Tamate R, Ueki T, Nakanishi J (2021) Design of azobenzene-bearing hydrogel with photoswitchable mechanics driven by photo-induced phase transition for in vitro disease modeling. Acta Biomater 132:103–113. https://doi.org/10.1016/j.actbio.2021.03.028
133. Coelho A, Sousa N (2022) Magnetic resonance elastography of the ageing brain in normal and demented populations: a systematic review. Hum Brain Mapp 43(13):4207–4218. https://doi.org/10.1002/hbm.25891
134. Arani A, Murphy MC, Glaser KJ, Manduca A, Lake DS, Kruse SA, Jack CR, Ehman RL, Huston J (2015) Measuring the effects of aging and sex on regional brain stiffness with MR elastography in healthy older adults. Neuroimage 111:59–64. https://doi.org/10.1016/j.neuroimage.2015.02.016
135. Pavon N, Diep K, Yang F, Sebastian R, Martinez-Martin B, Ranjan R, Sun Y, Pak CH (2024) Patterning ganglionic eminences in developing human brain organoids using a morphogen-gradient-inducing device. Cell Reports Methods 4(1):100689. https://doi.org/10.1016/j.crmeth.2023.100689
136. Qazi TH, Blatchley MR, Davidson MD, Yavitt FM, Cooke ME, Anseth KS, Burdick JA (2022) Programming hydrogels to probe spatiotemporal cell biology. Cell Stem Cell 29(5):678–691. https://doi.org/10.1016/j.stem.2022.03.013
137. Yan Y, Li X, Gao Y, Mathivanan S, Kong L, Tao Y, Dong Y, Li X, Bhattacharyya A, Zhao X, Zhang SC (2024) 3D bioprinting of human neural tissues with functional connectivity. Cell Stem Cell 31(2):260–274.e7. https://doi.org/10.1016/j.stem.2023.12.009
138. Jin Y, Mikhailova E, Lei M, Cowley SA, Sun T, Yang X, Zhang Y, Liu K, Catarino da Silva D, Campos Soares L, Bandiera S, Szele FG, Molnár Z, Zhou L, Bayley H (2023) Integration of 3D-printed cerebral cortical tissue into an ex vivo lesioned brain slice. Nat Commun 14(1):5986. https://doi.org/10.1038/s41467-023-41356-w
139. Pavlou G, Spitz S, Pramotton FM, Tsai A, Li BM, Wang X, Barr OM, Ko EC, Zhang S, Ashley SJ, Maaser-Hecker A, Choi SH, Jorfi M, Tanzi RE, Kamm RD (2025) Engineered 3D human neurovascular model of Alzheimer's disease to study vascular dysfunction. Biomaterials 314:122864. https://doi.org/10.1016/j.biomaterials.2024.122864
140. Abdelrahman S, Alsanie WF, Khan ZN, Albalawi HI, Felimban RI, Moretti M, Steiner N, Chaudhary AG, Hauser CAE (2022) A Parkinson's disease model composed of 3D bioprinted dopaminergic neurons within a biomimetic peptide scaffold. Biofabrication 14(4). https://doi.org/10.1088/1758-5090/ac7eec

141. Pavlou G, Spitz S, Pramotton FM, Tsai A, Li BM, Wang X, Barr OM, Ko EC, Zhang S, Ashley SJ, Tanzi RE, Kamm RD (2025) Engineered 3D human neurovascular model of Alzheimer's disease to study vascular dysfunction. Biomaterials 314:122864. https://doi.org/10.1016/j.biomaterials.2024.122864
142. Park J, Lee BK, Jeong GS, Hyun JK, Lee CJ, Lee S-H (2015) Three-dimensional brain-on-a-chip with an interstitial level of flow and its application as an in vitro model of Alzheimer's disease. Lab Chip 15(1):141–150. https://doi.org/10.1039/c4lc00962b
143. Fernandes JTS, Chutna O, Chu V, Conde JP, Outeiro TF (2016) A novel microfluidic cell co-culture platform for the study of the molecular mechanisms of Parkinson's disease and other synucleinopathies. Front Neurosci 10:511. https://doi.org/10.3389/fnins.2016.00511
144. Ruiz A, Joshi P, Mastrangelo R, Francolini M, Verderio C, Matteoli M (2014) Testing Aβ toxicity on primary CNS cultures using drug-screening microfluidic chips. Lab Chip 14(15):2860–2866. https://doi.org/10.1039/c4lc00174e
145. Liang CC, Chen PY, Liu NC, Lee IC (2024) Comparison between dynamic versus static models and real-time monitoring of neuronal dysfunction in an Amyloid-ß induced neuronal toxic model on a chip platform. Lab Chip 24(7):1887–1902. https://doi.org/10.1039/d3lc00507k
146. Zhi Y, Zhu Y, Wang J, Zhao J, Zhao Y (2023) Cortical organoid-on-a-chip with physiological hypoxia for investigating Tanshinone IIA-induced neural differentiation. Research 6:0273. https://doi.org/10.34133/research.0273
147. Tan SY, Feng X, Cheng LKW, Wu AR (2023) Vascularized human brain organoid on-chip. Lab Chip 23(12):2693–2709. https://doi.org/10.1039/D2LC01109C
148. Quintard C, Tubbs E, Jonsson G, Jiao J, Wang J, Werschler N, Laporte C, Pitaval A, Bah TS, Pomeranz G, Bissardon C, Kaal J, Leopoldi A, Long DA, Blandin P, Achard JL, Battail C, Hagelkruys A, Navarro F, Fouillet Y, Penninger JM, Gidrol X (2024) A microfluidic platform integrating functional vascularized organoids-on-chip. Nat Commun 15(1):1452. https://doi.org/10.1038/s41467-024-45710-4
149. Zhang S, Wan Z, Kamm RD (2021) Vascularized organoids on a chip: strategies for engineering organoids with functional vasculature. Lab Chip 21(3):473–488. https://doi.org/10.1039/D0LC01186J
150. Kook MG, Lee SE, Shin N, Kong D, Kim DH, Kim MS, Kang HK, Choi SW, Kang KS (2022) Generation of cortical brain organoid with vascularization by assembling with vascular spheroid. Int J Stem Cells 15(1):85–94. https://doi.org/10.15283/IJSC21157
151. Soden PA, Henderson AR, Lee E (2022) A microfluidic model of AQP4 polarization dynamics and fluid transport in the healthy and inflamed human brain: the first step towards glymphatics-on-a-chip. Adv Biol 6(12):e2200027. https://doi.org/10.1002/adbi.202200027
152. Pan X, Li J, Li W, Wang H, Durisic N, Li Z, Feng Y, Liu Y, Zhao CX, Wang T (2022) Axons-on-a-chip for mimicking non-disruptive diffuse axonal injury underlying traumatic brain injury. Lab Chip 22(23):4541–4555. https://doi.org/10.1039/d2lc00730d
153. Kim JW, Choi YY, Park S-H, Ha JH, Lee HU, Kang T, Sun W, Chung BG (2022) Microfluidic electrode array chip for electrical stimulation-mediated axonal regeneration. Lab Chip 22(11):2122–2130. https://doi.org/10.1039/d1lc01158h
154. Krauss R, Bosanac T, Devraj R, Engber T, Hughes RO (2020) Axons matter: the promise of treating neurodegenerative disorders by targeting SARM1-mediated axonal degeneration. Trends Pharmacol Sci 41(4):281–293. https://doi.org/10.1016/j.tips.2020.01.006
155. Ristola M, Fedele C, Hagman S, Sukki L, Kapucu FE, Mzezewa R, Hyvärinen T, Kallio P, Priimagi A, Narkilahti S (2021) Directional growth of human neuronal axons in a microfluidic device with nanotopography on azobenzene-based material. Adv Mater Interfaces 8(11):2100048. https://doi.org/10.1002/admi.202100048
156. Courte J, Renault R, Jan A, Viovy J-L, Peyrin J-M, Villard C (2018) Reconstruction of directed neuronal networks in a microfluidic device with asymmetric microchannels. Method Cell Biol 148:71–95. https://doi.org/10.1016/bs.mcb.2018.07.002
157. Gao C, Jiang J, Tan Y, Chen S (2023) Microglia in neurodegenerative diseases: mechanism and potential therapeutic targets. Signal Transduct Target Ther 8(1):359. https://doi.org/10.1038/s41392-023-01588-0

158. Zhang Y, Savvidou M, Liaudanskaya V, Singh P, Fu Y, Nasreen A, Coe M, Kelly M, Snapper D, Wagner C, Beheshti A, Georgakoudi I (2024) Synergistic label-free fluorescence imaging and MiRNA studies reveal dynamic human neuron-glial metabolic interactions following injury. Sci Adv 10(50):eadp1980. https://doi.org/10.1126/sciadv.adp1980
159. Robertson G, Bushell TJ, Zagnoni M (2014) Chemically induced synaptic activity between mixed primary hippocampal co-cultures in a microfluidic system. Integr Biol (U K) 6(6):636–644. https://doi.org/10.1039/c3ib40221e
160. Park J, Wetzel I, Marriott I, Dréau D, D'Avanzo C, Kim DY, Tanzi RE, Cho H (2018) A 3D human triculture system modeling neurodegeneration and neuroinflammation in Alzheimer's disease. Nat Neurosci 21(7):941–951. https://doi.org/10.1038/s41593-018-0175-4
161. Mulay AR, Hwang J, Kim D (2024) Microphysiological blood-brain barrier systems for disease modeling and drug development. Adv Healthc Mater 13(21):e2303180. https://doi.org/10.1002/adhm.202303180
162. Garcia-Polite F, Martorell J, Del Rey-Puech P, Melgar-Lesmes P, O'Brien CC, Roquer J, Ois A, Principe A, Edelman ER, Balcells M (2017) Pulsatility and high shear stress deteriorate barrier phenotypein brain microvascular endothelium. J Cereb Blood Flow Metab 37(7):2614–2625. https://doi.org/10.1177/0271678X16672482
163. Nawara TJ, Yuan J, Seeley LD, Sztul E, Mattheyses AL (2025) Fluidic shear stress alters clathrin dynamics and vesicle formation in endothelial cells. Biophys J 124(11):1763–1770. https://doi.org/10.1016/j.bpj.2024.06.007
164. Seo D, Holtzman DM (2024) Current understanding of the Alzheimer's disease-associated microbiome and therapeutic strategies. Exp Mol Med 56(1):86–94. https://doi.org/10.1038/s12276-023-01146-2
165. Das TK, Ganesh BP (2023) Interlink between the gut microbiota and inflammation in the context of oxidative stress in Alzheimer's disease progression. Gut Microbes 15(1):2206504. https://doi.org/10.1080/19490976.2023.2206504
166. Wang X, Sun G, Feng T, Zhang J, Huang X, Wang T, Xie Z, Chu X, Yang J, Wang H, Ding J, Geng M (2019) Sodium oligomannate therapeutically remodels gut microbiota and suppresses gut bacterial amino acids-shaped neuroinflammation to inhibit Alzheimer's disease progression. Cell Res 29(10):787–803. https://doi.org/10.1038/s41422-019-0216-x
167. Shabbir U, Tyagi A, Elahi F, Aloo SO, Oh DH (2021) The potential role of polyphenols in oxidative stress and inflammation induced by gut microbiota in Alzheimer's disease. Antioxidants 10:1370. https://doi.org/10.3390/antiox10091370
168. Bello-Corral L, Alves-Gomes L, Fernández-Fernández JA, Fernández-García D, Casado-Verdejo I, Sánchez-Valdeón L (2023) Implications of gut and oral microbiota in neuroinflammatory responses in Alzheimer's disease. Life Sci 333:122132. https://doi.org/10.1016/j.lfs.2023.122132
169. Kim M-S, Kim Y, Choi H, Kim W, Park S, Lee D, Kim DK, Kim HJ, Choi H, Hyun D-W, Bae J-W, Mook-Jung I (2020) Transfer of a healthy microbiota reduces amyloid and tau pathology in an Alzheimer's disease animal model. Gut 69(2):283–294. https://doi.org/10.1136/gutjnl-2018-317431
170. Zhuang Z-Q, Shen L-L, Li W-W, Fu X, Zeng F, Gui L, Lü Y, Cai M, Zhu C, Tan Y-L, Bu X-L, Wang Y-J (2018) Gut microbiota is altered in patients with Alzheimer's disease. J Alzheimers Disease 63(4):1337–1346. https://doi.org/10.3233/JAD-180176
171. Kim MH, Kim D, Sung JH (2021) A gut-brain axis-on-a-chip for studying transport across epithelial and endothelial barriers. J Ind Eng Chem 101:126–134. https://doi.org/10.1016/j.jiec.2021.06.021
172. Seo GM, Lee H, Kang YJ, Kim D, Sung JH (2024) Development of in vitro model of exosome transport in microfluidic gut-brain axis-on-a-chip. Lab Chip 24(19):4581–4593. https://doi.org/10.1039/d4lc00490f

173. Uzel SGM, Platt RJ, Subramanian V, Pearl TM, Rowlands CJ, Chan V, Boyer LA, So PTC, Kamm RD (2016) Microfluidic device for the formation of optically excitable, three-dimensional, compartmentalized motor units. Sci Adv 2(8):e1501429. https://doi.org/10.1126/sciadv.1501429
174. Osaki T, Uzel SGM, Kamm RD (2020) On-chip 3D neuromuscular model for drug screening and precision medicine in neuromuscular disease. Nat Protoc 15(2):421–449. https://doi.org/10.1038/s41596-019-0248-1
175. Moshayedi P, Nih LR, Llorente IL, Berg AR, Cinkornpumin J, Lowry WE, Segura T, Carmichael ST (2016) Systematic optimization of an engineered hydrogel allows for selective control of human neural stem cell survival and differentiation after transplantation in the stroke brain. Biomaterials 105:145–155. https://doi.org/10.1016/j.biomaterials.2016.07.028
176. Wevers NR, Nair AL, Fowke TM, Pontier M, Kasi DG, Spijkers XM, Hallard C, Rabussier G, van Vught R, Vulto P, de Vries HE, Lanz HL (2021) Modeling ischemic stroke in a triculture neurovascular unit on-a-chip. Fluids Barriers CNS 18(1):59. https://doi.org/10.1186/s12987-021-00294-9
177. Sheets KT, Ewend MG, Mohiti-Asli M, Tuin SA, Loboa EG, Aboody KS, Hingtgen SD (2020) Developing implantable scaffolds to enhance neural stem cell therapy for post-operative glioblastoma. Mol Ther 28(4):1056–1067. https://doi.org/10.1016/j.ymthe.2020.02.008
178. Yi HG, Jeong YH, Kim Y, Choi YJ, Moon HE, Park SH, Kang KS, Bae M, Jang J, Youn H, Paek SH, Cho DW (2019) A bioprinted human-glioblastoma-on-a-chip for the identification of patient-specific responses to chemoradiotherapy. Nat Biomed Eng 3(7):509–519. https://doi.org/10.1038/s41551-019-0363-x
179. Gao K, Han H, Cranick MG, Zhao S, Xu S, Yin B, Song H, Hu Y, Clarke MT, Wang D, Wong JM, Zhao Z, Burgstone BW, Farmer DL, Murthy N, Wang A (2024) Widespread gene editing in the brain via in utero delivery of MRNA using acid-degradable lipid nanoparticles. ACS Nano 18(44):30293–30306. https://doi.org/10.1021/acsnano.4c05169
180. Vatine GD, Barrile R, Workman MJ, Sances S, Barriga BK, Rahnama M, Barthakur S, Kasendra M, Lucchesi C, Kerns J, Wen N, Spivia WR, Chen Z, Van Eyk J, Svendsen CN (2019) Human IPSC-derived blood-brain barrier chips enable disease modeling and personalized medicine applications. Cell Stem Cell 24(6):995–1005.e6. https://doi.org/10.1016/j.stem.2019.05.011
181. Kamte YS, Chandwani MN, Michaels AC, O'donnell LA (2021) Neural stem cells: what happens when they go viral? Viruses 13(8):1468. https://doi.org/10.3390/v13081468
182. Selvan ST, Padmanabhan P, Gulyás BZ (2020) Nanotechnology-based diagnostics and therapy for pathogen-related infections in the CNS. ACS Chem Neurosci 11(16):2371–2377. https://doi.org/10.1021/acschemneuro.9b00470
183. Kim J, Lee KT, Lee JS, Shin J, Cui B, Yang K, Choi YS, Choi N, Lee SH, Lee JH, Bahn YS, Cho SW (2021) Fungal brain infection modelled in a human-neurovascular-unit-on-a-chip with a functional blood–brain barrier. Nat Biomed Eng 5(8):830–846. https://doi.org/10.1038/s41551-021-00743-8
184. Hu N, Shi J-X, Chen C, Xu H-H, Chang Z-H, Hu P-F, Guo D, Zhang X-W, Shao W-W, Fan X, Zuo J-C, Ming D, Li X-H (2024) Constructing organoid-brain-computer interfaces for neurofunctional repair after brain injury. Nat Commun 15(1):9580. https://doi.org/10.1038/s41467-024-53858-2
185. Huang Q, Tang B, Romero JC, Yang Y, Elsayed SK, Pahapale G, Lee TJ, Pantoja IEM, Han F, Berlinicke C, Xiang T, Solazzo M, Hartung T, Qin Z, Caffo BS, Smirnova L, Gracias DH (2022) Shell microelectrode arrays (MEAs) for brain organoids. Sci Adv 8(33):eabq5031. https://doi.org/10.1126/sciadv.abq5031
186. Martinelli E, Akouissi O, Liebi L, Furfaro I, Maulà D, Savoia N, Remy A, Nikles L, Roux A, Stoppini L, Lacour SP (2025) The E-flower: a hydrogel-actuated 3D MEA for brain spheroid electrophysiology. Sci Adv 10(42):eadp8054. https://doi.org/10.1126/sciadv.adp8054

187. Prudon N, Cordero-Espinoza L, Abarkan M, Gurchenkov B, Morel C, Lepleux M, De Luca V, Lartigue M, Cabanas H, Pujol N, Faggiani E, Feyeux M (2024) Bioreactor-produced IPSCs-derived dopaminergic neuron-containing neural microtissues innervate and normalize rotational bias in a dose-dependent manner in a Parkinson rat model. Neurotherapeutics 21(5):e00436. https://doi.org/10.1016/j.neurot.2024.e00436
188. Chen X, Sun G, Tian E, Zhang M, Davtyan H, Beach TG, Reiman EM, Blurton-Jones M, Holtzman DM, Shi Y (2021) Modeling sporadic Alzheimer's disease in human brain organoids under serum exposure. Adv Sci 8(18):e2101462. https://doi.org/10.1002/advs.202101462

Chapter 5
Stem Cell-Derived 3D Models to Study Alzheimer's Disease

Jose Inzunza, Aphrodite Demetriou, Patricia Muñoz, Jonathan Arias, and Ivan Nalvarte

Abstract Alzheimer's disease (AD) is a devastating age-related neurodegenerative disease. Despite significant research efforts, there is currently no general established disease-modifying therapy that can halt or reverse the progression of AD. A key challenge lies in the lack of relevant experimental models to effectively study disease progression. In recent years, complex stem cell-derived 3D brain models have emerged as promising new tools to study human brain development, brain aging, and neurodegeneration. These innovative models offer renewed hope for identifying disease-modifying therapies. However, significant challenges must be addressed before these 3D models can achieve broader application. This chapter provides an overview of the latest advancements in 3D brain in vitro models, their relevance in AD research, and their advantages and limitations.

Keywords Alzheimer's disease · Brain organoids · Assembloids · 3D models · Differentiation · Microglia · Bioreactors · Animal replacement · Aβ aggregates

---

J. Inzunza
Department of Laboratory Medicine, Division of Clinical Microbiology, Karolinska Institutet Stem Cell Organoid Facility (KISCO), Karolinska Institutet, Huddinge, Sweden

A. Demetriou · P. Muñoz · I. Nalvarte (✉)
Department of Neurobiology Care Sciences and Society, Division of Neurogeriatrics, Karolinska Institutet, Solna, Sweden
e-mail: ivan.nalvarte@ki.se

J. Arias
Department of Laboratory Medicine, Division of Clinical Microbiology, Karolinska Institutet Stem Cell Organoid Facility (KISCO), Karolinska Institutet, Huddinge, Sweden

VU LSC - EMBL Partnership Institute for Gene Editing Technologies, Laboratory of Nuclease-Enabled Cell Therapies, Vilnius University, Vilnius, Lithuania

R. C. B. Wong and Y. Hayashi (eds.), *Advances in Using Stem Cells to Treat Neural Diseases*, Stem Cell Biology and Regenerative Medicine 77,
https://doi.org/10.1007/978-3-031-94101-6_5

5.1 Introduction

Alzheimer's disease (AD) is an age-related neurodegenerative disease and the most common type of dementia. Today, there are around 55 million diagnosed cases of AD world-wide, and given the growing elderly population, this number is expected to reach 150 million by 2050 [1, 2]. The hallmarks of AD are accumulation and deposition of insoluble neurotoxic amyloid beta peptides ($A\beta_{42}$) in and around synapses, followed by neuroinflammation and intracellular accumulation of hyperphosphorylated tau protein that form insoluble neurofibrillary tangles in neurons, all leading to synaptic and neuronal loss, memory and cognitive deterioration, and ultimately to death. Familial AD (fAD) displays mutations in AD-risk genes, such as *APP*, *PSEN1*, *PSEN2*, and *TREM2*, which actively participate in the pathological processes [3, 4]. However, fAD only explains approximately 5% of all AD cases. In contrast, sporadic AD (sAD) comprises 95% of cases, but the etiology of sAD is not well known. Clearly, a complex interaction between genetics, environment, and disease history participates in disease risks [5]. Furthermore, women have a higher risk of being diagnosed with AD, suggesting that sex-hormonal factors are also involved [6].

Recently, there has been some success with antibody-based therapy directed against soluble Aβ protofibrils in early AD, which activates microglial cells to clear Aβ aggregates, thereby providing the first disease-modifiable treatment [7]. However, concerns about the efficacy and feasibility of this treatment have been raised [7, 8], and there is still no cure for AD. In fact, over 400 clinical trials that have shown preclinical promise have failed when reaching human testing. A major reason for this failure is the lack of appropriate human-relevant experimental models of AD.

5.2 Experimental Models to Study AD

5.2.1 Animal Models

Mouse models are the most common in vivo system of AD, encompassing dozens of different genetically modified strains that, to different extents, mimic AD. The majority of these overexpress mutated (i) amyloid precursor protein (*APP*) that promotes its cleavage to neurotoxic Aβ, (ii) gamma-secretase subunit presenilin 1 (*PSEN1*) that participates in APP cleavage, and (iii) Tau protein (*MAPT*) that directs it to hyperphosphorylation and aggregation. There are combinations of these mutations that result in different degrees of AD pathology in the mouse brain. Late-onset AD mouse models have also been developed, such as the *App-NL-F* [9] and *APOE4* strains [10]. The mouse models have contributed significantly to our understanding of the disease, and they are very popular since they can rapidly reproduce amyloid and tau pathology with accompanying increases in neuroinflammation [11, 12].

However, there are detrimental species-specific differences between the AD mouse models and human AD. Perhaps the most striking difference is that mice present very limited synaptic loss compared to humans, and there is almost no neuronal loss [12, 13], which leads to limited memory performance deterioration. In addition, the induction of amyloidosis in the mouse brain does not lead to tau hyperphosphorylation, which is seen in human AD, suggesting that the overall AD progression is fundamentally different between mice and humans [13]. Another drawback of the transgenic AD mouse models is that the mutated genes are overexpressed giving rise to artifacts that interfere with pathology. More recent AD models utilize knock-in strategies to avoid these artifacts and have shown promise in modeling amyloidosis [9] and tau pathology [14] more accurately. However, their limited synaptic and neuronal loss still results in modest effects on memory decline [13]. Transgenic rat models of AD have been proposed as a better alternative to mice since rats have more complex cognitive and social behaviors and can, in contrast to similar AD mouse models, show immunoreactivity to phosphorylated tau after amyloidosis, which is more similar to human AD [15, 16]. Additionally, non-human primates can mimic human AD very well, including sAD [17]. Nevertheless, rats and non-human primates are rarely used in AD research due to the lack of genetically modified strains, as well as their longer timespan to pathology. These animal models require larger facilities to house them, which limits throughput and increases costs [12, 17]. The use of non-human primates in research also comes with important ethical aspects [17]. These limitations in the experimental modeling of AD are major causes for the limited success rate in finding disease-modifying treatments. Therefore, great hope is placed on the recent advances in human 3D brain culture systems to provide experimental models that can more accurately mimic human disease progression and enable preclinical drug testing.

5.2.2 3D Models and Organoids

The discovery that adult somatic cells can be reprogrammed to pluripotent fate has revolutionized disease modeling [18]. Such induced pluripotent stem cells (iPSCs) can be derived from AD patients carrying AD-driving mutations and differentiated to neurons in 2D culture to study the impact of AD mutations on their functionality. The same approach can be used for iPSCs from sporadic AD patients. However, a hallmark of AD is the deposition of amyloid aggregates around the synapses. Such aggregations rarely form in 2D cultures due to the constant wash-out of culture medium, resulting in the absence of this pathological hallmark. Additionally, 2D cultured neurons do not form mature and accurate functional interactions compared to the in vivo context [19, 20]. In this sense, iPSC-derived 3D brain models and organoids may be a better option, trapping amyloid peptides in the 3D environment. To this end, different 3D culture methods and approaches have been laid out: (i) brain organoids, (ii) neurospheroids and assembloids, (iii) microfluidic organ-on-chips, and (iv) primary brain slices. In this chapter, we will not discuss microfluidic organ-on-chips or primary brain slices, but focus on the brain organoid and neurospheroids/assembloids approaches, highlighting the advantages and limitations of these methods.

5.3 Guided and Unguided Differentiation of Brain Organoids

In the last decade, brain organoids have attracted attention as promising models to study the brain, reflected by the sharp increase in number of research articles and citations since the first publication on their use in 2008 [21], and since the first use of the term cerebral organoids in 2013 [22] (Fig. 5.1a). Similarly, brain organoids have been used to study AD—with the first publication in 2016 (Fig. 5.1b). Although brain organoids have proven useful in studying neurodevelopmental diseases, their use in modeling neurodegenerative diseases has been more challenging, which will be discussed in this chapter. But first, we will introduce the two main types of organoid differentiation protocols, unguided and guided differentiation protocols, and highlight some advantages and limitations of these protocols.

5.3.1 Unguided Differentiation of Brain Organoids

The first 3D cell culture model of the brain was developed by Eiraku *et al.* [21]. They used a method that relaied on the self-organized formation of apico-basally polarized cortical tissues from human embryonic stem cells (ESCs) using an efficient three-dimensional aggregation culture (SFEBq culture). SFEBq organoid displayed the formation of polarized cortical neuroepithelial tissue, Cajal-Retzius cells, and cortical-plate neurons. Emerging studies used a similar method without the addition of specific patterning factors (besides progesterone in some protocols),

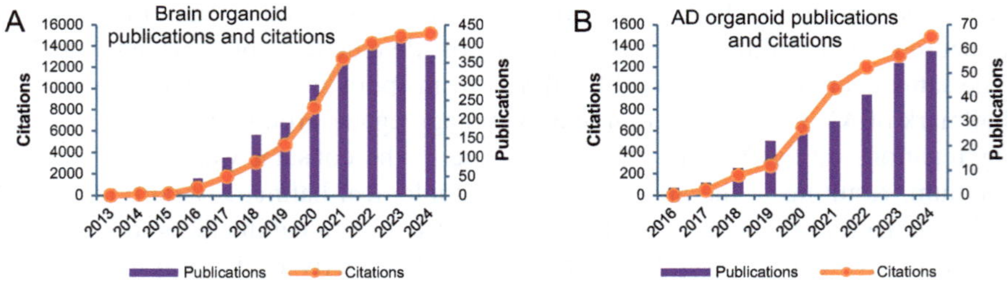

Fig. 5.1 Number of publications and citations of (**a**) brain organoid and (**b**) Alzheimer's organoid publications between 2013–2024. A literature search was conducted on Web of Science using the search terms "brain organoid(s)" or "Cerebral organoid(s)" in (**a**), and "Alzheimer's disease" and "organoid(s)" in (**b**)

relying on the self-organizing potential of iPSCs to form different tissues [18, 22–25] which led to a cerebral organoid containing different brain regions such as hippocampal and retinal structures, and different cortical regions, as well as ventricles [22, 26]. These cerebral organoids have the advantage of randomly forming brain structures, and are derived by a relatively simple protocol that guides to neuroectodermal differentiation and maturation [22]. Indeed, the protocol also allows differentiation of astrocytes after more extended culture periods [27], mimicking human brain development. Furthermore, the addition of oligodendrocyte patterning cues (PDGF-AA and IGF1, followed by T3) has been demonstrated to induce oligodendrocyte precursor cell (OPC) expansion and differentiation into mature myelinating oligodendrocytes [28]. But this is a slow process, requiring more than 200 days in culture to produce minimum adequate myelination [28]. However, Ng and coworkers [29] identified that iPSC differentiation into myelinating oligodendrocytes can be accelerated by inducing SOX9 expression at the beginning of organoid production, which reduced the timeline for myelin formation from 103–210 days to just 40 days. A limitation of unguided protocols is the inherent variability in organoid formation, which can lead to batch-to-batch and intra-batch inconsistencies. This variability arises from the lack of precise control over the differentiation process, resulting in heterogeneous cellular compositions and developmental outcomes across different organoid batches [30–32]. Another limitation is that the unguided protocols are based on expelling non-neuronal cell types during Matrigel embedding, which poses a risk of incomplete expelling, resulting in contamination of the organoid with unwanted cell types. Indeed, reports have shown that mesodermal progenitors may reside in the organoid [33, 34]. But this could also be an advantage as demonstrated by Ormel and colleagues [33] that modified the protocol by reducing the levels of neuroectodermal stimulants and delayed Matrigel embedding, leading to the development of microglia-like cells from trapped mesodermal progenitor cells. Microglia are, as we discuss below, an important cell type in the pathophysiology of brain diseases.

5.3.2 Guided Differentiation of Brain Organoids

Unguided organoid differentiation has the disadvantage that it generates heterogeneous organoids containing multiple brain regions, which may complicate the analysis of specific brain functions and pathologies, and thereby limit their applicability [35]. For this reason, guided protocols have been developed to direct the iPSCs into distinct brain regions. These protocols enhance the efficiency of neural induction and promote the formation of distinct neural identities by using exogenous signals such as WNT, BMP, and SHH [36, 37]. Two of the most affected regions in AD are the forebrain and the hippocampus. Although attempts to generate hippocampal organoids or spheroids have been made (e.g., [36, 38, 39]), the typical hippocampal regional and cellular architecture is not recreated, which may question their relevance as models of the hippocampus. In addition, the lack of good hippocampus-specific markers makes it hard to exclude the presence of non-hippocampal regions in these 3D models. In contrast, forebrain organoid protocols have been developed, based on temporal modulation of WNT, BMP, and SHH. This can accurately guide differentiation to specific functional forebrain regions, including cortical and subcortical regions, as well as dorsal and ventral forebrain regions [40–42]. Forebrain organoids closely mimic the complex architecture and cellular diversity of the developing human brain, replicating key aspects of early cortical development, including the formation of distinct forebrain structures, and the generation of various neuronal subtypes, such as glutamatergic pyramidal neurons and GABAergic interneurons [43, 44]. Thus, forebrain organoids have accelerated in popularity as a model of neurodevelopment.

However, the use of brain organoids to model the aging brain and neurodegenerative diseases such as AD is still controversial. Although brain organoids can recapitulate key aspects of neurodevelopment, they do not represent an aging brain, and they normally lack microglia, which are a key cell type in neurodegenerative diseases. In addition, full maturation of neurons, astrocytes, and oligodendrocytes requires lengthy protocols, limiting the use of high-throughput platforms, while increasing the risk of necrotic core formation or microbiological contamination.

5.3.3 Bioreactors and Other Methods to Minimize Necrosis

Miniaturized spinning bioreactors provide a low-shear environment that facilitates the efficient exchange of nutrients and gases, reducing the risk of necrotic cores and thereby allowing growth over extended times without affecting organoid health negatively and allowing growth of larger organoids [40, 45]. Similarly, orbital shakers also provide efficient exchange of nutrients and gases [26, 46] and have the advantage of low cost and easy maintenance but have been reported to not be as efficient in preventing necrotic cores [40]. Importantly, since brain organoids lack vasculature, keeping organoids in constant low-shear suspension movement, either in a bioreactor or on an orbital shaker, is a critical step in generating healthy organoids.

Although in vitro vascularization of tissue-engineered constructs is a major goal in the field of bioengineering, it has not yet been achieved in a way that preserves self-organization and tissue architecture. To enhance oxygen supply while preserving the accessibility and scalability of in vitro models, researchers explored a novel strategy by adapting a classic organotypic slice culture technique using the air-liquid interface. Cerebral organoids cultured at the air–liquid interface (referred to as ALI-COs) exhibit significantly improved survival, enhanced morphology, and extensive axon outgrowths resembling nerve tracts [47]. Future microfluidic applications to generate vascularized organoids could improve the long-term culturing of organoids even more and could thus aid in acquiring more "aged" phenotypes, which is needed to model neurodegeneration.

5.3.4 Microglia Co-cultures

Microglia cell is an important neuroinflammatory cell type in the brain that participates in synaptic pruning, debris clearance, and immune responses. Fagerlund *et al.* showed that microglia-like cells promote neuronal functions and network maturation in cerebral organoids [48]. But since microglia originate from the mesoderm, and most brain organoid protocols rely on neuroectodermal differentiation, microglia are not generated in most brain organoid protocols, which is a limitation and can result in missing out on crucial physiological processes. This is especially important when studying neurological diseases where the neuroimmune system is a key player in pathology, such as in AD. Microglia become activated in response to Aβ aggregation, working to clear the aggregates by secreting enzymes and cytokines that facilitate their elimination [49]. However, excessive or chronic activation can result in the release of neurotoxic factors, contributing to neuronal damage and cognitive decline [49, 50]. Although Ormel *et al.* [33] showed that the unguided organoid protocol could give rise to microglia-like cells, Kälvälä [51] noted that this process introduces variability in cell composition, making guided patterning a more reliable method for generating homogeneous organoids that through co-culturing methods (e.g., assembloid techniques, see below) can include microglia [52]. Additionally, Xu *et al.* [53] highlighted the potential to control the ratio of microglia in organoids by co-culturing iPSC-derived primitive macrophage progenitors (PMPs) with neural progenitor cells (NPCs), thereby enhancing the study of microglial functions in both development and disease contexts. Xu *et al.* [53] demonstrated that a ratio of 7:3 of NPCs to PMPs resulted in ca 8% microglia in the mature cell population of the brain organoid, which is comparable to the human brain setting in which microglia are abundant during development, participating in developmental processes including synaptic pruning and neuronal survival, but drop at more mature stages to ca 5–10% of the total brain cell population [54, 55]. Furthermore, a recent discovery that overexpression of PU.1, a transcription factor involved in hematopoiesis, in organoid cultures results in microglia differentiation [56] and may be an amenable way to introduce microglia into the system.

It can be noted that controlling neuroinflammation in an organoid setting can be challenging, since external cues such as growth factors, cytokines, pH, and osmosis are known to modulate microglia activation, morphology, and inflammatory responses [52, 57, 58]. Unless controlled, neuroinflammation can destroy the organoid.

5.4 3D Models of Alzheimer's Disease

The limitations of animal models in Alzheimer's disease (AD) research have turned focus to the potential of human 3D brain cultures. This is an emerging and innovative field of research, with different models being developed to encompass all the

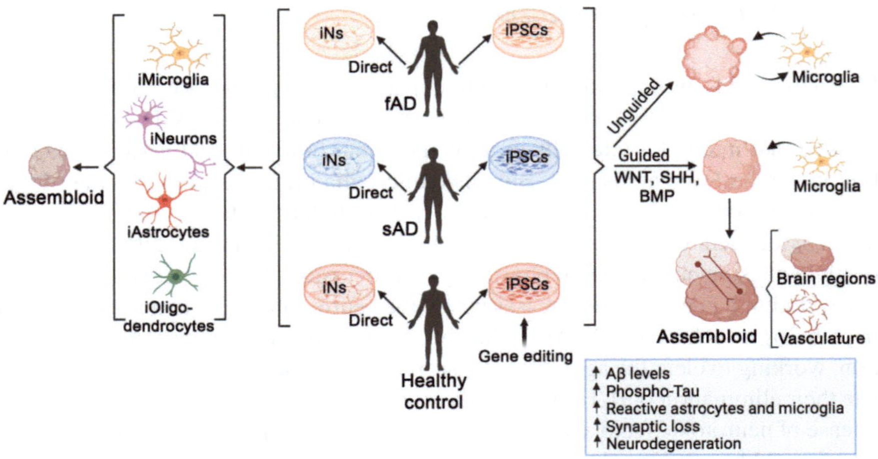

Fig. 5.2 Schematic illustration of the main organoid models in AD research. Human-induced pluripotent stem cells (hiPSCs) can be derived from familial (fAD) and sporadic AD (sAD) patients or from healthy control individuals. AD-driving mutations can be introduced at the iPSC stage. Unguided differentiation, relying on the self-organizing potential of iPSCs, yields cerebral organoids, while guided differentiation (e.g., by modulating WNT, SHH, and BMP pathways) can give rise to, e.g., forebrain, thalamic, and midbrain organoids. Co-culture and merging of different brain region-specific organoids results in assembloids. These assembloids can also be merged with cells from other lineages at various differentiation stages, such as vascular endothelial cells. Microglia, derived from mesodermal lineage can be incorporated into organoids using various protocols [52]. It was also shown that microglia can be present in cerebral organoids from unguided differentiation [33]. In addition to iPSC-derived organoids, it is possible to acquire microglia, neurons, astrocytes, and oligodendrocytes (iNeural cells (iNs)) from direct reprogramming of somatic cells. These directly reprogrammed cells retain the aging characteristic of the somatic cells and can be grown in 3D cultures alone or in combination to form assembloids. Organoids from sAD and fAD iPSCs present hallmarks of AD pathology, including increased Aβ and phospho-tau levels, increased neuroinflammation, and prominent synaptic and neuronal loss. There are advantages and limitations with all these protocols as discussed in the main text

molecular hallmarks of AD in vitro (Fig. 5.2). In the following sections, we will highlight some of the major achievements in this area.

5.4.1 AD Brain Organoids

As shown in Fig. 5.1b, there has been a sharp increase in AD brain organoid publications. This is not surprising due to the severe species-specific differences between rodent models and humans, and due to the difficulties in reproducing amyloid pathology in 2D cultures. The latter problem was exemplified when wild-type neuroepithelial stem cells were cultured within a self-assembling 3D peptide scaffolds that upon Aβ oligomer exposure produced in vivo-like responses associated with AD (e.g., p21-activated kinase-mediated Aβ sensing). These were not observed in traditional 2D cultures [59]. Thus, the 3D cultures mimic the cellular organization in the brain better than 2D cultures, giving more physiological relevance and possibilities to study the interaction of extracellular dynamics with pathology [60]. They also allow Aβ aggregation, which is not possible in 2D culture systems [61]. Similarly, brain organoids have shown great promise in modeling several of the hallmarks of AD in vitro. For example, it has been demonstrated that cerebral organoids derived from fAD patient iPSCs spontaneously develop Aβ and tau pathologies over time [62] and that fAD mutations lead to premature neurogenesis by reducing Notch signaling [63]. Similarly, Choe *et al.* [64] demonstrated that introducing fAD mutations generated AD pathology in 70-day old cerebral organoids. Another study by Ghatak *et al.* [65] showed that cerebral organoids can be utilized to investigate the hyperexcitability of neurons associated with AD, revealing altered electrophysiological neuronal activity in organoids carrying mutations linked to AD. They later also showed that 2-month-old cerebral fAD organoids placed on multielectrode arrays (MEAs) display hypersynchronous network activity compared to WT organoids [66].

The majority of AD 3D and organoid models are derived from fAD patients or are modified to carry AD-driving mutations (e.g., mutations in *APP*, *PSEN1*, and/or *PSEN2*) to mimic fAD [60]. As an example, cerebral organoids derived from AD patients with *PSEN1/2* mutations showed defective tissue patterning and altered development [67]. However, sporadic AD models have also been studied, of which organoids derived from *APOE-ε4* allele carriers (the most prevalent sporadic gene variant in AD) are the most common. These *APOE-ε4* organoids show elevated Aβ and phospho-tau levels, as well as synapse loss, but may require up to 6 months in culture before pathology can be studied [32, 68–70]. Other sAD models include incubation of 3D spheres and organoids with Aβ oligomers to study Aβ aggregation and/or tau-phosphorylation [71, 72]. Furthermore, the incubation of cerebral organoids with the small molecule Aftin-5 that shifts APP processing toward neurotoxic $A\beta_{42}$ has been demonstrated as a model of sAD [73]. However, extracellular Aβ aggregations and elevation in Tau-phosphorylation were missing in this model.

Although microglia play an integral part in both neurodevelopment and neurodegeneration, most organoid and 3D-spheroid models lack the microglial population since it originates from the mesoderm as opposed to neurons, astrocytes, and oligodendrocytes that originate from the ectoderm as discussed above. This means that studying AD pathology in organoids without microglia can miss important neuroprotective and neurotoxic processes.

iPSC-derived microglia have been extensively studied in 2D culture. As an example, Konttinen *et al.* [74] found that *PSEN1*ΔE9, *APPswe*, and *APOE-ε4* microglia-like cells confer disparate phenotypes. While *APOE-ε4* genotype had a profound impact on several functions of inflammatory responses, reduced metabolism, phagocytosis, and migration, the familial mutations *APPswe* and *PSEN1ΔE9* had only minor effects on functionality [74]. Recently, a few studies have successfully introduced microglia in AD organoids. Lin *et al.* [75] showed that fAD organoids in co-culture with *APOE-ε4* microglia had more extracellular Aβ aggregates compared to fAD organoids in co-culture with WT (*APOE-ε3*) microglia. Cai *et al.*

[72] used an acoustofluidic high-throughput assembly method to generate 3D spheres of AD neurons, astrocytes and microglia, and could demonstrate the localization of microglia to Aβ aggregates. Furthermore, Park *et al.* [76] generated a 3D microfluidic platform to generate a tri-culture organoid system containing fAD neurons, astrocytes, and microglia and could model microglia recruitment to Aβ aggregates as well as their neurotoxic functions such as axonal cleavage and nitric oxide (NO) release. It was also shown that microglia that develop innately in an unguided cerebral organoid differentiation protocol could model early microglial protective functions upon Aβ exposure, whereas co-culture protocols with mature microglia resulted in Aβ oligomerization and overactive microglia [77]. Such results are perhaps not surprising since co-culture of mature cells does not recreate their cooperative developmental trajectories, which may give rise to artifacts and excessive neuroinflammation. Thus, the methods by Xu *et al.* [37], using a controlled ratio of PMPs and NPCs in co-culture may be a good option for integrating microglia into AD organoids. As will be discussed below, the immature neuronal state of organoids is a limitation in modeling age-related diseases. Different strategies have been proposed to increase the biological age of the organoids [78], of which one promising method may be through telomere shortening [79]. Another method that has gained in popularity is direct reprogramming of aged somatic cells to neurons and glia.

5.4.2 Direct Reprogramming in AD Modeling

Brain organoids can mimic human brain development surprisingly well, including establishment of neurogenic ventricular zones with dividing radial glial cells that give rise to neurons and intermediate progenitors [22, 64, 80]. However, in contrast to the adult human brain, aged organoids maintain regions of neuronal stem cell renewal and increased neuronal differentiation, as illustrated by their *SOX2* and *PAX6* expression [22, 80]. This may mean that even aged organoids can maintain neurogenesis and that they may cope better with neuronal damage than the adult human brain, simply by replenishing damaged or dying neurons. Indeed, the aged phenotype is generally missing in brain organoid models, which can be controversial when using brain organoids to model neurodegenerative diseases such as AD. An alternative approach to iPSC-derived brain organoids is to use direct reprogramming of adult somatic cells, e.g., from an AD patient, which retains the aged cell phenotype.

Direct reprogramming, or transdifferentiation, is a process by which differentiated cells are converted directly into another cell type without passing through a pluripotent stem cell stage. This mechanism involves the introduction of specific transcription factors that drive the conversion of one cell type to another, effectively altering the cell's identity and function [81–84]. Unlike neurons derived from iPSCs, neurons directly reprogrammed from somatic cells retain aging characteristics, including significant DNA damage and disrupted nuclear organization [85], as well as aging-associated epigenetic and transcriptomic signatures, and age-related nucleocytoplasmic defects [86, 87]. For this reason, direct reprogramming has been proposed as a good method to more accurately obtain an aged or neurodegenerative phenotype in vitro compared to using iPSC-derived neuronal cultures that retain neural stem cell properties [88, 89]. However, using direct reprogramming comes with significant challenges. Perhaps the biggest challenge is that reprogrammed terminally differentiated neurons cannot replicate, which means that one would need large numbers of somatic cells to reprogram. This is not feasible for large-scale and high-throughput disease modeling since it would require a large number of donors and would risk of severe batch-to-batch variability because each experiment would also require its own transdifferentiation step. However, the applicability and value of direct reprogramming in AD modeling were demonstrated in a recent study by Sun *et al.* [90]. In particular, they demonstrated that direct reprogramming may be of notable importance when modeling late-onset sAD that is heavily dependent on the age signature [86, 91]. They used a micro-RNA-based direct reprogramming method converting fibroblasts from fAD and sAD patients to cortical neurons, and observed Aβ accumulation, seed-competent and insoluble tau, as well as neurodegeneration in both fAD and sAD models already after 30 days post-reprogramming. Chemical inhibition of APP amyloidogenic processing was effective in the early stages but not in the later timepoints [90]. Notably, they also demonstrate that age-associated retrotransposons may have a direct role in AD neuropathology [90]. This study underscores the importance of taking biological age into consideration when studying neurodegenerative diseases such as AD, and that direct reprogramming of aged somatic cells to aged neurons may be the most relevant experimental model of both fAD and sAD today. But obvious limitations exist that need to be addressed before wider use of directly reprogrammed 3D neuronal disease models will be a reality.

5.4.3 Assembloids as Models of AD

A limitation with current 3D and organoid models is that it is challenging to mimic interaction between distinct brain regions. Another limitation is that most organoid methods lack non-ectodermal derived cells, like microglia and endothelial cells. To address these limitations, assembloids that combine different brain region organoids or different cell populations stemming from different developmental lineages have been established [92]. For example, Xiang *et al.* [93] fused two organoids representing the thalamus and the cortex to study how the projections between these regions form during development. Furthermore, Miura *et al.* [94] fused cortical and striatal organoids, and could identify cortical-striatal deficits in connectivity in assembloids from patients with the neurodevelopmental 22q13.3 deletion syndrome. Another study of Miura *et al.* [95] generated functional loop assembloids that resemble the key elements of the cortico-striatal-thalamic-cortical circuit and reveal the emergence of synchronized patterns of neuronal activity. Similarly, Wu *et al.* [96] used striato-nigral assembloids to model medium spiny neuronal projection defects in Huntington's disease.

Assembloids have also been used to model neurodegenerative diseases. For example, Rickner *et al.* [97] developed an astrocyte-neuron assembloid to study cell-specific tau pathology and could demonstrate its use for rapid drug screening applications. Furthermore, age-induced midbrain-striatum assembloids have been used to model early phenotypes of Parkinson's disease in an age-inducible manner through progerin overexpression, highlighting the interconnected relationship between aging and neurodegeneration [98]. The lack of vasculature has also been addressed in assembloid models. Sun *et al.* [99] developed an assembloid from myeloid progenitors, umbilical vein endothelial cells, and neuroepithelial cells and could demonstrate vascular structures within the assembloid. These assembloids displayed early microglia and astrocyte reactivity when modified to carry a neurotoxic tau mutation, providing proof-of-concept as a working neuroinflammation model [99]. Another study by Kong [100] and coworkers generated a cortical-blood vessel assembloid system by fusing a cortical organoid with a blood-vessel organoid. They showed that SARS-Cov2 viral infection in this assembloid induced microvascular injury, and since both microglia and vascular system originate from the mesoderm, they could observe microglial activation in response to SARS-CoV2 infection, which surprisingly led to hallmarks of AD with increased Aβ and phospho-tau levels [100]. This furthermore illustrates the power of complex 3D systems in modeling the aging brain.

An interesting application of the assembloid technology is that different brain regions or cell types can be modified to carry disease-specific mutations or originate from healthy or AD patients. This means that a hybrid organoid can be established containing, for example, healthy neural cells in combination with isogenic gene-modified fAD microglia, or vice versa. This could enable the identification of cell type specific contributions to pathology, and can provide more precise therapeutic targeting.

5.4.4 Considerations on Modeling Disease

Brain organoid and assembloid models face limitations mostly due to the absence of functional vascularization as well as the lack of significant sensory and physiological inputs essential for shaping circuit activity during development [101]. One should keep in mind that bridging the gap between cellular models and clinical manifestations is challenging, and careful interpretation is needed when connecting cellular phenotypes to complex cognitive and behavioral processes. There is substantial genetic heterogeneity caused by somatic polymorphisms carried by the different iPSCs donors, which far exceeds the genetic variance observed in inbred animal models. It is crucial to always include appropriate genetically related controls and track detailed donor information, as well as to match clinical features, sex, and age to reduce variability. In addition, in vitro differentiation offers multiple alternative protocols. It is well known that differentiation factors and culture conditions can influence the epigenetic regulation such as methylation, histone modification and chromatin structure of iPSCs and therefore their differentiation properties. Thus, by accepting that modeling a disease with brain organoids might represent only specific aspects of said disease, we can conclude that it is important to incorporate validation of findings through independent experimental systems or clinical data.

5.5 Conclusions and Perspectives

Despite significant efforts to combat neurodegenerative diseases such as AD, there has been limited success in developing effective disease-modifying therapies. A major reason for this is the heavy reliance on rodent models in AD research. Therapies that seem to work in an AD mouse model most often show little efficacy in human clinical trials. Therefore, the implementation of new more human-relevant experimental models is urgently needed to combat the growing threat of AD and other neurodegenerative diseases, especially in the context of an ever-growing aging population. The recent US FDA ruling permitting evidence from human microphysiological systems and organoids to support clinical trial applications underscores the growing importance of deeply phenotyped, highly reproducible, and scalable brain organoid models. However, there are obvious challenges in using organoids to model age-related neurodegeneration, and to be aware of these challenges and limitations is important for designing studies using such microphysiological systems and for interpreting the data generated by these. Nevertheless, the recent advancements in brain organoids as human disease-relevant experimental models hold great promise to not only unmask disease-relevant pathways but also to provide deeper molecular understanding of the interaction between the aging phenotype and neurodegenerative processes, as well as serve as platforms for drug screening and personalized medicine. Clearly, although further refinements are needed, brain organoids have already demonstrated their utility in advancing our understanding of neurological disorders and accelerating the development of novel therapies.

Acknowledgments The Nalvarte lab is supported by the Swedish Research Council, 2023-02366, 2024-02689, Forska Utan Djurförsök, F2023-004, and the Karolinska Institutet KID Funds FUS-2021:0004. This work was conducted as part of the execution of project "Mission-driven Implementation of Science and Innovation Program" 02-002-P-0001, funded by the Economic Revitalization and Resilience Enhancement Plan "New Generation Lithuania" supporting the Arias Lab.

References

1. GBD 2019 Dementia Forecasting Collaborators (2022) Estimation of the global prevalence of dementia in 2019 and forecasted prevalence in 2050: an analysis for the global burden of disease study 2019. Lancet Public Health 7(2):e105–e125
2. Alzheimer's disease facts and figures. (2023). Alzheimers Dement 19(4):1598–1695
3. Bekris LM, Yu CE, Bird TD, Tsuang DW (2010) Genetics of Alzheimer disease. J Geriatr Psychiatry Neurol 23(4):213–227
4. Jónsson T, Stefánsson H, Steinberg S, Jónsdóttir I, Jónsson PV, Snædal J et al (2013) Variant of *TREM2* associated with the risk of Alzheimer's disease. N Engl J Med 368(2):107–116
5. Hogan R, Flamier A, Nardini E, Bernier G (2020) The role of BMI1 in late-onset sporadic Alzheimer's disease. Genes 11(7):825
6. Andrew MK, Tierney MC (2018) The puzzle of sex, gender and Alzheimer's disease: why are women more often affected than men? Womens Health (Lond) 14:1745506518817995
7. CHV D, Swanson CJ, Aisen P, Bateman RJ, Chen C, Gee M et al (2023) Lecanemab in early Alzheimer's disease. N Engl J Med 388(1):9–21
8. Nguyen HV, Mital S, Knopman DS, Alexander GC (2024) Cost-effectiveness of Lecanemab for individuals with early-stage Alzheimer disease. Neurology 102(7):e209218
9. Saito T, Matsuba Y, Mihira N, Takano J, Nilsson P, Itohara S et al (2014) Single app knock-in mouse models of Alzheimer's disease. Nat Neurosci 17(5):661–663
10. Balu D, Karstens AJ, Loukenas E, Maldonado Weng J, York JM, Valencia-Olvera AC et al (2019) The role of APOE in transgenic mouse models of AD. Neurosci Lett 707:134285
11. Zhong MZ, Peng T, Duarte ML, Wang M, Cai D (2024) Updates on mouse models of Alzheimer's disease. Mol Neurodegener 19(1):23
12. Drummond E, Wisniewski T (2017) Alzheimer's disease: experimental models and reality. Acta Neuropathol 133(2):155–175
13. Granzotto A, Vissel B, Sensi SL (2024) Lost in translation: inconvenient truths on the utility of mouse models in Alzheimer's disease research. elife 13:e90633
14. Saito T, Mihira N, Matsuba Y, Sasaguri H, Hashimoto S, Narasimhan S et al (2019) Humanization of the entire murine Mapt gene provides a murine model of pathological human tau propagation. J Biol Chem 294(34):12754–12765
15. Liu L, Orozco IJ, Planel E, Wen Y, Bretteville A, Krishnamurthy P et al (2008) A transgenic rat that develops Alzheimer's disease-like amyloid pathology, deficits in synaptic plasticity and cognitive impairment. Neurobiol Dis 31(1):46–57
16. Pang K, Jiang R, Zhang W, Yang Z, Li L-L, Shimozawa M et al (2022) An app knock-in rat model for Alzheimer's disease exhibiting aβ and tau pathologies, neuronal death and cognitive impairments. Cell Res 32(2):157–175
17. Li HW, Zhang L, Qin C (2019) Current state of research on non-human primate models of Alzheimer's disease. Animal Model Exp Med 2(4):227–238
18. Takahashi K, Yamanaka S (2006) Induction of pluripotent stem cells from mouse embryonic and adult fibroblast cultures by defined factors. Cell 126(4):663–676
19. Dehaene-Lambertz G, Spelke ES (2015) The infancy of the human brain. Neuron 88(1):93–109
20. Silbereis John C, Pochareddy S, Zhu Y, Li M, Sestan N (2016) The cellular and molecular landscapes of the developing human central nervous system. Neuron 89(2):248–268
21. Eiraku M, Watanabe K, Matsuo-Takasaki M, Kawada M, Yonemura S, Matsumura M et al (2008) Self-organized formation of polarized cortical tissues from ESCs and its active manipulation by extrinsic signals. Cell Stem Cell 3(5):519–532
22. Lancaster MA, Renner M, Martin CA, Wenzel D, Bicknell LS, Hurles ME et al (2013) Cerebral organoids model human brain development and microcephaly. Nature 501(7467):373–379
23. Chiaradia I, Lancaster MA (2020) Brain organoids for the study of human neurobiology at the interface of in vitro and in vivo. Nat Neurosci 23(12):1496–1508

24. Eiraku M, Takata N, Ishibashi H, Kawada M, Sakakura E, Okuda S et al (2011) Self-organizing optic-cup morphogenesis in three-dimensional culture. Nature 472(7341):51–56
25. Paşca AM, Sloan SA, Clarke LE, Tian Y, Makinson CD, Huber N et al (2015) Functional cortical neurons and astrocytes from human pluripotent stem cells in 3D culture. Nat Methods 12(7):671–678
26. Lancaster MA, Knoblich JA (2014) Generation of cerebral organoids from human pluripotent stem cells. Nat Protoc 9(10):2329–2340
27. Lanjewar SN, Sloan SA (2021) Growing glia: cultivating human stem cell models of gliogenesis in health and disease. Front Cell Dev Biol 9:649538
28. Madhavan M, Nevin ZS, Shick HE, Garrison E, Clarkson-Paredes C, Karl M et al (2018) Induction of myelinating oligodendrocytes in human cortical spheroids. Nat Methods 15(9):700–706
29. Ng AHM, Khoshakhlagh P, Rojo Arias JE, Pasquini G, Wang K, Swiersy A et al (2021) A comprehensive library of human transcription factors for cell fate engineering. Nat Biotechnol 39(4):510–519
30. Chiaradia I, Imaz-Rosshandler I, Nilges BS, Boulanger J, Pellegrini L, Das R et al (2023) Tissue morphology influences the temporal program of human brain organoid development. Cell Stem Cell 30(10):1351–67.e10
31. Qian X, Song H, Ming GL (2019) Brain organoids: advances, applications and challenges. Development 146(8):dev166074
32. Hernández D, Rooney LA, Daniszewski M, Gulluyan L, Liang HH, Cook AL et al (2022) Culture variabilities of human iPSC-derived cerebral organoids are a major issue for the modelling of phenotypes observed in Alzheimer's disease. Stem Cell Rev Rep 18(2):718–731
33. Ormel PR, Vieira de Sá R, van Bodegraven EJ, Karst H, Harschnitz O, Sneeboer MAM et al (2018) Microglia innately develop within cerebral organoids. Nat Commun 9(1):4167
34. Quadrato G, Nguyen T, Macosko EZ, Sherwood JL, Min Yang S, Berger DR et al (2017) Cell diversity and network dynamics in photosensitive human brain organoids. Nature 545(7652):48–53
35. Luo D, Xu J, Liu F, Gu Z (2024) Advances and challenges in cerebral organoids research. Adv Nanobiomed Res 4(5):2300126
36. Susaimanickam PJ, Kiral FR, Park IH (2022) Region specific brain organoids to study neurodevelopmental disorders. Int J Stem Cells 15(1):26–40
37. Xu R, Boreland AJ, Li X, Erickson C, Jin M, Atkins C et al (2021) Developing human pluripotent stem cell-based cerebral organoids with a controllable microglia ratio for Modeling brain development and pathology. Stem Cell Reports 16(8):1923–1937
38. Pomeshchik Y, Klementieva O, Gil J, Martinsson I, Hansen MG, de Vries T et al (2020) Human iPSC-derived hippocampal spheroids: an innovative tool for stratifying Alzheimer disease patient-specific cellular phenotypes and developing therapies. Stem Cell Reports 15(1):256–273
39. Ciarpella F, Zamfir RG, Campanelli A, Ren E, Pedrotti G, Bottani E et al (2021) Murine cerebral organoids develop network of functional neurons and hippocampal brain region identity. iScience 24(12):103438
40. Qian X, Nguyen Ha N, Song Mingxi M, Hadiono C, Ogden Sarah C, Hammack C et al (2016) Brain-region-specific organoids using mini-bioreactors for modeling ZIKV exposure. Cell 165(5):1238–1254
41. Cederquist GY, Asciolla JJ, Tchieu J, Walsh R, Cornacchia D, Resh MD et al (2019) Specification of positional identity in forebrain organoids. Nat Biotechnol 37(4):436–444
42. Mulder L, Depla J, Sridhar A, Wolthers KC, Sá RV (2023) A beginner's guide on the use of brain organoids for neuroscientists: a systematic review. Stem Cell Res Ther 14(1):87
43. Bagley JA, Reumann D, Bian S, Lévi-Strauss J, Knoblich JA (2017) Fused cerebral organoids model interactions between brain regions. Nat Methods 14(7):743–751

44. Velasco S, Kedaigle AJ, Simmons SK, Nash A, Rocha M, Quadrato G et al (2019) Individual brain organoids reproducibly form cell diversity of the human cerebral cortex. Nature 570(7762):523–527
45. Goto-Silva L, Ayad NME, Herzog IL, Silva NP, Lamien B, Orlande HRB et al (2019) Computational fluid dynamic analysis of physical forces playing a role in brain organoid cultures in two different multiplex platforms. BMC Dev Biol 19(1):3
46. Koch LS, Choy Buentello D, Broersen K (2023) Robust tissue fabrication for long-term culture of iPSC-derived brain organoids for aging research. J Vis Exp 195. https://doi.org/10.3791/64586
47. Giandomenico SL, Mierau SB, Gibbons GM, Wenger LMD, Masullo L, Sit T et al (2019) Cerebral organoids at the air–liquid interface generate diverse nerve tracts with functional output. Nat Neurosci 22(4):669–679
48. Fagerlund I, Dougalis A, Shakirzyanova A, Gómez-Budia M, Pelkonen A, Konttinen H et al (2022) Microglia-like cells promote neuronal functions in cerebral organoids. Cells 11(1):124
49. Leng F, Edison P (2021) Neuroinflammation and microglial activation in Alzheimer disease: where do we go from here? Nat Rev Neurol 17(3):157–172
50. Spangenberg EE, Lee RJ, Najafi AR, Rice R, Elmore MRP, Blurton-Jones M et al (2016) Eliminating microglia in Alzheimer's mice prevents neuronal loss without modulating amyloid-B pathology. Brain 139(4):1265–1281
51. Kälvälä S, Abushik P, Dougalis A, Malm T, Pelkonen A, Lehtonen Š (2023) Air-Liquid interface culture of midbrain organoids improves neuronal functionality and integration of microglia. BioRxiv. https://doi.org/10.1101/2023.10.10.561672
52. Zhang W, Jiang J, Xu Z, Yan H, Tang B, Liu C et al (2023) Microglia-containing human brain organoids for the study of brain development and pathology. Mol Psychiatry 28(1):96–107
53. Xu R, Boreland AJ, Li X, Erickson C, Jin M, Atkins C et al (2020) Developing human pluripotent stem cell-based cerebral organoids with a controllable microglia ratio for modeling brain development and pathology. Stem Cell Reports 16(8):1923–1937
54. Frost JL, Schafer DP (2016) Microglia: architects of the developing nervous system. Trends Cell Biol 26(8):587–597
55. Nikodémová M, Kimyon RS, De I, Small AL, Collier LS, Watters JJ (2015) Microglial numbers attain adult levels after undergoing a rapid decrease in cell number in the third postnatal week. J Neuroimmunol 278:280–288
56. Cakir B, Tanaka Y, Kiral FR, Xiang Y, Dagliyan O, Wang J et al (2022) Expression of the transcription factor PU.1 induces the generation of microglia-like cells in human cortical organoids. Nat Commun 13(1):430
57. Pasternak O, Westin CF, Bouix S, Seidman LJ, Goldstein JM, Woo T-UW et al (2012) Excessive extracellular volume reveals a neurodegenerative pattern in schizophrenia onset. J Neurosci 32(48):17365–17372
58. Aktories P, Petry P, Kierdorf K (2022) Microglia in a dish—which techniques are on the menu for functional studies? Front Cell Neurosci 16:908315
59. Zhang D, Pekkanen-Mattila M, Shahsavani M, Falk A, Teixeira AI, Herland A (2014) A 3D Alzheimer's disease culture model and the induction of P21-activated kinase mediated sensing in iPSC derived neurons. Biomaterials 35(5):1420–1428
60. Bubnys A, Tsai L-H (2022) Harnessing cerebral organoids for Alzheimer's disease research. Curr Opin Neurobiol 72:120–130
61. Centeno EGZ, Cimarosti H, Bithell A (2018) 2D versus 3D human induced pluripotent stem cell-derived cultures for neurodegenerative disease modelling. Mol Neurodegener 13(1):27
62. Gonzalez C, Armijo E, Bravo-Alegria J, Becerra-Calixto A, Mays CE, Soto C (2018) Modeling amyloid beta and tau pathology in human cerebral organoids. Mol Psychiatry 23(12):2363–2374
63. Arber C, Lovejoy C, Harris L, Willumsen N, Alatza A, Casey JM et al (2021) Familial Alzheimer's disease mutations in PSEN1 Lead to premature human stem cell neurogenesis. Cell Rep 34(2):108615

64. Choe MS, Yeo HC, Kim JS, Lee J, Lee HJ, Kim H-R et al (2024) Simple modeling of familial Alzheimer's disease using human pluripotent stem cell-derived cerebral organoid technology. Stem Cell Res Ther 15(1):118
65. Ghatak S, Dolatabadi N, Trudler D, Zhang X, Wu Y, Mohata M et al (2019) Mechanisms of hyperexcitability in Alzheimer's disease hiPSC-derived neurons and cerebral organoids vs isogenic controls. elife 8:e50333
66. Ghatak S, Dolatabadi N, Gao R, Wu Y, Scott H, Trudler D et al (2021) NitroSynapsin ameliorates hypersynchronous neural network activity in Alzheimer hiPSC models. Mol Psychiatry 26(10):5751–5765
67. Vanova T, Sedmik J, Raska J, Amruz Cerna K, Taus P, Pospisilova V et al (2023) Cerebral organoids derived from patients with Alzheimer's disease with PSEN1/2 mutations have defective tissue patterning and altered development. Cell Rep 42(11):113310
68. Blanchard JW, Bula M, Davila-Velderrain J, Akay LA, Zhu L, Frank A et al (2020) Reconstruction of the human blood–brain barrier in vitro reveals a pathogenic mechanism of APOE4 in pericytes. Nat Med 26(6):952–963
69. Zhao J, Fu Y, Yamazaki Y, Ren Y, Davis MD, Liu C-C et al (2020) APOE4 exacerbates synapse loss and neurodegeneration in Alzheimer's disease patient iPSC-derived cerebral organoids. Nat Commun 11(1):5540
70. Park J-C, Jang S-Y, Lee D, Lee J, Kang U, Chang H et al (2021) A logical network-based drug-screening platform for Alzheimer's disease representing pathological features of human brain organoids. Nat Commun 12(1):280
71. Bejoy J, Song L, Wang Z, Sang Q-X, Zhou Y, Li Y (2018) Neuroprotective activities of heparin, Heparinase III, and hyaluronic acid on the Aβ42-treated forebrain spheroids derived from human stem cells. ACS Biomater Sci Eng 4(8):2922–2933
72. Cai H, Ao Z, Hu L, Moon Y, Wu Z, Lu H-C et al (2020) Acoustofluidic assembly of 3D neurospheroids to model Alzheimer's disease. Analyst 145(19):6243–6253
73. Pavoni S, Jarray R, Nassor F, Guyot AC, Cottin S, Rontard J et al (2018) Small-molecule induction of Aβ-42 peptide production in human cerebral organoids to model Alzheimer's disease associated phenotypes. PLoS One 13(12):e0209150
74. Konttinen H, Cabral-da-Silva MC, Ohtonen S, Wojciechowski S, Shakirzyanova A, Caligola S et al (2019) PSEN1ΔE9, APPswe, and APOE4 confer disparate phenotypes in human iPSC-derived microglia. Stem Cell Reports 13(4):669–683
75. Lin YT, Seo J, Gao F, Feldman HM, Wen HL, Penney J et al (2018) APOE4 causes widespread molecular and cellular alterations associated with Alzheimer's disease phenotypes in human iPSC-derived brain cell types. Neuron 98(6):1141–54.e7
76. Park J, Wetzel I, Marriott I, Dreau D, D'Avanzo C, Kim DY et al (2018) A 3D human triculture system modeling neurodegeneration and neuroinflammation in Alzheimer's disease. Nat Neurosci 21(7):941–951
77. Wenzel TJ, Desjarlais JD, Mousseau DD (2024) Human brain organoids containing microglia that have arisen innately adapt to a β-amyloid challenge better than those in which microglia are integrated by co-culture. Stem Cell Res Ther 15(1):258
78. Hossain MK, Kim H-R, Chae HJ (2024) Aging phenotype in AD brain organoids: track to success and challenges. Ageing Res Rev 96:102256
79. Harley J, Santosa MM, Ng CY, Grinchuk OV, Hor J-H, Liang Y et al (2024) Telomere shortening induces aging-associated phenotypes in hiPSC-derived neurons and astrocytes. Biogerontology 25(2):341–360
80. Ferdaos N, Lowell S, Mason JO (2022) Pax6 mutant cerebral organoids partially recapitulate phenotypes of Pax6 mutant mouse strains. PLoS One 17(11):e0278147
81. Richard JP, Zuryn S, Fischer N, Pavet V, Vaucamps N, Jarriault S (2011) Direct in vivo cellular reprogramming involves transition through discrete, non-pluripotent steps. Development 138(8):1483–1492
82. Kelaini S, Cochrane A, Margariti A (2014) Direct reprogramming of adult cells: avoiding the pluripotent state. Stem Cells Cloning 7:19–29

83. Pfisterer U, Kirkeby A, Torper O, Wood J, Nelander J, Dufour A et al (2011) Direct conversion of human fibroblasts to dopaminergic neurons. Proc Natl Acad Sci U S A 108(25):10343–10348
84. Vierbuchen T, Ostermeier A, Pang ZP, Kokubu Y, Südhof TC, Wernig M (2010) Direct conversion of fibroblasts to functional neurons by defined factors. Nature 463(7284):1035–1041
85. Tang Y, Liu ML, Zang T, Zhang CL (2017) Direct reprogramming rather than iPSC-based reprogramming maintains aging hallmarks in human motor neurons. Front Mol Neurosci 10:359
86. Mertens J, Paquola ACM, Ku M, Hatch EM, Böhnke L, Ladjevardi S et al (2015) Directly reprogrammed human neurons retain aging-associated transcriptomic signatures and reveal age-related nucleocytoplasmic defects. Cell Stem Cell 17(6):705–718
87. Huh CJ, Zhang B, Victor MB, Dahiya S, Batista LFZ, Horvath S et al (2016) Maintenance of age in human neurons generated by microRNA-based neuronal conversion of fibroblasts. elife 5:e18648
88. Zhang Y, Xie X, Hu J, Afreen KS, Zhang CL, Zhuge Q et al (2020) Prospects of directly reprogrammed adult human neurons for neurodegenerative disease modeling and drug discovery: iN vs. iPSCs models. Front Neurosci 14:546484
89. Drouin-Ouellet J, Pircs K, Barker RA, Jakobsson J, Parmar M (2017) Direct neuronal reprogramming for disease modeling studies using patient-derived neurons: what have we learned? Front Neurosci 11:530
90. Sun Z, Kwon J-S, Ren Y, Chen S, Walker CK, Lu X et al (2024) Modeling late-onset Alzheimer's disease neuropathology via direct neuronal reprogramming. Science 385(6708):adl2992
91. Mertens J, Herdy JR, Traxler L, Schafer ST, Schlachetzki JCM, Böhnke L et al (2021) Age-dependent instability of mature neuronal fate in induced neurons from Alzheimer's patients. Cell Stem Cell 28(9):1533–48.e6
92. Paşca SP (2019) Assembling human brain organoids. Science 363(6423):126
93. Xiang Y, Tanaka Y, Çakır B, Patterson B, Kim K-Y, Sun P et al (2019) hESC-derived thalamic organoids form reciprocal projections when fused with cortical organoids. Cell Stem Cell 24(3):487–97.e7
94. Miura Y, Li M-Y, Birey F, Ikeda K, Revah O, Thete MV et al (2020) Generation of human striatal organoids and cortico-striatal assembloids from human pluripotent stem cells. Nat Biotechnol 38(12):1421–1430
95. Miura Y, Kim JI, Jurjuţ O, Kelley KW, Yang X, Chen X et al (2024) Assembloid model to study loop circuits of the human nervous system. Bio Rxiv
96. Wu S, Hong Y, Chu C, Gan Y, Li X, Tao M et al (2024) Construction of human 3D striatonigral assembloids to recapitulate medium spiny neuronal projection defects in Huntington's disease. Proc Natl Acad Sci U S A 121(22):e2316176121
97. Rickner HD, Jiang L, Hong R, O'Neill NK, Mojica CA, Snyder BJ et al (2022) Single cell transcriptomic profiling of a neuron-astrocyte assembloid tauopathy model. Nat Commun 13(1):6275
98. Barmpa K, Saraiva C, Lopez-Pigozzi D, Gomez-Giro G, Gabassi E, Spitz S et al (2024) Modeling early phenotypes of Parkinson's disease by age-induced midbrain-striatum assembloids. Commun Biol 7(1):1561
99. Sun X, Kofman S, Ogbolu VC, Karch CM, Ibric L, Qiang L (2023) Vascularized brain assembloids with enhanced cellular complexity provide insights into the cellular deficits of tauopathy. Int J Cell Cloning 42(2):107–115
100. Kong D, Park KH, Kim D-H, Kim NG, Lee S-E, Shin N et al (2023) Cortical-blood vessel assembloids exhibit Alzheimer's disease phenotypes by activating glia after SARS-CoV-2 infection. Cell Death Discovery 9(1):32
101. Paşca SP, Arlotta P, Bateup HS, Camp JG, Cappello S, Gage FH et al (2025) A framework for neural organoids, assembloids and transplantation studies. Nature 639(8054):315–320

Chapter 6
The Application of Directly Induced Neurons to Neurodegenerative Disease Research

Sumihiro Maeda

Abstract Traditionally, research on neurodegenerative diseases using human cells has primarily relied on the application of human-induced pluripotent stem (iPS) cells-derived neural cells. However, since neurodegenerative diseases typically manifest in older individuals, studies using human iPS cells, which reset cellular age to zero, have inherent limitations. To overcome these challenges, a new approach has emerged: directly induced neurons (iNs), which are converted directly from somatic cells. This paper provides an overview of how these cells are induced and their applications in studying various diseases.

Keywords Alzheimer's disease · Aβ · Tau · Tauopathy · Neurodegenerative disease · Aggregation · Directly induced neuron · iN · Tau isoform · hiPSC-derived neurons

6.1 Necessity of Human-Cell-Based Models for Neurodegenerative Diseases and Limitations of hiPSC Models

With the global advent of super-aged societies, developing treatments, preventive measures, and diagnostic tools for neurodegenerative diseases prevalent among the elderly has become an urgent issue. Alzheimer's disease and frontotemporal lobar degeneration are prime examples, significantly burdening caregivers and contributing to socioeconomic challenges due to lost labor opportunities.

In neurodegenerative diseases, familial cases account for approximately 5–10%, and shared pathological features are observed between familial and sporadic cases. This has spurred research into familial neurodegeneration mechanisms caused by

S. Maeda (✉)
Department of Physiology, Keio University School of Medicine, Shinjuku-ku, Tokyo, Japan
e-mail: sumihiro.maeda@keio.jp

R. C. B. Wong and Y. Hayashi (eds.), *Advances in Using Stem Cells to Treat Neural Diseases*, Stem Cell Biology and Regenerative Medicine 77,
https://doi.org/10.1007/978-3-031-94101-6_6

genetic mutations, with the expectation that such studies could illuminate mechanisms underlying sporadic cases. Transgenic mice expressing human genes with familial neurodegenerative mutations have partially replicated pathological features seen in patients, including cognitive impairments. However, key pathological features, such as neuronal cell death, often fail to manifest in mouse models. This gap raises questions about whether the shorter lifespan of mice (approximately 2 years) precludes sufficient observation time or if intrinsic differences between human and mouse cells explain this discrepancy.

Recent advances in human-induced pluripotent stem (hiPS) cells, pioneered by Professor Shinya Yamanaka's group at Kyoto University, have enabled comparative studies of human and mouse cellular differences [1]. Notably, embedding hiPSC-derived human neuronal cells into mouse brains has demonstrated neuronal cell death in human cells, but not in mouse neurons, indicating differential vulnerability to pathogenic proteins [2]. These findings underscore the necessity of using human-cell-based models for studying human neurodegenerative diseases.

However, a critical limitation of hiPS cells is the "resetting" of cellular age to zero during reprogramming [3], which poses challenges for modeling diseases that manifest in aged populations. To overcome this limitation, recent advancements have focused on directly converting somatic cells into neurons (Directly induced neurons, iNs) without passing through an iPS cell stage, preserving the donor's cellular age (Fig. 6.1). This direct induction method represents a promising tool for studying age-dependent neurodegenerative diseases.

6.2 Generation of Human Neuronal Models via Direct Induction

As mentioned earlier, using hiPS cells results in the generation of immature neurons with a "zero-aged" cellular state. While approaches such as introducing progeria-causing genes (Miller et al. 2013) or enhancing aging-related gene expression identified through comprehensive transcriptomics of healthy brains [4] have been explored to artificially induce aging in hiPSC-derived neurons, it remains unclear whether these methods effectively replicate the aged cellular state.

In contrast, direct induction of neurons bypasses the reprogramming process, preserving the donor's cellular age. This makes directly induced neurons (iNs) more suitable for modeling neurodegenerative diseases that are known to manifest in an age-dependent manner [4]. Fibroblasts and blood-derived cells have been the primary somatic cells used for this direct induction.

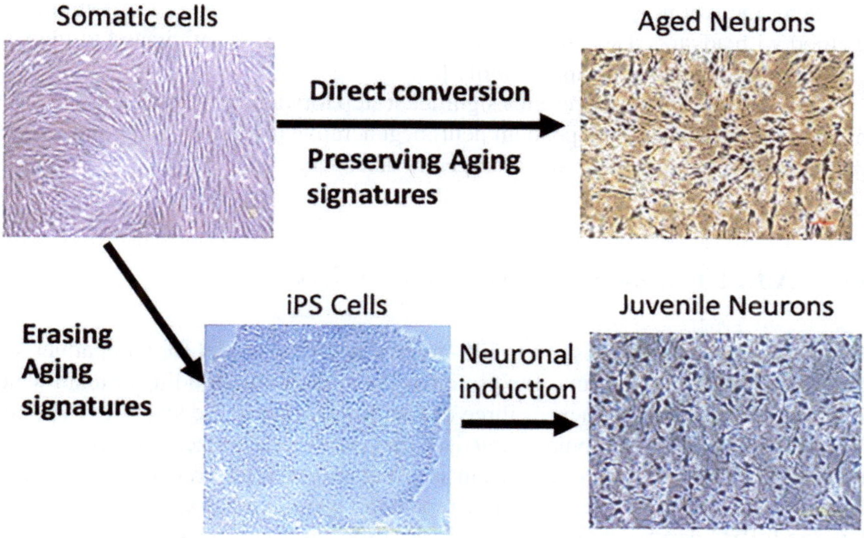

Fig. 6.1 Comparison between neurons directly induced and those derived via iPS cells

6.2.1 Methods of Direct Neuronal Induction

Two main strategies for direct neuronal induction from fibroblasts have been developed:

1. **MicroRNA-Mediated Induction**: This approach involves using neuron-specific microRNAs such as miR9 to reprogram the cells, followed by the introduction of neuronal transcription factors to drive differentiation into the desired neuronal subtype [5].
2. **Transcription Factor-Driven Induction**: Neuronal induction is achieved through the transient overexpression of transcription factors such as Ngn2 and Ascl1. Subsequently, neurons are isolated using methods like fluorescence-activated cell sorting (FACS) [6].

While fibroblast-based induction has been extensively studied, obtaining fibroblasts is invasive and requires explicit consent. This has shifted focus to peripheral blood cells, offering a less invasive alternative.

6.2.2 Challenges and Future Directions

Direct neuronal induction from blood cells utilizes transient overexpression of transcription factors like Ngn2 and Ascl1, similar to fibroblast-based protocols. However, due to the inherent diversity of blood-derived cells, the efficiency of neuronal induction is lower—typically around a few percent. Combining this method with advanced sorting techniques, such as FACS, is anticipated to improve efficiency and broaden its applicability [7].

These developments represent a significant step forward in creating age-reflective models for studying age-dependent neurodegenerative diseases, paving the way for improved disease modeling and therapeutic discovery.

6.3 Adult Tau Isoform Expression in iNs

One of the main limitations to use iPS-derived neuronal model for tauopathies was the lack of adult tau isoforms. Tau protein has a microtubule-binding domain which is composed repeated sequences three or four times. Due to the alternative splicing at microtubule-binding domain, tau isoforms were categorized to three or four repeats tau. Only the three-repeat tau is expressed in fetal brains, while in adult human brains, both of three- and four-repeat tau were expressed. Therefore, to utilize the hiPSC-derived neuronal model for adult-onset neurodegenerative diseases like AD. However, in iPS-derived neuron express only the fetus form of tau, three-repeat tau [8]. Longer culture period like 1 year or MAPT mutations enhance the expression of adult form of tau, four-repeat tau [9]. However, the requirement of long-culture period to drive the expression of four-repeat tau limits the utility of hiPSC-derived neuronal culture for neurodegenerative research. On the other hand, MAPT mutations are known to enhance four-repeat tau expression in the patient's brain. Thus, it is possible to drive four-repeat tau expression by introducing MAPT mutations into iPSCs. However, most MAPT mutations trigger frontotemporal dementia but not AD. Therefore, the application of MAPT mutations make it difficult to extrapolate the obtained data to AD pathogenesis. Among the more than 40 MAPT mutations, only one MAPT mutation, the A152T mutation, is reported to increase the risk to develop AD. However, the A152T mutation does not promote the expression of four-repeat tau in hiPS-derived neuronal cultures. Thus, driving four-repeat tau expression was one of the biggest obstacles for tau researchers to apply human cellular model for AD researches. Because the iNs maintain the cellular age of the original somatic cells, the iNs were predicted to express four-repeat tau. Actually, four-repeat tau was found to be expressed well in iNs.

6.4 Application of iNs in Neurodegenerative Disease Research

Among the direct induction methods, fibroblast-derived iNs are most commonly employed in disease research. The miRNA-based approach, involving miR9, has been used primarily for studying Huntington's disease (HD) [5], while transcription factor-only methods using Ngn2 and similar factors are applied to familial and sporadic Alzheimer's disease (AD) [6].

6.4.1 Huntington's Disease

HD, a hereditary neurodegenerative disease, is characterized by intracellular protein aggregates leading to neurodegeneration. This makes it well-suited for modeling using patient-derived neuronal cultures. In HD, aggregates of huntingtin protein form in the brains of patients but were not reproduced in hiPSC-derived neurons. In contrast, fibroblast-derived iNs from pre- and post-symptomatic HD patients successfully recapitulated intracellular huntingtin aggregation. Moreover, fibroblasts first reverted to hiPSCs and then re-induced into fibroblasts for iN induction failed to form aggregates. These findings highlight the utility of aged neurons in modeling age-related neurodegenerative diseases [5].

6.4.2 Alzheimer's Disease

For AD, direct neuronal induction without miRNA overexpression, relying instead on factors like Ngn2, was used to generate iNs from fibroblasts of familial and sporadic AD patients. Increased Aβ secretion was observed only in familial AD-derived fibroblasts. While the extracellular formation of amyloid plaques—a hallmark of AD—cannot be replicated in two-dimensional cultures, sporadic AD-derived iNs displayed aberrations such as reactivation of cell cycle-related gene expression and dedifferentiation tendencies. These findings align with neuropathological reports of cell cycle reactivation in AD patient brains, suggesting that iNs can partially model AD pathology even in sporadic cases [6].

Mertens et al. also reported a three-dimensional (3D) culture model of cortical spheroids composed of iNs induced by miR9/9* and 124 method [10]. The 3D culture method enabled the retention of secreted Aβ within the spheroids, facilitating endogenous Aβ aggregation. However, while they successfully induced the expression of the adult tau isoform (four-repeat tau) in iNs, they did not observe endogenous tau aggregation. To investigate tau–tau interactions, they relied on the overexpression of exogenous mutant tau, as demonstrated by a proximity ligation assay (PLA). In iNs derived from AD patients, PLA signals of tau–tau interactions were increased compared to those from healthy controls. Although these signals did not indicate tau aggregation, the findings highlighted that the environment of AD patient-derived iNs could promote tau aggregation. Further research will be required to achieve endogenous tau aggregation in iNs.

These studies underscore the advantages of direct induction for maintaining the cellular age of donor cells, making iNs a valuable tool for investigating the pathophysiology and therapeutic strategies for age-related neurodegenerative diseases.

6.5 Future Prospects for Neurological Disease Research Using Direct Induction Methods

The significance of iNs in neurodegenerative disease research lies in their ability to retain the donor's cellular age, thereby faithfully recapitulating disease phenotypes. However, challenges remain, particularly regarding the invasive nature of obtaining fibroblasts, the primary somatic cells used for iN generation, and the low efficiency of iN induction from blood-derived cells, which are easier to collect. These limitations highlight the need for improved methodologies.

Developing techniques to achieve high-efficiency iN induction from easily accessible and non-invasive cell sources would greatly advance the field. For example, if repeatable cell sampling from the same individual were made feasible, researchers could map longitudinal changes in neuronal cell aging and correlate them with observed phenotypic alterations. This would allow for the construction of personalized neural aging curves and, by aggregating data, the derivation of standardized neural aging models.

By aligning artificially aged cellular phenotypes with these standardized models, it would become possible to predict individual disease onset, track disease progression, and evaluate drug responsiveness. Such advancements could pave the way for a transformative era in personalized medicine for neurodegenerative diseases (see Fig. 6.2).

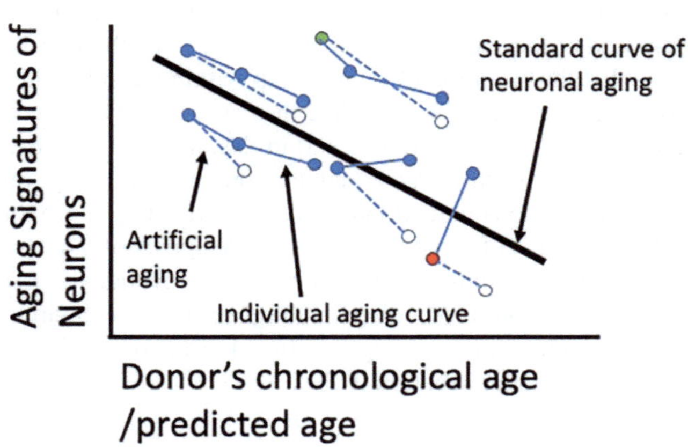

Fig. 6.2 Construction of neuronal aging curves using directly induced neurons

References

1. Supakul S, Okano H, Maeda S (2021) Utilization of human induced pluripotent stem cells-derived in vitro models for the future study of sex differences in Alzheimer's disease. Front Aging Neurosci 13:768948. https://doi.org/10.3389/fnagi.2021.768948
2. Espuny-Camacho I, Arranz AM, Fiers M, Snellinx A, Ando K, Munck S, Bonnefont J, Lambot L, Corthout N, Omodho L et al (2017) Hallmarks of Alzheimer's disease in stem-cell-derived human neurons transplanted into mouse brain. Neuron 93(5):1066–1081.e8. https://doi.org/10.1016/j.neuron.2017.02.001
3. Horvath S (2013) DNA methylation age of human tissues and cell types. Genome Biol 14(10):R115. https://doi.org/10.1186/gb-2013-14-10-r115
4. Mertens J, Paquola ACM, Ku M, Hatch E, Böhnke L, Ladjevardi S, McGrath S, Campbell B, Lee H, Herdy JR et al (2015) Directly reprogrammed human neurons retain aging-associated transcriptomic signatures and reveal age-related nucleocytoplasmic defects. Cell Stem Cell 17:705–718. https://doi.org/10.1016/j.stem.2015.09.001
5. Victor MB, Richner M, Olsen HE, Lee SW, Monteys AM, Ma C, Huh CJ, Zhang B, Davidson BL, Yang XW et al (2018) Striatal neurons directly converted from Huntington's disease patient fibroblasts recapitulate age-associated disease phenotypes. Nat Neurosci 21:341–352. https://doi.org/10.1038/s41593-018-0075-7
6. Mertens J, Herdy JR, Traxler L, Schafer ST, Schlachetzki JCM, Böhnke L, Reid DA, Lee H, Zangwill D, Fernandes DP et al (2021) Age-dependent instability of mature neuronal fate in induced neurons from Alzheimer's patients. Cell Stem Cell 28:1533–1548.e6. https://doi.org/10.1016/j.stem.2021.04.004
7. Tanabe K, Ang CE, Chanda S, Olmos VH, Haag D, Levinson DF, Südhof TC, Wernig M (2018) Transdifferentiation of human adult peripheral blood T cells into neurons. Proc Natl Acad Sci USA 115:6470–6475. https://doi.org/10.1073/pnas.1720273115
8. Sato C, Barthélemy NR, Mawuenyega KG, Patterson BW, Gordon BA, Jockel-Balsarotti J, Sullivan M, Crisp MJ, Kasten T, Kirmess KM et al (2018) Tau kinetics in neurons and the human central nervous system. Neuron 97:1284–1298.e7. https://doi.org/10.1016/j.neuron.2018.02.015
9. Sposito T, Preza E, Mahoney CJ, Setó-Salvia N, Ryan NS, Morris HR, Arber C, Devine MJ, Houlden H, Warner TT et al (2015) Developmental regulation of tau splicing is disrupted in stem cell-derived neurons from frontotemporal dementia patients with the 10 + 16 splice-site mutation in MAPT. Hum Mol Genet 24:5260–5269. https://doi.org/10.1093/hmg/ddv246
10. Sun Z, Kwon JS, Ren Y, Chen S, Walker CK, Lu X, Cates K, Karahan H, Sviben S, Fitzpatrick JAJ et al (2024) Modeling late-onset Alzheimer's disease neuropathology via direct neuronal reprogramming. Science 385:adl2992. https://doi.org/10.1126/science.adl2992
11. Justine D Miller 1, Yosif M Ganat, Sarah Kishinevsky, Robert L Bowman, Becky Liu, Edmund Y Tu, Pankaj K Mandal, Elsa Vera, Jae-won Shim, Sonja Kriks, Tony Taldone, Noemi Fusaki, Mark J Tomishima, Dimitri Krainc, Teresa A Milner, Derrick J Rossi, Lorenz Studer. Cell Stem Cell 2013 Dec 5;13(6):691–705. https://doi.org/10.1016/j.stem.2013.11.006

Chapter 7
Breakthroughs in Stem Cell-Based Treatments for ALS and FTD: Promises and Challenges Ahead

Henriette Reventlow Frederiksen and Kristine Freude

Abstract Amyotrophic lateral sclerosis (ALS) and frontotemporal dementia (FTD) are now recognized as a disease spectrum with overlapping genetic, molecular, and pathological features, including TDP-43 dysfunction, RNA processing abnormalities, and neuroinflammation. Biomarker advancements, such as T cell subtypes in ALS and HDGFL2 in both diseases, are improving diagnosis and treatment monitoring, while personalized therapies like antisense oligonucleotides (ASOs) and CRISPR-based approaches show promise in modifying disease progression. Neuroinflammation, driven by microglial activation and cytokine dysregulation, presents key therapeutic targets for immune-modulating and anti-inflammatory interventions. Current ALS treatments provide limited benefits, and no FDA-approved therapies exist for FTD, highlighting the need for novel approaches. Stem cell-based therapies offer potential neuroprotection and immune regulation, with allogeneic transplants holding more promise than autologous ones, despite immune rejection challenges. Advances in gene editing and immune-evasive strategies could enhance transplantation outcomes, paving the way for personalized, off-the-shelf treatments for ALS and FTD.

Keywords Neuroinflammation · FTD · ALS · ALS-FTD spectrum · Biomarkers · Genetic mutations · Stem cell therapy · Cell engineering · Gene editing · Personalized treatment · Neurodegeneration

H. R. Frederiksen · K. Freude (✉)
Department of Veterinary and Animal Sciences, Faculty of Health and Medical Sciences, University of Copenhagen, Copenhagen, Denmark
e-mail: kkf@sund.ku.dk

R. C. B. Wong and Y. Hayashi (eds.), *Advances in Using Stem Cells to Treat Neural Diseases*, Stem Cell Biology and Regenerative Medicine 77,
https://doi.org/10.1007/978-3-031-94101-6_7

7.1 Common Mechanisms and Pathophysiology in Neurodegenerative Disorders: Insights from FTD and ALS

Neurodegenerative disorders are a broad group of diseases characterized by the progressive degeneration of structural components and function of the nervous system, specifically the neurons and glial cells. Despite the diversity of these diseases in terms of clinical presentation and affected brain regions, they share several common features at the cellular and molecular levels. These hallmarks include protein aggregation, neuronal and synaptic dysfunction, cell loss, altered metabolism and neuroinflammation. This congruence in pathophysiology points toward common underlying mechanisms driving these diseases. As such, shared dysregulated pathways may comprise potential targets for future therapy. Common therapeutic interventions for multiple neurodegenerative diseases might include modulation of protein aggregation, enhancement of neuronal survival, metabolic regulation, and/or suppression of inflammation. Frontotemporal dementia (FTD) and amyotrophic lateral sclerosis (ALS) are examples of such neurodegenerative disorders with divergent disease phenotypes, but which share many cellular, mechanistic and genetic hallmarks, including clinical commonalities [1] (Fig. 7.1).

FTD primarily affects cognitive functions, with patients typically experiencing changes in personality, behavior, and language abilities. It predominantly targets the frontal and temporal lobes of the brain, leading to alterations in executive function, decision-making, and social behavior. FTD can be classified by clinical syndromes into behavioral-variant frontotemporal dementia (bvFTD), semantic dementia (SVPPA), and progressive nonfluent aphasia (NFVPPA) with unique regional neuroanatomical deficits in both gray matter and white matter in the frontal and temporal lobes of the brain [2]. While variable, age of onset is usually between 45 and 65 years of age. This variability has been attributed to genetic modifiers such as TMEM106B in GRN [3] and a polymorphism at a locus containing two overlapping genes *LOC101929163* and *C6orf10* in C9orf72 carriers [4]. The progression of FTD leads to significant cognitive, emotional, and functional decline with there being no cure at the present time. Around 30% of FTD cases are believed to be familial, with common mutations in *MAPT*, *GRN*, or *C9orf72* and rare mutations in *FUS*, *VCP*, *CHMP2B*, or *TBK1* [5]. Molecular phenotypes of the disease are dependent on FTD type, with the most common phenotype being referred to as TDP-43 pathology (FTD-TDP), associated with mutations in *GRN*, *VCP*, *FUS*, and *C9orf72* [6]. TDP-43 pathology is characterized by insoluble, hyperphosphorylated, and ubiquitinated aggregates of TAR DNA-binding protein 43 (TDP-43) in the cytoplasm, nucleus, and cell processes of neurons and glia. Additional less frequent phenotypes include ubiquitin-immunoreactive neuronal inclusions (FTD-FET) linked to mutations in *FUS*, *EWS*, and *TAF15* [7], and the presence of hyperphosphorylated tau (FTD-tau) associated with MAPT [6].

ALS is a motor neuron disease leading to progressive muscle weakness and paralysis, as it primarily affects the motor neurons in both brain and spinal cord

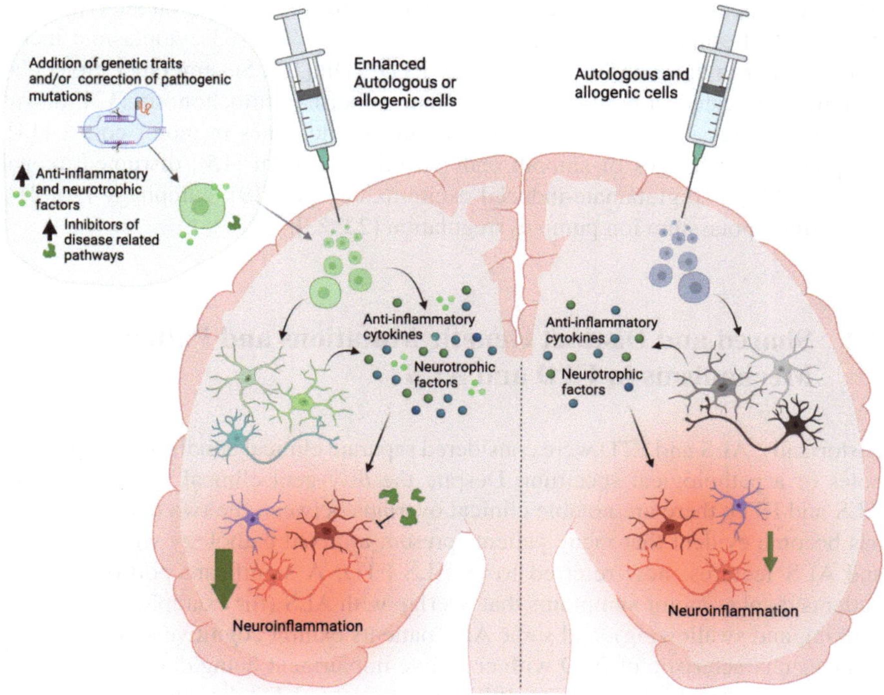

Fig. 7.1 Schematic representation of stem cell-based therapeutic strategies for ALS and FTD. *Right-Current approaches:* Injection of autologous or allogeneic stem cell products which differentiate in vivo, replace lost cells, and secrete neurotrophic and anti-inflammatory factors to reduce neuroinflammation. *Left-Potential future approach:* Genome editing is used to correct pathogenic mutations in patient cells or introduce genes encoding neurotrophic and anti-inflammatory factors as well as inhibitors of disease-associated pathways

responsible for controlling voluntary muscle movements. ALS progresses rapidly with respiratory failure typically manifesting within a few years after symptom onset, due to the involvement of motor neurons that control breathing. The disease is classified into four distinct clinical subtypes: (1). Limb onset: the most prevalent form, characterized by the involvement of both upper and lower motor neurons affecting the limbs; (2). Bulbar onset: this subtype primarily manifests as difficulties in swallowing and breathing; (3). Progressive muscular atrophy (PMA): a rare variant involving only lower motor neuron degeneration, and (4). Primary lateral sclerosis (PLS): an exceedingly rare form of ALS limited to upper motor neuron involvement [8]. While most cases of ALS are sporadic with unknown cause, approximately 10% are familial with pathogenic mutations in disease-related genes. The most common of these is *C9orf72* accounting for up to 40% of familial cases and 6% of sporadic cases [9]. Other genes frequently associated with ALS include *SOD1*, *FUS,* and *TARDBP*, amongst more than 40 other ALS-associated genes identified to date [10]. At the molecular level, the major pathological hallmarks of

ALS are atrophy and gliosis, targeting different areas dependent on subtype [11]. Another well-established hallmark is the presence of TDP-43 cytoplasmic inclusions, similar to those described above for FTD-TDP [12]. Several other molecular findings implicated in both disorders include abnormal mitochondria [13], downregulated expression of mitochondrial and glycolytic genes in motor cortex [14], morphological changes of smooth endoplasmic reticulum [15], disrupted axonal transport [16, 17], glutamate-induced excitotoxicity [18, 19], autophagy [20, 21], and sodium potassium ion pump dysregulation [22, 23].

7.2 Shared and Distinct Genetic Mutations and Pathological Mechanisms in FTD and ALS

Historically, ALS and FTD were considered separate clinical conditions at opposing poles of a pathological spectrum. Despite the divergent clinical presentations of ALS and FTD, there are notable clinical overlaps between the two diseases, and it has become evident that many patients present a mixed pathology with both FTD and ALS features, now referred to as ALS-FTD. A significant portion of FTD patients display motor symptoms that overlap with ALS (for example, difficulty in walking and swallowing), and some ALS patients exhibit cognitive and behavioral changes characteristic of FTD with cognitive impairment being diagnosed in 30% to 75% of cases of which 45% fulfill the criteria for FTD diagnosis [19, 24–26]. Beyond these symptomatic overlaps, shared genetic mutations and disrupted biological pathways are now well recognized. Similar transcriptional changes in familial and sporadic cases support the concept of common pathophysiological mechanisms [27] and mutations in *TARDBP, C9orf72, VCP, UBQLN2, SQSTM1, OPTN, FUS, CHMP2B, CHCHD10, CCCNF, TIA1*, and *TBK1* are associated with an increased risk of developing either disease [28, 29]. Amongst these, *C9orf72* is the most prevalent shared genetic variant and has been extensively studied, highlighting its critical role in ALS and FTD pathogenesis [29]. The proteins encoded by these genes all play key roles in biological pathways and are integral to biological pathways dysregulated in both diseases. These pathways include regulation of RNA processing [30, 31], autophagy [32], vesicle trafficking [33, 34], DNA repair [35], nucleocytoplasmic transport [36], stress granule dynamics [37], protein folding [38], and immune processes [39]. A key hallmark of sporadic and familial ALS, as well as some FTD subtypes, is the accumulation of abnormally phosphorylated and ubiquitinated TDP-43 [40–42]. Many shared risk genes (*C9orf72, TARDBP, VCP, UBQLN2, SQSTM1*) are implicated in similar cellular processes as TDP-43 [28]. This interconnectivity has been shown to be even more complex by the recent finding that microglia lacking GRN enhance TDP-43 accumulation in neurons [43]. A similar exacerbation of neuron phenotypes due to increased aggregation of endogenous TDP-43 results from expression of glycine-arginine repeat protein [poly(GR)] translated from expanded G4C2 repeats in mutant C9orf72 [44]. These examples showcase the intricate interdependencies of cellular phenotypes in both ALS and FTD, highlighting the significance of cell type-specific responses in ALS and FTD pathogenesis.

Some of the mutations associated with both ALS and FTD, such as in *TIA1* [45], occur with low prevalence [46, 47]. Consequently, it remains unclear whether these rare mutations are disease-causing or serve as risk factors that contribute to disease in combination with other genes. As most ALS and FTD cases are sporadic, many risk factors remain unidentified. A large study comparing the genetics of over 6000 sporadic ALS cases identified pathogenic or likely pathogenic variants in 5.89% of cases in 90 risk genes, with the most common mutations occurring in *SOD1, TARDBP, FUS, NEK1, OPTN,* and *TBK1*. *C9orf72* repeat expansions were present in 5.21% of sporadic ALS patients. However, 71.83% of cases showed no identifiable genetic markers, while 17.33% exhibited isolated risk factors [48]. The absence of identifiable genetic markers in these cases of ALS, as in most neurodegenerative disorders, indicates the involvement of other factors besides genetics. One such factor is the role of epigenetics, particularly in relation to aging. One study of an ALS/FTD mouse model supports this notion, revealing decreased gene repression via H3K9me3 in astrocytes and neurons, consistent with disease-associated transcriptomic changes [49]. The impact of aging on disease progression is further supported by the longer overall survival rates observed in ALS patients with early-onset symptoms [50].

Both ALS and FTD are multisystem disorders with heterogeneous molecular findings, complicating their study, characterization, and treatment. Diagnosis primarily relies on clinical examination and postmortem histological analysis, which only provide insights into the later stages of disease. However, pathological changes typically occur years before symptom onset, and for ALS, the short lifespan of patients underscores the importance of early therapeutic intervention. In vivo and in vitro data from early disease stages is therefore crucial for tailoring therapy. To better understand early molecular changes, several models have been established, including induced pluripotent stem cells from ALS and FTD patients, enabling the modeling and comparison of molecular alterations across different cell types and genetic backgrounds. Increased understanding of early pathology can facilitate biomarker discovery, potentially reducing the average diagnostic delay for ALS, which currently takes around a year due to the need for multiple tests [28]. Furthermore, the identification of biomarkers will be essential not only for diagnosing ALS at an earlier stage but also for distinguishing between ALS and FTD with their overlapping clinical and pathological features. The ability to differentiate between these two disorders in their early phases would allow for more tailored and accurate therapeutic strategies, improving patient outcomes. Advancing biomarker discovery is key to shortening diagnostic timelines and enhancing the precision of both ALS and FTD diagnosis, ultimately supporting the development of targeted treatments and interventions.

7.3 Biomarkers and Their Role in Differentiating Between ALS and FTD

As outlined above, ALS and FTD exist on a disease spectrum with significant clinical, molecular, and genetic overlap. More than 50 genes have been identified as being associated with FTD or ALS [51]. The implementation of rodent and human induced pluripotent stem cell models has enabled great strides to be made in deciphering the molecular and cell type-specific pathologies manifesting from specific mutations [52–57]. Clinical genetic assessments have substantiated that mutations in two genes are solely responsible for causing FTD (*MAPT* and *GRN*) [52, 58]. Currently, mutations in 33 genes have been demonstrated to cause ALS, with 11 of these genes also being causative for FTD [51]. These genetic associations with the disease phenotype spectrum underline the necessity and suitability of gene-based biomarkers. Besides genetic profiling, such biomarkers include TDP-43 encoded by *TARDBP*, tau encoded by *MAPT*, *C9orf72*, and *SOD1*.

TDP-43 accumulation pathology can be observed in up to 50% and up to 80% of FTD and ALS patients, respectively [59]. TDP-43 has been investigated as a potential biomarker in serum, plasma, and cerebrospinal fluid (CSF). Phosphorylated TDP-43 (p-TDP-43) is a useful marker for distinguishing *C9orf72* and *GRN* mutation carriers from those with other causative mutations but does not differentiate between *C9orf72* and *GRN* carriers, as both exhibit upregulated p-TDP-43 [60]. Total CSF TDP-43 (tTDP-43) is elevated in ALS and can help distinguish ALS from bv-FTD [61]. Nevertheless, tTDP-43 is similarly increased in the CSF of Parkinson's Disease (PD), Multiple Sclerosis and Guillain-Barre syndrome patients and it is highly debated if truncated or full-length forms of TDP-43 are relevant in addition to the concern of measuring peripheral tTDP-43 over brain-derived tTDP-43 [62]. Another contributor to TDP-43 pathology supports the notion that instead of a gain-of-function phenotype, TDP-43 mutations create a loss-of-function phenotype associated with the activation of a cryptic splice site. This leads to expression of cryptic hepatoma-derived growth factor-like protein 2 (HDGFL2) in FTD and ALS-FTD prior to the traditional neurofilament light and phosphorylated neurofilament heavy chain protein becoming detectable. This renders HDGFL2 a promising biomarker, especially since the upregulation of HDGFL2 can be detected in presymptomatic carriers, making it one of the earliest ALS biomarkers [63].

Hyperphosphorylation of tau protein is characteristic of FTD, particularly bvFTD and progressive nonfluent aphasia (nfvPPA) with the ratio between phosphorylated (p-tau181) and total (t-tau) tau being implemented in diagnostics. Increased p-tau181 CSF levels are strongly linked to the genetic cause of MAPT mutations, while this is absent in other forms of FTD and ALS. It remains challenging to use p-tau181 CSF levels for FTD diagnosis due to a similar increase observed in Alzheimer's disease (AD) [64].

C9orf72 repeat expansion is a causative mutation observed in 40% of ALS and 20% of FTD cases. This mutation leads to the formation of dipeptide repeat proteins, resulting in the production of poly-GA, poly-PR, and poly-GR peptides through a mechanism known as repeat-associated non-AUG (RAN) translation [65]. Poly-GA, poly-PR, and poly-GR peptides are therefore considered useful biomarkers, since they can be detected in the CSF of patients carrying *C9orf72* mutations. Of particular interest is the emerging possibility of treating the abnormally extended *C9orf72* mRNA via targeted delivery of CRISPR effectors. The genome editing tool CRISPR/Cas13d provides a promising approach with which to specifically target GGGGCC repeat RNA [66]. While CRISPR/Cas13d-based delivery to target cells and tissues in vivo remains challenging and will require delivery via adeno-associated virus (AAV), CSF based readout of poly-GA, poly-PR, and poly-GR will enable assessment of gene-editing efficacy.

Mutations in the *SOD1* gene are the second most common cause of familial ALS, accounting for approximately 10–20% of cases, and are also detected in 1–2% of sporadic ALS cases. SOD1 is an antioxidant enzyme which protects the cell from reactive oxygen species toxicity with SOD1 malfunction therefore manifesting in neurodegeneration. CSF SOD1 activity levels are decreased in patients with *SOD1* mutations compared to non-carriers of the mutation leading to its use as a biomarker [67]. SOD1 has therefore received attention as an ALS therapeutic target with recent advances in antisense oligonucleotide (ASO) treatments as seen with the phase three clinical trial for BIIB067 (Tofersen—NCT04856982), an ASO targeting mutated *SOD1*. Assaying for changes in CSF SOD1 activity will be more relevant in determining Tofersen efficacy than the assessment of total SOD1 levels.

Besides diagnosis based on the above gene-based biomarkers, classification and distinction between ALS and FTD are supported by the assessment of neurofilament light chain (NFL) in the CSF. NFL has been reported to be upregulated in patients with ALS and, to a lesser extent, FTD, compared to healthy controls [68]. Other studies have confirmed that plasma NFL is valuable for prognosis and distinguishing between clinical and genetic ALS subgroups [69]. Like NFL, phosphorylated neurofilament heavy chain (pNFH) is similarly elevated in ALS patients [70]. Both NFL and pNFH are markers of neuronal injury and are associated with inflammation in ALS and FTD. The inflammatory component is common across various neurodegenerative diseases, making inflammatory proteins promising candidates as distinctive biomarkers for ALS and FTD.

Pro-inflammatory cytokines can be detected in both plasma and CSF. Elevated levels of several inflammatory cytokines and chemokines have been observed in the blood and CSF of ALS patients compared to controls, including interleukin (IL)-1β, IL-4, IL-6, IL-8, IL-17, tumor necrosis factor (TNF), and granulocyte colony-stimulating factor (GCSF) [71], while no clear pattern is reported in FTD. YKL-40 (also known as Chitinase 3-like 1) is a glycoprotein, with roles in both cell survival and proliferation. YKL-40 is elevated in the brain and CSF for both ALS and FTD patients, rendering it a non-discriminative biomarker [72]. Similarly, an upregulation of C-reactive protein (CRP) as a marker of systemic inflammation is upregulated in both FTD and ALS. The adaptive immune system and response plays a key role in the inflammatory processes associated with ALS and can be exploited via assessment of infiltrating T cells into the central nervous system (CNS). A recent study in ALS patients found that high $CD4^{+}FOXP3^{-}$ effector T cells in blood and

CSF predict poorer survival, while increased activated Tregs and a higher activated/resting Treg ratio in blood correlate with better survival [73]. $CD8^+$ T cell infiltration was observed in FTD patients with p-tau pathology. This was accompanied by profound astrocyte activation and only moderate microglia activation [74]. These differential T cell activation patterns could potentially be exploited for differential diagnosis and, importantly, be predictive of severity and progression of the diseases.

Cell-free DNA (cfDNA) has recently emerged as a potential biomarker for monitoring disease progression, although its utility in neurodegenerative diseases remains underexplored. cfDNAs are degraded DNA released from apoptotic or necrotic cells into the extracellular space and subsequently detectable in the CSF and plasma. These cfDNAs are differentially methylated contingent upon cellular origin, making it possible to determine whether the cfDNA originated from degradation of brain cells or from cells in peripheral organs. A recent study introduced a cell-free DNA methylation capture panel targeting ALS-relevant DNA sites, combined with prediction algorithms. This panel both accurately predicted ALS status and exhibited reproducibility across different cohorts [75].

Extracellular vesicles (EVs) and their contents constitute a further promising source of biomarkers. EVs are universal carriers of intercellular communication, serving as immune mediators and facilitating signaling in juxtracrine, autocrine, paracrine, and endocrine interactions between cells [76]. Besides transmission of misfolded proteins, cytokines, and chemokines, EVs contain miRNAs. These miRNAs can impact surrounding tissues via endocytic uptake or release of the EV content to the extracellular space. Several miRNAs have been implicated in ALS and are differentially expressed between patients and healthy controls. These include miR-205-5p, miR-206, miR-376a-5p, miR-412-5p, miR-3927-3p, miR-4701-3p, miR-6763-5p, and miR-6801-3 [77]. Similarly, many miRNAs correlated with FTD have been identified as being dysregulated in the various gene mutation contexts [78]. Therefore, it could potentially be beneficial to include miRNA profiling as a complementary diagnostic tool with which to supplement clinical presentation and aid in the early recognition and differentiation of ALS and FTD.

In summary, many biomarkers have been developed to discriminate between ALS and FTD and their respective subforms. While a clear distinction remains challenging, such biomarkers are evolving to become predictive for the trajectory of disease development, as shown for the T cell subtypes in ALS and presence of HDGFL2 in both FTD and ALS. Moreover, they can be used to determine whether personalized medicine is efficient, as demonstrated in the ASO treatment targeting mutated SOD1 and the CRISPR/Cas13d targeting expansion repeats in *C9orf72* leading to reduced poly-GA, poly-PR, and poly-GR biomarker readouts. A comprehensive understanding of disease pathologies and biomarkers as readouts for successful treatments is imperative for the development of stem cell-based treatments for ALS and FTD.

7.4 Neuroinflammation in FTD and ALS: Mechanisms and Implications

Shared genetic factors between ALS and FTD are reflected in their similar pathology, with increased expression of neuroinflammatory markers observed across the ALS-FTD spectrum [79]. Neuroinflammation, driven by innate immune responses of glial cells, is present in end-stage ALS and FTD pathology and begins early in disease progression, supported by human imaging studies and rodent models. Perturbations in immune cell populations and cytokine changes in ALS and FTD patients suggest an autoinflammatory state [80]. Regarding FTD, there is a strong evidence of inappropriate microglia activation and responses across the divergent mutations. It has been shown that higher levels of microglia activation, measured via ([^{11}C]PK11195 PET) and gray matter volume (structural MRI) are correlated with predictions of cognitive decline in the three major variants of FTD (bvFTD, nfvPPA, and svPPA), which is relevant for discerning between fast and slow progressors [81]. This study also concludes that microglia activation in FTD is a key pathological feature and driver of both tau and TDP-43 pathology.

Progranulin (PGRN) is a secreted immune-regulating glycoprotein released by myeloid cells and various cells in the CNS [82]. PGRN is essential for lysosomal function, neuronal survival, and the regulation of inflammation. Mutations leading to a loss of function in the *GRN* gene manifest in neuronal ceroid lipofuscinosis in FTD in a manner dependent on gene dosage [83]. It appears that microgliosis predominantly damages the white matter in FTD patients with *GRN* haploinsufficiency, based on heterozygous loss-of-function mutations. In this condition, microglia display enlarged endosomes containing myelin accompanied with TDP-43 accumulation [84]. In contrast, overexpression of PGRN is protective in animal models for several neurodegenerative and inflammatory diseases such as AD, PD, FTD, ALS, stroke, Gaucher disease, and arthritis [83], suggesting that the promotion of PGRN expression could potentially be beneficial not only for patients with FTD and PGRN mutations, but also more generally, to counteract neurodegeneration and the inflammatory response in various diseases.

For rare variants of genetic FTD, such as FTD3 associated with *CHMP2B* mutation, roles of the inflammatory profile and microglia reactivity in their pathophysiology is not as well-defined as for the three major variants. This is partly due to the very small cohort size of patients with ultra-rare mutations in *CHMP2B*. Therefore, a clear intrathecal inflammatory profile of FTD3 could not be established in patient CSF samples, even though TREM2 levels were found to increase with disease progression [85]. *CHMP2B* mutations lead to ubiquitinated neuronal cytoplasmic inclusions that do not contain tau, TDP-43, or FUS. Notably, the unique phenotype of iPSC-derived CHMP2B mutant microglia shows that in the heterozygous condition, reflecting the patient situation, microglia are rather unresponsive and not activated [56]. Consequently, it comes as no surprise that subforms of FTD, other than the three major variants, might be associated with divergent microglial responses resulting in lower immune reactivity. In contrast, for FTD3 and *CHMP2B* mutation, it appears that microglia stimulation with IFN-γ might serve as a potentially beneficial therapeutic intervention in the early phases of disease [56].

SOD1 mutations, strictly relevant for ALS, have been demonstrated to yield relevant microglia phenotypes in model organisms. Microglia are activated by misfolded proteins such as mutant (mut)SOD1 originating from diseased motor neurons via CD14/TLR2/TLR4 and scavenger receptor pathways [86, 87]. Once activated, mutSOD1-expressing microglia overproduce extracellular superoxide, leading to neuronal damage [88]. Microglia play a key role in non-cell-autonomous motor neuron degeneration, with mutSOD1 expression in these cells driving inflammatory toxicity and neuronal death. This neurodegeneration is exacerbated by NF-κB activation and increased production of ROS, NO, and TNF-α [89]. Notably, removing mutSOD1 from Cd11b$^+$ microglial cells has been shown to extend lifespan in ALS models, highlighting the critical impact of microglial function on disease progression [90].

As a major genetic driver of ALS and FTD, C9orf72 is abundantly expressed in myeloid cells, which play a key role in antigen presentation and the development of autoimmune responses. *C9orf72* repeat expansion mutations have been identified in such myeloid cells. Loss of C9orf72 has been reported to result in lysosomal accumulation and heightened immune activity in bone marrow macrophages and microglia [91, 92], implicating neuroinflammation in ALS and FTD disease pathology. Effects of heterozygous C9orf72 loss-of-function mutation are more subtle in microglia, showing lysosomal accumulation and a pro-inflammatory phenotype with increased expression of IL-6 and IL-1β [91]. In contrast, intrathecal ablation of C9orf72 in wild-type mice results in upregulated expression of TREM2, C1QA, and TYRO protein tyrosine kinase binding protein (Tyrobp) but without neurodegeneration [93]. These findings are equally applicable for ALS and FTD and indicate a more profound role for C9orf72 in myeloid cells, relevant for later stages of the disease with impaired blood brain barrier integrity and infiltration of myeloid cells.

Haploinsufficiency for *TBK1*—encoding ANK-binding kinase 1—has been demonstrated to cause both ALS and FTD. TBK1 primarily functions to stimulate the production of type I interferons (IFN-α/β) in innate immune cells by interacting with, and phosphorylating, IRF3 and IRF7 [94]. Upstream of TBK1, cytosolic TDP-43 activates the cGAS/STING pathway: STING directly binds to, and activates, TBK1, triggering the transcription of genes comprising the IFN-dependent response in microglia [95–97].

Like microglia and macrophages, T cells are also gaining recognition as disease modifying cell types and contributors to the neuroinflammation observed both in FTD and ALS. In bvFTP, T cell recruitment via chemokine signaling has been reported to play a role in the immune response to tau pathology, with a specific expansion of CD8$^+$ T cells [74]. CD8$^+$ T cells are mostly known for their activation and response to viral infections. CD8$^+$ T cells persist in the brain as long-term tissue-resident memory cells which locally damage neurons and activate glial cells, as described in AD pathophysiology [98]. Intriguingly, single-cell immune profiling of ALS patients versus healthy control CSF revealed activated macrophages and clonally expanded $CD8^+$ T cells in the former, suggesting potential interactions between the CNS and peripheral circulation [99]. As mentioned above, elevated levels of $CD4^+$ T cells are positively correlated with disease progression and negatively correlated with survival time. $CD4^+$ T cells play multiple critical roles in the suppression of the immune reaction by activating cells within the innate immune system, B-lymphocytes, and cytotoxic T cells, as well as non-immune cells [100]. Clearly, more research should be dedicated to deciphering the role of peripheral immune cells and the involvement of adaptive immune cells in both the pathology and progression of disease. Nevertheless, inflammatory responses are largely conserved between both ALS and FTD, rendering them attractive targets for disease modifying intervention strategies. Traditionally, these intervention strategies are based on drug development, but promising new strategies could potentially comprise cell transplantations and/or delivery of key components of disease modifiers.

7.5 Current Cell Therapy Strategies for ALS and FTD

The majority of studies investigating ALS and FTD etiology have been conducted on postmortem human tissue or mouse models. By their nature, postmortem samples reflect the late stages of these disorders, when neuronal damage and cell death have already occurred, limiting treatment options to cell replacement. However, replacing lost neurons, particularly motor neurons with axonal projections of up to 1 m in length, remains unfeasible in humans and has only been achieved in animal models to date [101–103].

Currently only a few drugs have been approved by the FDA for treatment of ALS, which have been shown to prolong the patient's lifetime, while none have been approved for treatment of FTD. These include Riluzole (1995), which decreases the amount of glutamate released [104], Edaravone (2017), which scavenges free radicals, so reducing oxidative stress [105], and Tofersen (2023) which is an antisense oligonucleotide that has shown a beneficial effect in ALS patients carrying mutations in *SOD1* but not *C9orf72* [106]. Treatment with this drug is restricted to cases involving gain-of-function mutations in the gene; however, it holds significant promise for the development of other therapeutics targeting molecular mechanisms.

Instead of transplanting a differentiated cell type, such as has been trialed for the treatment of PD [107], transplantation of multipotent stem cells is expected to prevent disease progression by in vivo differentiation and secretion of protective factors, protecting the remaining cells [108]. Several transplantation studies using multipotent stem cells have been conducted for the treatment of ALS. Most of these studies, which transplanted autologous mesenchymal stem cells (MSCs) from ALS patient bone marrow, consistently demonstrate treatment safety [109–114]. Other studies have explored the transplantation of allogeneic neural stem cells as a promising therapeutic approach. One significant advantage of using allogeneic stem cells is that they lack the aged epigenetic footprint and, more importantly, the disease-specific molecular and cellular abnormalities present in autologous cells. Consequently, allogeneic stem cells may offer greater therapeutic benefits compared to autologous cells. However, a notable drawback of allogeneic transplantation is the requirement for lifelong immunosuppressive therapy to prevent graft rejection or graft-versus-host disease [115]. Immunosuppressive medications are associated with a wide range of side effects, but these may be considered acceptable if the transplants lead to increased survival and slower disease progression. In contrast, autologous stem cell transplants, while not subject to rejection, have so far failed to demonstrate clinically significant outcomes in trials. For instance, MUSE cells, a subpopulation of stem cells in MSC associated with superior tissue repair abilities [116], were proven to be safe when transplanted into patients [117]. Their application was, however, discontinued in 2023. Additionally, a phase III clinical trial investigated the ability of MSCs cultured in vitro to enhance neurotrophic factor (NTF) secretion (NurOwn/Debamestrocel). While biomarker analyzes indicated some biological activity [118], the results did not meet the FDA's threshold for clinical significance, leading to rejection of the therapy in 2023 (BrainStorm Cell Therapeutics Provides Update on FDA Advisory Committee Meeting to Review NurOwn for the Treatment of ALS—Sep 27, 2023). Despite this, the company reported an average survival increase of 5.5 months at a conference in 2024. The concept of enhancing the intrinsic properties of the transplanted stem cells has been investigated by other researchers, such as in a study using fetal neural stem cells, transduced to express GDNF [119]. This NTF supports dopaminergic, motor, and sensory neurons and enhances neuronal resistance to oxidative stress, excitotoxicity, and apoptosis. GDNF further enhances axonal regeneration and synaptic repair, modulates microglial activation and suppresses pro-inflammatory cytokines. Consequently, GDNF overexpression holds significant promise for neuroprotection and regeneration in neurodegenerative diseases, but optimizing its delivery, dosage, and long-term effects is essential for clinical application. While similar genetic modifications could be applied to autologous cells, timing of treatment is critical and the time required to generate an edited autologous cell line delays treatment, potentially diminishing its effectiveness as the disease and neurodegenerative damage progresses. In contrast, allogeneic cells enable quality-controlled, readily available progenitor stem cells, allowing for more immediate intervention, especially favorable for rapidly progressing cases. However, the major challenge of immune rejection remains a significant obstacle.

The field of allogeneic transplantations is constantly advancing, with recent animal studies demonstrating the development of cell lines that evade an immune response in allogeneic or even xenogeneic recipients [120]. This paves the way for the generation of cell lines that can be transplanted to treat neurological disorders without the need for immunosuppression [121].

For FTD, no clinical trials have been conducted to date. However, studies transplanting multipotent stem cells into FTD mouse models have demonstrated partial symptom alleviation [122]. In the context of FTD, stem cells are primarily employed to generate iPSC-based models to acquire a better understanding of disease mechanisms and elucidate patient-specific differences [108].

Ultimately, patient selection is crucial for achieving meaningful treatment effects. Ensuring comparability in disease stage, symptoms, and subtype is essential, as variability in these factors can influence outcomes. For autologous treatments, minimizing the time required for cell preparation is particularly important to mitigate the effects of disease progression during this period. For allogeneic approaches, improving immune tolerance and leveraging gene-editing technologies offer exciting avenues for future research and clinical applications.

7.6 Future Directions for Stem Cell Therapies in ALS and FTD

The promising outcomes of stem cell transplantation studies to date pave the way for the development of more advanced therapeutic strategies for ALS and FTD. However, multiple factors influence the efficacy of stem cell transplantation, necessitating careful optimization to ensure the best possible treatment outcomes for ALS, FTD, and potentially other neurodegenerative disorders.

A critical step in optimizing stem cell treatment is accurately determining the patient's disease subtype and stage. This classification is essential for tailoring the most appropriate intervention. In the context of cell replacement therapies, which are particularly relevant for late-stage disease and rapidly progressing cases, ensuring full differentiation and successful integration of transplanted cells remains a key challenge. Various strategies have been explored to enhance graft survival and integration. For instance, the administration of co-factors such as granulocyte colony-stimulating factor [123] and progranulin [124], in combination with transplantation, has been shown to improve cell mobilization. Additionally, delivering stem cells within a hydrogel or other biomaterial scaffolds has been investigated as a means to enhance graft survival [125].

Another factor in stem cell transplantation is the selection of the appropriate cell type and transplantation site, as transplanted cells exhibit different behaviors dependent upon their microenvironment [126]. Neural stem cell transplantation has been shown to result in differentiation primarily into glial-like cells, as well as other neuronal subtypes such as neurons and oligodendrocytes [127]. The transplantation of neural stem cells into the anterior horn of SOD1 mutant rats showed graft differentiation, survival, and migration, associated with slowing of disease progression [128], highlighting the benefit of using progenitor cells with differentiation capabilities. However, transplanting differentiated cell populations might prove a more strategic approach for the tighter control of grafted cell behavior. For ALS, transplantation of oligodendrocyte progenitor cells may be particularly advantageous, as these cells support neuronal survival, promote myelination, and facilitate axonal growth by secreting NTFs [129].

Beyond direct stem cell replacement, stem cells and their derivatives also hold promise for cell-based therapy, where they may modulate disease progression by secreting neuroprotective and anti-inflammatory factors that help restore cellular homeostasis [130]. Additionally, transplanted cells may benefit from extended survival and superior tissue integration compared to pharmacological treatments, thereby prolonging the localized release of beneficial factors and enhancing therapeutic efficacy.

To maximize therapeutic efficacy, transplanted cells must be delivered to the affected region, which varies depending on the disease type and stage. In ALS, stem cells have been administered into the CSF to mitigate motor neuron degeneration. However, since cells cannot cross an intact blood brain barrier, this approach may yield only limited effects on neuroinflammation within the brain. Consequently, while CSF injection may extend patient lifespan, repeated administrations might be necessary, as persistent neuroinflammation in the brain continues to release inflammatory mediators into the CSF. In diseases such as FTD, direct brain injection may enhance therapeutic outcomes by targeting the affected region rather than a single-cell type [131].

A key concern with both CSF and intracerebral stem cell transplantation is safety. Current clinical trials have demonstrated long-term safety in patients, even with multiple injections [110, 132, 133] and high cell dosages [134]. However, the transplantation of self-renewing cells carries an inherent risk of uncontrolled proliferation. Such risks have been observed in long-term studies of other conditions, such as ataxia-telangiectasia, where adverse effects emerged 4 years post-transplantation [135]. If future advancements in stem cell therapy significantly extend ALS patient survival, it may become advantageous to develop a cell line incorporating a built-in safety mechanism, such as an inducible "suicide" system. The incorporation of such a system in transplanted cells would necessitate genetic modifications, achieved using precise genome editing techniques such as CRISPR-Cas9, first identified in 2012 [136]. Compared to conventional transduction methods used for gene overexpression, precise genome editing provides a safer alternative by enabling controlled gene insertion and expression under the control of endogenous promoters [137, 138]. In contrast, lentiviral induction allows for the simultaneous integration of genetic elements at multiple genomic loci, thereby enhancing the expression of therapeutic genes.

CRISPR-Cas9 is already being applied therapeutically for the ex vivo treatment of blood disorders [139] and is currently being evaluated in clinical trials for in vivo therapy (YOLT-204). For familial cases of both FTD and ALS, similar strategies could be employed to correct pathogenic mutations in MSCs prior to transplantation. This approach would generate a genetically "healthy" cell line, eliminating the need for immunosuppression. However, since the majority of ALS and FTD cases are sporadic and highly heterogeneous, a more effective therapeutic strategy may comprise the development of an enhanced cell line designed for targeting key molecular pathways implicated in disease progression.

Researchers have explored various strategies to optimize transplanted cells by enhancing their intrinsic properties. These approaches include differentiating autologous cells in specialized media to increase their endogenous expression of NTFs [140], genetically modifying cells to overexpress GDNF via viral transduction

[119] (clinical trial: NCT05306457), and engineering hamster cells to overexpress human NTFs [141].

Genetic modifications may play a critical role in enhancing transplanted cells, ensuring they do not contribute to pathology over time due to influences from the diseased microenvironment. Postmortem analyzes of ALS patient brains have revealed increased glial cell populations [142] and elevated inflammatory markers in these cells [79], suggesting that unprotected glial transplants might exacerbate neuroinflammation and disease progression in case they become pro-inflammatory.

To prevent transplanted microglia from contributing to disease progression, an inducible system to trigger cell death could be integrated, activated only if the cells transition from an anti-inflammatory to a pro-inflammatory state. Additionally, to minimize immune system sensitization, transplanted cells should secrete anti-inflammatory factors selectively in response to inflammatory signals in the micro-environment. This could be achieved by designing an inducible system that is only activated when pro-inflammatory molecule concentrations exceed a predefined threshold.

In the context of ALS and related neurodegenerative diseases, several potential genetic modifications could enhance the efficacy of stem cell-based transplantation therapies beyond those currently being tested in humans. One promising approach is that of engineering transplanted cells to secrete poly-dipeptide-neutralizing agents targeting poly-GA, poly-PR and poly-GR, present in patients with *C9orf72* repeat expansion. Neutralization of these has been shown to reduce oxidative stress, improve mitochondrial function [143], rescue ALS and FTD phenotypes [141, 144], and decrease inflammation in ALS mouse models [145].

Several other promising molecular targets for therapeutic intervention have been explored such as inhibition of Glycoprotein nonmetastatic melanoma protein B (GPNMB), which is elevated in CSF of ALS patients with *SOD1* mutations [146]. Other approaches include genetic depletion of NLRP3 and general suppression of the NLRP3 inflammasome [147] as well as direct suppression of TDP-43 [148]. Inhibition of microglial release of matrix metalloproteinase-9 (MMP-9) has also been explored, preventing the release of pro-inflammatory cytokines such as TNF-α and IL-1β [95]. Other approaches trialed include upregulation of the expression of ciliary neurotrophic factor (CNTF) [141], ablation of TBK1 [149] or inhibition of the downstream target of TBK1, RIPK1, which controls cell survival and apoptosis in response to inflammatory and stress signals [150].

More broadly, targeting immune response pathways implicated in ALS and FTD might provide therapeutic benefits. By engineering transplanted cells to locally secrete immunomodulatory factors, the intervention could be restricted to affected regions, minimizing systemic side effects associated with generalized anti-inflammatory treatments.

Inflammatory pathways implicated in ALS and FTD pathophysiology that may serve as therapeutic targets via their inhibition include the STING pathway [151], which is known to be activated by TDP-43 pathology [151], as well as the NF-κB pathway and interferon signaling [152].

While these strategies may also provide therapeutic benefits for FTD patients, a specific approach under investigation for FTD involves developing transplanted cells engineered to secrete progranulin for patients with GRN mutations [153–155]. This strategy is currently being evaluated in clinical trials using AAV-mediated progranulin gene therapy [156]. Notably, progranulin supplementation may also be beneficial in ALS patients [119], as progranulin deficiency has been linked with TDP-43 mislocalization in ALS organoid models [155].

Finally, an emerging approach for prospective stem cell-based therapies in ALS and FTD is to use transplanted cells as delivery vehicles for therapeutic agents [157]. These cells could be engineered to transport and release treatments such as antisense oligonucleotides, small interfering RNA (siRNA), or other neuroprotective molecules, providing a localized and sustained therapeutic effect.

In conclusion, while stem cell-based therapies hold great promise for ALS and FTD, their success depends on precise tailoring to individual disease phenotypes, genetic backgrounds, and affected biological pathways. Emerging biomarkers will be critical for early diagnosis, for distinguishing FTD from ALS and other neurodegenerative diseases, and for the evaluation of therapeutic efficacy in clinical trials. Additionally, ensuring patient safety in transplantation studies remains paramount. The development of immune-evasive, off-the-shelf iPSC-derived therapies holds great potential to enhance accessibility and affordability, ultimately broadening patient reach and optimizing clinical outcomes [121].

References

1. Ferrari R, Kapogiannis D, Huey ED, Momeni P (2011) FTD and ALS: a tale of two diseases. Curr Alzheimer Res 8(3):273–294
2. Chen Y, Landin-Romero R, Kumfor F, Irish M, Hodges JR, Piguet O (2020) Cerebellar structural connectivity and contributions to cognition in frontotemporal dementias. Cortex 129:57–67
3. Pottier C, Zhou X, Perkerson RB 3rd, Baker M, Jenkins GD, Serie DJ et al (2018) Potential genetic modifiers of disease risk and age at onset in patients with frontotemporal lobar degeneration and GRN mutations: a genome-wide association study. Lancet Neurol 17(6):548–558
4. Kocoglu C, Gossye H, Dillen L, Van Mossevelde S, De Bleecker JL, Vandenberghe R et al (2021) No association of CpG SNP rs9357140 with onset age in Belgian C9orf72 repeat expansion carriers. Neurobiol Aging 97(145):e1–e4
5. Greaves CV, Rohrer JD (2019) An update on genetic frontotemporal dementia. J Neurol 266(8):2075–2086
6. Rademakers R, Neumann M, Mackenzie IR (2012) Advances in understanding the molecular basis of frontotemporal dementia. Nat Rev Neurol 8(8):423–434
7. Mackenzie IR, Neumann M (2016) Molecular neuropathology of frontotemporal dementia: insights into disease mechanisms from postmortem studies. J Neurochem 138(Suppl 1):54–70
8. Verma A (2021) Clinical manifestation and management of amyotrophic lateral sclerosis. In: Araki T (ed) Amyotrophic lateral sclerosis. Exon, Brisbane
9. Majounie E, Renton AE, Mok K, Dopper EG, Waite A, Rollinson S et al (2012) Frequency of the C9orf72 hexanucleotide repeat expansion in patients with amyotrophic lateral sclerosis and frontotemporal dementia: a cross-sectional study. Lancet Neurol 11(4):323–330

10. Wang H, Guan L, Deng M (2023) Recent progress of the genetics of amyotrophic lateral sclerosis and challenges of gene therapy. Front Neurosci 17:1170996
11. Jankovska N, Matej R (2021) Molecular pathology of ALS: what we currently know and what important information is still missing. Diagnostics (Basel) 11(8):1365
12. Mackenzie IR, Rademakers R (2008) The role of transactive response DNA-binding protein-43 in amyotrophic lateral sclerosis and frontotemporal dementia. Curr Opin Neurol 21(6):693–700
13. Smith EF, Shaw PJ, De Vos KJ (2019) The role of mitochondria in amyotrophic lateral sclerosis. Neurosci Lett 710:132933
14. Lederer CW, Torrisi A, Pantelidou M, Santama N, Cavallaro S (2007) Pathways and genes differentially expressed in the motor cortex of patients with sporadic amyotrophic lateral sclerosis. BMC Genomics 8:26
15. Sasaki S, Iwata M (1996) Impairment of fast axonal transport in the proximal axons of anterior horn neurons in amyotrophic lateral sclerosis. Neurology 47(2):535–540
16. Yang X, Ma Z, Lian P, Xu Y, Cao X (2023) Common mechanisms underlying axonal transport deficits in neurodegenerative diseases: a mini review. Front Mol Neurosci 16:1172197
17. De Vos KJ, Hafezparast M (2017) Neurobiology of axonal transport defects in motor neuron diseases: opportunities for translational research? Neurobiol Dis 105:283–299
18. Van Den Bosch L, Van Damme P, Bogaert E, Robberecht W (2006) The role of excitotoxicity in the pathogenesis of amyotrophic lateral sclerosis. Biochim Biophys Acta 1762(11–12):1068–1082
19. Benussi A, Alberici A, Buratti E, Ghidoni R, Gardoni F, Di Luca M et al (2019) Toward a glutamate hypothesis of frontotemporal dementia. Front Neurosci 13:304
20. Chua JP, De Calbiac H, Kabashi E, Barmada SJ (2022) Autophagy and ALS: mechanistic insights and therapeutic implications. Autophagy 18(2):254–282
21. Hung C, Patani R (2024) 4R tau drives endolysosomal and autophagy dysfunction in frontotemporal dementia. Autophagy 20(5):1201–1202
22. Ragagnin AMG, Shadfar S, Vidal M, Jamali MS, Atkin JD (2019) Motor neuron susceptibility in ALS/FTD. Front Neurosci 13:532
23. Sirabella R, Valsecchi V, Anzilotti S, Cuomo O, Vinciguerra A, Cepparulo P et al (2018) Ionic homeostasis maintenance in ALS: focus on new therapeutic targets. Front Neurosci 12:510
24. Crockford C, Newton J, Lonergan K, Chiwera T, Booth T, Chandran S et al (2018) ALS-specific cognitive and behavior changes associated with advancing disease stage in ALS. Neurology 91(15):e1370–e1e80
25. Burkhardt C, Neuwirth C, Weber M (2017) Longitudinal assessment of the Edinburgh cognitive and behavioural amyotrophic lateral sclerosis screen (ECAS): lack of practice effect in ALS patients? Amyotroph Lateral Scler Frontotemporal Degener 18(3–4):202–209
26. Beeldman E, Raaphorst J, Klein Twennaar M, de Visser M, Schmand BA, de Haan RJ (2016) The cognitive profile of ALS: a systematic review and meta-analysis update. J Neurol Neurosurg Psychiatry 87(6):611–619
27. Pineda SS, Lee H, Ulloa-Navas MJ, Linville RM, Garcia FJ, Galani K et al (2024) Single-cell dissection of the human motor and prefrontal cortices in ALS and FTLD. Cell 187(8):1971–89 e16
28. Abramzon YA, Fratta P, Traynor BJ, Chia R (2020) The overlapping genetics of amyotrophic lateral sclerosis and frontotemporal dementia. Front Neurosci 14:42
29. Gao FB, Almeida S, Lopez-Gonzalez R (2017) Dysregulated molecular pathways in amyotrophic lateral sclerosis-frontotemporal dementia spectrum disorder. EMBO J 36(20):2931–2950
30. Ling SC, Polymenidou M, Cleveland DW (2013) Converging mechanisms in ALS and FTD: disrupted RNA and protein homeostasis. Neuron 79(3):416–438
31. Nedelsky NB, Taylor JP (2022) Pathological phase transitions in ALS-FTD impair dynamic RNA-protein granules. RNA 28(1):97–113

32. Beckers J, Tharkeshwar AK, Fumagalli L, Contardo M, Van Schoor E, Fazal R et al (2023) A toxic gain-of-function mechanism in C9orf72 ALS impairs the autophagy-lysosome pathway in neurons. Acta Neuropathol Commun 11(1):151
33. Ragagnin AMG, Sundaramoorthy V, Farzana F, Gautam S, Saravanabavan S, Takalloo Z et al (2023) ALS/FTD-associated mutation in cyclin F inhibits ER-Golgi trafficking, inducing ER stress, ERAD and Golgi fragmentation. Sci Rep 13(1):20467
34. Shi Y, Lin S, Staats KA, Li Y, Chang WH, Hung ST et al (2018) Haploinsufficiency leads to neurodegeneration in C9ORF72 ALS/FTD human induced motor neurons. Nat Med 24(3):313–325
35. Farg MA, Konopka A, Soo KY, Ito D, Atkin JD (2017) The DNA damage response (DDR) is induced by the C9orf72 repeat expansion in amyotrophic lateral sclerosis. Hum Mol Genet 26(15):2882–2896
36. Chou CC, Zhang Y, Umoh ME, Vaughan SW, Lorenzini I, Liu F et al (2018) TDP-43 pathology disrupts nuclear pore complexes and nucleocytoplasmic transport in ALS/FTD. Nat Neurosci 21(2):228–239
37. Baradaran-Heravi Y, Van Broeckhoven C, van der Zee J (2020) Stress granule mediated protein aggregation and underlying gene defects in the FTD-ALS spectrum. Neurobiol Dis 134:104639
38. San Gil R, Pascovici D, Venturato J, Brown-Wright H, Mehta P, Madrid San Martin L et al (2024) A transient protein folding response targets aggregation in the early phase of TDP-43-mediated neurodegeneration. Nat Commun 15(1):1508
39. Garcia-Garcia R, Martin-Herrero L, Blanca-Pariente L, Perez-Cabello J, Roodveldt C (2021) Immune signaling kinases in amyotrophic lateral sclerosis (ALS) and frontotemporal dementia (FTD). Int J Mol Sci 22(24):13280
40. Arai T, Hasegawa M, Akiyama H, Ikeda K, Nonaka T, Mori H et al (2006) TDP-43 is a component of ubiquitin-positive tau-negative inclusions in frontotemporal lobar degeneration and amyotrophic lateral sclerosis. Biochem Biophys Res Commun 351(3):602–611
41. Cairns NJ, Neumann M, Bigio EH, Holm IE, Troost D, Hatanpaa KJ et al (2007) TDP-43 in familial and sporadic frontotemporal lobar degeneration with ubiquitin inclusions. Am J Pathol 171(1):227–240
42. Brettschneider J, Del Tredici K, Toledo JB, Robinson JL, Irwin DJ, Grossman M et al (2013) Stages of pTDP-43 pathology in amyotrophic lateral sclerosis. Ann Neurol 74(1):20–38
43. Zhang J, Velmeshev D, Hashimoto K, Huang YH, Hofmann JW, Shi X et al (2020) Neurotoxic microglia promote TDP-43 proteinopathy in progranulin deficiency. Nature 588(7838):459–465
44. Cook CN, Wu Y, Odeh HM, Gendron TF, Jansen-West K, Del Rosso G et al (2020) C9orf72 poly(GR) aggregation induces TDP-43 proteinopathy. Sci Transl Med 12(559):eabb3774
45. Mackenzie IR, Nicholson AM, Sarkar M, Messing J, Purice MD, Pottier C et al (2017) TIA1 mutations in amyotrophic lateral sclerosis and frontotemporal dementia promote phase separation and Alter stress granule dynamics. Neuron 95(4):808–16 e9
46. Gu X, Chen Y, Wei Q, Cao B, Ou R, Yuan X et al (2018) Mutation screening of the TIA1 gene in Chinese patients with amyotrophic lateral sclerosis/frontotemporal dementia. Neurobiol Aging 68(161):e1–e3
47. Baradaran-Heravi Y, Dillen L, Nguyen HP, Van Mossevelde S, Baets J, De Jonghe P et al (2018) No supportive evidence for TIA1 gene mutations in a European cohort of ALS-FTD spectrum patients. Neurobiol Aging 69(293):e9–e11
48. Van Daele SH, Moisse M, van Vugt J, Zwamborn RAJ, van der Spek R, van Rheenen W et al (2023) Genetic variability in sporadic amyotrophic lateral sclerosis. Brain 146(9):3760–3769
49. Jury N, Abarzua S, Diaz I, Guerra MV, Ampuero E, Cubillos P et al (2020) Widespread loss of the silencing epigenetic mark H3K9me3 in astrocytes and neurons along with hippocampal-dependent cognitive impairment in C9orf72 BAC transgenic mice. Clin Epigenetics 12(1):32
50. del Aguila MA, Longstreth WT Jr, McGuire V, Koepsell TD, van Belle G (2003) Prognosis in amyotrophic lateral sclerosis: a population-based study. Neurology 60(5):813–819

51. Liscic RM, Alberici A, Cairns NJ, Romano M, Buratti E (2020) From basic research to the clinic: innovative therapies for ALS and FTD in the pipeline. Mol Neurodegener 15(1):31
52. Hutton M, Lendon CL, Rizzu P, Baker M, Froelich S, Houlden H et al (1998) Association of missense and 5′-splice-site mutations in tau with the inherited dementia FTDP-17. Nature 393(6686):702–705
53. Meijboom KE, Abdallah A, Fordham NP, Nagase H, Rodriguez T, Kraus C et al (2022) CRISPR/Cas9-mediated excision of ALS/FTD-causing hexanucleotide repeat expansion in C9ORF72 rescues major disease mechanisms in vivo and in vitro. Nat Commun 13(1):6286
54. Szebenyi K, Wenger LMD, Sun Y, Dunn AWE, Limegrover CA, Gibbons GM et al (2021) Human ALS/FTD brain organoid slice cultures display distinct early astrocyte and targetable neuronal pathology. Nat Neurosci 24(11):1542–1554
55. Lee C, Frew J, Weilinger NL, Wendt S, Cai W, Sorrentino S et al (2023) hiPSC-derived GRN-deficient astrocytes delay spiking activity of developing neurons. Neurobiol Dis 181:106124
56. Haukedal H, Syshoj Lorenzen S, Winther Westi E, Corsi GI, Gadekar VP, McQuade A et al (2023) Alteration of microglial metabolism and inflammatory profile contributes to neurotoxicity in a hiPSC-derived microglia model of frontotemporal dementia 3. Brain Behav Immun 113:353–373
57. Marques C, Held A, Dorfman K, Sung J, Song C, Kavuturu AS et al (2024) Neuronal STING activation in amyotrophic lateral sclerosis and frontotemporal dementia. Acta Neuropathol 147(1):56
58. Sirkis DW, Geier EG, Bonham LW, Karch CM, Yokoyama JS (2019) Recent advances in the genetics of frontotemporal dementia. Curr Genet Med Rep 7(1):41–52
59. Neumann M, Sampathu DM, Kwong LK, Truax AC, Micsenyi MC, Chou TT et al (2006) Ubiquitinated TDP-43 in frontotemporal lobar degeneration and amyotrophic lateral sclerosis. Science 314(5796):130–133
60. Suarez-Calvet M, Dols-Icardo O, Llado A, Sanchez-Valle R, Hernandez I, Amer G et al (2014) Plasma phosphorylated TDP-43 levels are elevated in patients with frontotemporal dementia carrying a C9orf72 repeat expansion or a GRN mutation. J Neurol Neurosurg Psychiatry 85(6):684–691
61. Junttila A, Kuvaja M, Hartikainen P, Siloaho M, Helisalmi S, Moilanen V et al (2016) Cerebrospinal fluid TDP-43 in frontotemporal lobar degeneration and amyotrophic lateral sclerosis patients with and without the C9ORF72 Hexanucleotide expansion. Dement Geriatr Cogn Dis Extra 6(1):142–149
62. Feneberg E, Steinacker P, Lehnert S, Schneider A, Walther P, Thal DR et al (2014) Limited role of free TDP-43 as a diagnostic tool in neurodegenerative diseases. Amyotroph Lateral Scler Frontotemporal Degener 15(5–6):351–356
63. Irwin KE, Jasin P, Braunstein KE, Sinha IR, Garret MA, Bowden KD et al (2024) A fluid biomarker reveals loss of TDP-43 splicing repression in presymptomatic ALS-FTD. Nat Med 30(2):382–393
64. Thijssen EH, La Joie R, Strom A, Fonseca C, Iaccarino L, Wolf A et al (2021) Plasma phosphorylated tau 217 and phosphorylated tau 181 as biomarkers in Alzheimer's disease and frontotemporal lobar degeneration: a retrospective diagnostic performance study. Lancet Neurol 20(9):739–752
65. Lee YB, Baskaran P, Gomez-Deza J, Chen HJ, Nishimura AL, Smith BN et al (2017) C9orf72 poly GA RAN-translated protein plays a key role in amyotrophic lateral sclerosis via aggregation and toxicity. Hum Mol Genet 26(24):4765–4777
66. Liu H, Zhao XF, Lu YN, Hayes LR, Wang J (2024) CRISPR/Cas13d targeting suppresses repeat-associated non-AUG translation of C9orf72 hexanucleotide repeat RNA. J Clin Invest 134(21):e179016
67. Leykam L, Forsberg KME, Nordstrom U, Hjertkvist K, Oberg A, Jonsson E et al (2024) Specific analysis of SOD1 enzymatic activity in CSF from ALS patients with and without SOD1 mutations. Neurobiol Dis 202:106718

68. Verde F, Licaj S, Soranna D, Ticozzi N, Silani V, Zambon A (2024) Cerebrospinal fluid and blood neurofilament light chain levels in amyotrophic lateral sclerosis and frontotemporal degeneration: a meta-analysis. Eur J Neurol 31(9):e16371
69. Behzadi A, Pujol-Calderon F, Tjust AE, Wuolikainen A, Hoglund K, Forsberg K et al (2021) Neurofilaments can differentiate ALS subgroups and ALS from common diagnostic mimics. Sci Rep 11(1):22128
70. Zucchi E, Bonetto V, Soraru G, Martinelli I, Parchi P, Liguori R et al (2020) Neurofilaments in motor neuron disorders: towards promising diagnostic and prognostic biomarkers. Mol Neurodegener 15(1):58
71. Femiano C, Bruno A, Gilio L, Buttari F, Dolcetti E, Galifi G et al (2024) Inflammatory signature in amyotrophic lateral sclerosis predicting disease progression. Sci Rep 14(1):19796
72. Llorens F, Thune K, Tahir W, Kanata E, Diaz-Lucena D, Xanthopoulos K et al (2017) YKL-40 in the brain and cerebrospinal fluid of neurodegenerative dementias. Mol Neurodegener 12(1):83
73. Yazdani S, Seitz C, Cui C, Lovik A, Pan L, Piehl F et al (2022) T cell responses at diagnosis of amyotrophic lateral sclerosis predict disease progression. Nat Commun 13(1):6733
74. Hartnell IJ, Woodhouse D, Jasper W, Mason L, Marwaha P, Graffeuil M et al (2024) Glial reactivity and T cell infiltration in frontotemporal lobar degeneration with tau pathology. Brain 147(2):590–606
75. Caggiano C, Morselli M, Qian X, Celona B, Thompson M, Wani S et al (2024) Tissue informative cell-free DNA methylation sites in amyotrophic lateral sclerosis. medRxiv
76. Buzas EI (2023) The roles of extracellular vesicles in the immune system. Nat Rev Immunol 23(4):236–250
77. Cheng J, Ho WK, Wu BT, Liu HP, Lin WY (2023) miRNA profiling as a complementary diagnostic tool for amyotrophic lateral sclerosis. Sci Rep 13(1):13805
78. Kmetzsch V, Latouche M, Saracino D, Rinaldi D, Camuzat A, Gareau T et al (2022) MicroRNA signatures in genetic frontotemporal dementia and amyotrophic lateral sclerosis. Ann Clin Transl Neurol 9(11):1778–1791
79. Umoh ME, Dammer EB, Dai J, Duong DM, Lah JJ, Levey AI et al (2018) A proteomic network approach across the ALS-FTD disease spectrum resolves clinical phenotypes and genetic vulnerability in human brain. EMBO Mol Med 10(1):48–62
80. McCauley ME, Baloh RH (2019) Inflammation in ALS/FTD pathogenesis. Acta Neuropathol 137(5):715–730
81. Malpetti M, Cope TE, Street D, Jones PS, Hezemans FH, Mak E et al (2023) Microglial activation in the frontal cortex predicts cognitive decline in frontotemporal dementia. Brain 146(8):3221–3231
82. Daniel R, He Z, Carmichael KP, Halper J, Bateman A (2000) Cellular localization of gene expression for progranulin. J Histochem Cytochem 48(7):999–1009
83. Rhinn H, Tatton N, McCaughey S, Kurnellas M, Rosenthal A (2022) Progranulin as a therapeutic target in neurodegenerative diseases. Trends Pharmacol Sci 43(8):641–652
84. Wu Y, Shao W, Todd TW, Tong J, Yue M, Koga S et al (2021) Microglial lysosome dysfunction contributes to white matter pathology and TDP-43 proteinopathy in GRN-associated FTD. Cell Rep 36(8):109581
85. Roos P, von Essen MR, Nielsen TT, Johannsen P, Stokholm J, Bie AS et al (2018) Inflammatory markers of CHMP2B-mediated frontotemporal dementia. J Neuroimmunol 324:136–142
86. Yao L, Kan EM, Lu J, Hao A, Dheen ST, Kaur C et al (2013) Toll-like receptor 4 mediates microglial activation and production of inflammatory mediators in neonatal rat brain following hypoxia: role of TLR4 in hypoxic microglia. J Neuroinflammation 10:23
87. Zeineddine R, Pundavela JF, Corcoran L, Stewart EM, Do-Ha D, Bax M et al (2015) SOD1 protein aggregates stimulate macropinocytosis in neurons to facilitate their propagation. Mol Neurodegener 10:57
88. Zhao W, Beers DR, Henkel JS, Zhang W, Urushitani M, Julien JP et al (2010) Extracellular mutant SOD1 induces microglial-mediated motoneuron injury. Glia 58(2):231–243

89. Dimayuga FO, Wang C, Clark JM, Dimayuga ER, Dimayuga VM, Bruce-Keller AJ (2007) SOD1 overexpression alters ROS production and reduces neurotoxic inflammatory signaling in microglial cells. J Neuroimmunol 182(1–2):89–99
90. Clarke BE, Patani R (2020) The microglial component of amyotrophic lateral sclerosis. Brain 143(12):3526–3539
91. O'Rourke JG, Bogdanik L, Yanez A, Lall D, Wolf AJ, Muhammad AK et al (2016) C9orf72 is required for proper macrophage and microglial function in mice. Science 351(6279):1324–1329
92. Lall D, Baloh RH (2017) Microglia and C9orf72 in neuroinflammation and ALS and frontotemporal dementia. J Clin Invest 127(9):3250–3258
93. Lagier-Tourenne C, Baughn M, Rigo F, Sun S, Liu P, Li HR et al (2013) Targeted degradation of sense and antisense C9orf72 RNA foci as therapy for ALS and frontotemporal degeneration. Proc Natl Acad Sci USA 110(47):E4530–E4539
94. Runde AP, Mack R, S JP, Zhang J. (2022) The role of TBK1 in cancer pathogenesis and anticancer immunity. J Exp Clin Cancer Res 41(1):135
95. Vahsen BF, Nalluru S, Morgan GR, Farrimond L, Carroll E, Xu Y et al (2023) C9orf72-ALS human iPSC microglia are pro-inflammatory and toxic to co-cultured motor neurons via MMP9. Nat Commun 14(1):5898
96. Oakes JA, Davies MC, Collins MO (2017) TBK1: a new player in ALS linking autophagy and neuroinflammation. Mol Brain 10(1):5
97. Ahmad L, Zhang SY, Casanova JL, Sancho-Shimizu V (2016) Human TBK1: a gatekeeper of neuroinflammation. Trends Mol Med 22(6):511–527
98. Reagin KL, Funk KE (2022) The role of antiviral CD8(+) T cells in cognitive impairment. Curr Opin Neurobiol 76:102603
99. Kim HJ, Ban JJ, Kang J, Im HR, Ko SH, Sung JJ et al (2024) Single-cell analysis reveals expanded CD8(+) GZMK (high) T cells in CSF and shared peripheral clones in sporadic amyotrophic lateral sclerosis. Brain Commun 6(6):fcae428
100. Luckheeram RV, Zhou R, Verma AD, Xia B (2012) CD4(+)T cells: differentiation and functions. Clin Dev Immunol 2012:925135
101. Bryson JB, Kourgiantaki A, Jiang D, Demosthenous A, Greensmith L (2024) An optogenetic cell therapy to restore control of target muscles in an aggressive mouse model of amyotrophic lateral sclerosis. elife 12:RP88250
102. Wyatt TJ, Rossi SL, Siegenthaler MM, Frame J, Robles R, Nistor G et al (2011) Human motor neuron progenitor transplantation leads to endogenous neuronal sparing in 3 models of motor neuron loss. Stem Cells Int 2011:207230
103. Kato S, Kurimoto S, Nakano T, Yoneda H, Ishii H, Mita-Sugiura S et al (2015) Successful transplantation of motoneurons into the peripheral nerve depends on the number of transplanted cells. Nagoya J Med Sci 77(1–2):253–263
104. Bensimon G, Lacomblez L, Meininger V (1994) A controlled trial of Riluzole in amyotrophic lateral sclerosis. ALS/Riluzole Study Group. N Engl J Med 330(9):585–591
105. Writing G, Edaravone ALSSG (2017) Safety and efficacy of edaravone in well defined patients with amyotrophic lateral sclerosis: a randomised, double-blind, placebo-controlled trial. Lancet Neurol 16(7):505–512
106. Cantara S, Simoncelli G, Ricci C (2024) Antisense oligonucleotides (ASOs) in motor neuron diseases: a road to cure in light and shade. Int J Mol Sci 25(9):4809
107. Freed CR, Greene PE, Breeze RE, Tsai WY, DuMouchel W, Kao R et al (2001) Transplantation of embryonic dopamine neurons for severe Parkinson's disease. N Engl J Med 344(10):710–719
108. Sivandzade F, Cucullo L (2021) Regenerative stem cell therapy for neurodegenerative diseases: an overview. Int J Mol Sci 22(4):2153
109. Sykova E, Rychmach P, Drahoradova I, Konradova S, Ruzickova K, Vorisek I et al (2017) Transplantation of mesenchymal stromal cells in patients with amyotrophic lateral sclerosis: results of phase I/IIa clinical trial. Cell Transplant 26(4):647–658

110. Mazzini L, Ferrero I, Luparello V, Rustichelli D, Gunetti M, Mareschi K et al (2010) Mesenchymal stem cell transplantation in amyotrophic lateral sclerosis: a phase I clinical trial. Exp Neurol 223(1):229–237
111. Karussis D, Karageorgiou C, Vaknin-Dembinsky A, Gowda-Kurkalli B, Gomori JM, Kassis I et al (2010) Safety and immunological effects of mesenchymal stem cell transplantation in patients with multiple sclerosis and amyotrophic lateral sclerosis. Arch Neurol 67(10):1187–1194
112. Blanquer M, Perez Espejo MA, Iniesta F, Gomez Espuch J, Meca J, Villaverde R et al (2010) Bone marrow stem cell transplantation in amyotrophic lateral sclerosis: technical aspects and preliminary results from a clinical trial. Methods Find Exp Clin Pharmacol 32 Suppl A:31–37
113. Oh KW, Moon C, Kim HY, Oh SI, Park J, Lee JH et al (2015) Phase I trial of repeated intrathecal autologous bone marrow-derived mesenchymal stromal cells in amyotrophic lateral sclerosis. Stem Cells Transl Med 4(6):590–597
114. Berry JD, Cudkowicz ME, Windebank AJ, Staff N P, Owegi M, Nicholson K et al (2019) NurOwn, phase 2, randomized, clinical trial in patients with ALS: safety, clinical, and biomarker results. Neurology 93(24):e2294–ee305
115. Pidala J, Martens M, Anasetti C, Carreras J, Horowitz M, Lee SJ et al (2020) Factors associated with successful discontinuation of immune suppression after allogeneic hematopoietic cell transplantation. JAMA Oncol 6(1):e192974
116. Dezawa M (2016) Muse cells provide the pluripotency of mesenchymal stem cells: direct contribution of Muse cells to tissue regeneration. Cell Transplant 25(5):849–861
117. Yamashita T, Nakano Y, Sasaki R, Tadokoro K, Omote Y, Yunoki T et al (2023) Safety and clinical effects of a Muse cell-based product in patients with amyotrophic lateral sclerosis: results of a phase 2 clinical trial. Cell Transplant 32:9636897231214370
118. Lindborg SR, Goyal NA, Katz J, Burford M, Li J, Kaspi H et al (2024) Debamestrocel multimodal effects on biomarker pathways in amyotrophic lateral sclerosis are linked to clinical outcomes. Muscle Nerve 69(6):719–729
119. Baloh RH, Johnson JP, Avalos P, Allred P, Svendsen S, Gowing G et al (2022) Transplantation of human neural progenitor cells secreting GDNF into the spinal cord of patients with ALS: a phase 1/2a trial. Nat Med 28(9):1813–1822
120. Hu X, White K, Olroyd AG, DeJesus R, Dominguez AA, Dowdle WE et al (2024) Hypoimmune induced pluripotent stem cells survive long term in fully immunocompetent, allogeneic rhesus macaques. Nat Biotechnol 42(3):413–423
121. Frederiksen HR, Doehn U, Tveden-Nyborg P, Freude KK (2020) Non-immunogenic induced pluripotent stem cells, a promising way forward for allogenic transplantations for neurological disorders. Front Genome Ed 2:623717
122. Yang Y, Aloi MS, Cudaback E, Josephsen SR, Rice SJ, Jorstad NL et al (2014) Wild-type bone marrow transplant partially reverses neuroinflammation in progranulin-deficient mice. Lab Investig 94(11):1224–1236
123. Tarella C, Rutella S, Gualandi F, Melazzini M, Scime R, Petrini M et al (2010) Consistent bone marrow-derived cell mobilization following repeated short courses of granulocyte-colony-stimulating factor in patients with amyotrophic lateral sclerosis: results from a multicenter prospective trial. Cytotherapy 12(1):50–59
124. Yamagami K, Samata B, Doi D, Tsuchimochi R, Kikuchi T, Amimoto N et al (2024) Progranulin enhances the engraftment of transplanted human iPS cell-derived cerebral neurons. Stem Cells Transl Med 13(11):1113–1128
125. Wang Y, Zoneff ER, Thomas JW, Hong N, Tan LL, McGillivray DJ et al (2023) Hydrogel oxygen reservoirs increase functional integration of neural stem cell grafts by meeting metabolic demands. Nat Commun 14(1):457
126. Yuan Q, Zhang SC (2024) Circuit integration by transplanted human neurons. Curr Opin Genet Dev 89:102225

127. Lundberg C, Martinez-Serrano A, Cattaneo E, McKay RD, Bjorklund A (1997) Survival, integration, and differentiation of neural stem cell lines after transplantation to the adult rat striatum. Exp Neurol 145(2 Pt 1):342–360
128. Zalfa C, Rota Nodari L, Vacchi E, Gelati M, Profico D, Boido M et al (2019) Transplantation of clinical-grade human neural stem cells reduces neuroinflammation, prolongs survival and delays disease progression in the SOD1 rats. Cell Death Dis 10(5):345
129. Priest CA, Manley NC, Denham J, Wirth ED 3rd, Lebkowski JS (2015) Preclinical safety of human embryonic stem cell-derived oligodendrocyte progenitors supporting clinical trials in spinal cord injury. Regen Med 10(8):939–958
130. Tang X, Deng P, Li L, He Y, Wang J, Hao D et al (2024) Advances in genetically modified neural stem cell therapy for central nervous system injury and neurological diseases. Stem Cell Res Ther 15(1):482
131. Frisoni GB, Beltramello A, Geroldi C, Weiss C, Bianchetti A, Trabucchi M (1996) Brain atrophy in frontotemporal dementia. J Neurol Neurosurg Psychiatry 61(2):157–165
132. Glass JD, Boulis NM, Johe K, Rutkove SB, Federici T, Polak M et al (2012) Lumbar intraspinal injection of neural stem cells in patients with amyotrophic lateral sclerosis: results of a phase I trial in 12 patients. Stem Cells 30(6):1144–1151
133. Chen L, Chen D, Xi H, Wang Q, Liu Y, Zhang F et al (2012) Olfactory ensheathing cell neurorestorotherapy for amyotrophic lateral sclerosis patients: benefits from multiple transplantations. Cell Transplant 21(Suppl 1):S65–S77
134. Glass JD, Hertzberg VS, Boulis NM, Riley J, Federici T, Polak M et al (2016) Transplantation of spinal cord-derived neural stem cells for ALS: analysis of phase 1 and 2 trials. Neurology 87(4):392–400
135. Amariglio N, Hirshberg A, Scheithauer BW, Cohen Y, Loewenthal R, Trakhtenbrot L et al (2009) Donor-derived brain tumor following neural stem cell transplantation in an ataxia telangiectasia patient. PLoS Med 6(2):e1000029
136. Jinek M, Chylinski K, Fonfara I, Hauer M, Doudna JA, Charpentier E (2012) A programmable dual-RNA-guided DNA endonuclease in adaptive bacterial immunity. Science 337(6096):816–821
137. Sun C, Li C, Liu W, Schioth HB (2022) Generation of endogenous promoter-driven luciferase reporter system using CRISPR/Cas9 for investigating transcriptional regulation of the Core clock gene BMAL1. Biomedicines 10(12):3108
138. Rojas-Fernandez A, Herhaus L, Macartney T, Lachaud C, Hay RT, Sapkota GP (2015) Rapid generation of endogenously driven transcriptional reporters in cells through CRISPR/Cas9. Sci Rep 5:9811
139. Frangoul H, Altshuler D, Cappellini MD, Chen YS, Domm J, Eustace BK et al (2021) CRISPR-Cas9 gene editing for sickle cell disease and beta-thalassemia. N Engl J Med 384(3):252–260
140. Petrou P, Gothelf Y, Argov Z, Gotkine M, Levy YS, Kassis I et al (2016) Safety and clinical effects of mesenchymal stem cells secreting neurotrophic factor transplantation in patients with amyotrophic lateral sclerosis: results of phase 1/2 and 2a clinical trials. JAMA Neurol 73(3):337–344
141. Aebischer P, Schluep M, Deglon N, Joseph JM, Hirt L, Heyd B et al (1996) Intrathecal delivery of CNTF using encapsulated genetically modified xenogeneic cells in amyotrophic lateral sclerosis patients. Nat Med 2(6):696–699
142. Hasan R, Humphrey J, Bettencourt C, Newcombe J, Consortium NA, Lashley T et al (2022) Transcriptomic analysis of frontotemporal lobar degeneration with TDP-43 pathology reveals cellular alterations across multiple brain regions. Acta Neuropathol 143(3):383–401
143. Lopez-Gonzalez R, Lu Y, Gendron TF, Karydas A, Tran H, Yang D et al (2016) Poly(GR) in C9ORF72-related ALS/FTD compromises mitochondrial function and increases oxidative stress and DNA damage in iPSC-derived motor neurons. Neuron 92(2):383–391

144. Rojas F, Aguilar R, Almeida S, Fritz E, Corvalan D, Ampuero E et al (2023) Mature iPSC-derived astrocytes of an ALS/FTD patient carrying the TDP43(A90V) mutation display a mild reactive state and release polyP toxic to motoneurons. Front Cell Dev Biol 11:1226604
145. Zhou Q, Mareljic N, Michaelsen M, Parhizkar S, Heindl S, Nuscher B et al (2020) Active poly-GA vaccination prevents microglia activation and motor deficits in a C9orf72 mouse model. EMBO Mol Med 12(2):e10919
146. Clarke BE, Ziff OJ, Tyzack G, Petric Howe M, Wang Y, Klein P et al (2024) Human VCP mutant ALS/FTD microglia display immune and lysosomal phenotypes independently of GPNMB. Mol Neurodegener 19(1):90
147. Trageser KJ, Yang EJ, Smith C, Iban-Arias R, Oguchi T, Sebastian-Valverde M et al (2023) Inflammasome-mediated neuronal-microglial crosstalk: a therapeutic substrate for the familial C9orf72 variant of frontotemporal dementia/amyotrophic lateral sclerosis. Mol Neurobiol 60(7):4004–4016
148. Ke YD, van Hummel A, Stevens CH, Gladbach A, Ippati S, Bi M et al (2015) Short-term suppression of A315T mutant human TDP-43 expression improves functional deficits in a novel inducible transgenic mouse model of FTLD-TDP and ALS. Acta Neuropathol 130(5):661–678
149. Duan W, Guo M, Yi L, Zhang J, Bi Y, Liu Y et al (2019) Deletion of Tbk1 disrupts autophagy and reproduces behavioral and locomotor symptoms of FTD-ALS in mice. Aging (Albany NY) 11(8):2457–2476
150. Xu D, Jin T, Zhu H, Chen H, Ofengeim D, Zou C et al (2018) TBK1 suppresses RIPK1-driven apoptosis and inflammation during development and in aging. Cell 174(6):1477–91.e19
151. Yu CH, Davidson S, Harapas CR, Hilton JB, Mlodzianoski MJ, Laohamonthonkul P et al (2020) TDP-43 triggers mitochondrial DNA release via mPTP to activate cGAS/STING in ALS. Cell 183(3):636–49.e18
152. Wang R, Yang B, Zhang D (2011) Activation of interferon signaling pathways in spinal cord astrocytes from an ALS mouse model. Glia 59(6):946–958
153. Kurnellas M, Mitra A, Schwabe T, Paul R, Arrant AE, Roberson ED et al (2023) Latozinemab, a novel progranulin-elevating therapy for frontotemporal dementia. J Transl Med 21(1):387
154. Ward M, Carter LP, Huang JY, Maslyar D, Budda B, Paul R et al (2024) Phase 1 study of latozinemab in progranulin-associated frontotemporal dementia. Alzheimers Dement (N Y) 10(1):e12452
155. de Majo M, Koontz M, Marsan E, Salinas N, Ramsey A, Kuo YM et al (2023) Granulin loss of function in human mature brain organoids implicates astrocytes in TDP-43 pathology. Stem Cell Reports 18(3):706–719
156. Sevigny J, Uspenskaya O, Heckman LD, Wong LC, Hatch DA, Tewari A et al (2024) Progranulin AAV gene therapy for frontotemporal dementia: translational studies and phase 1/2 trial interim results. Nat Med 30(5):1406–1415
157. Luo EY, Sugimura RR (2024) Taming microglia: the promise of engineered microglia in treating neurological diseases. J Neuroinflammation 21(1):19

Chapter 8
Retinal Organoid Development: Disease Modeling and Therapeutic Strategies

Kritika Sharma, Julia S. Pawlick, and Volker Busskamp

Abstract The human retina is a multi-layered sensory neural tissue that forms the innermost part of the eye. It contains specialized cells, including photoreceptors (PRs), which convert light signals into electrochemical signals. These signals are processed by multiple retinal neurons and transmitted via the optic nerve to the brain, where visual perception occurs. The intricate structure and function of the retina is essential for vision, and its health is critical for maintaining function. Acquired and inherited forms of retinal degenerative diseases can lead to visual impairment and blindness. Animal-based models and retinal explants have increased our basic understanding of retinal diseases and potential therapeutic targets. However, more efficient and human-relevant in vitro models are needed to better translate therapeutic potential into clinical applications. In this context, stem cell-derived 3D human retinal organoids (hROs) recapitulate the structural and cellular complexity of the human retina, making them a valuable tool to study disease mechanisms, screen potential therapies, and explore the progression of retinal diseases in a controlled environment. In addition, hROs may pave the way for cell-based transplantation to restore vision.

Keywords Retina · Photoreceptors · Retinal diseases · Stem cell research · 3D in vitro model system · Human retinal organoids · Disease modeling · Therapeutic intervention · Retinogenesis · Retinitis pigmentosa · Age-related macular degeneration · Retinal ganglion cells · Retinal explants · Blindness · 3Rs principle

Kritika Sharma and Julia S. Pawlick contributed equally.

K. Sharma · J. S. Pawlick · V. Busskamp (✉)
Department of Ophthalmology, Medical Faculty, University Hospital Bonn, Bonn, Germany
e-mail: volker.busskamp@ukbonn.de

R. C. B. Wong and Y. Hayashi (eds.), *Advances in Using Stem Cells to Treat Neural Diseases*, Stem Cell Biology and Regenerative Medicine 77,
https://doi.org/10.1007/978-3-031-94101-6_8

8.1 Introduction

8.1.1 Introduction to the Human Retina

Vision is the most dominant and complex of the human senses, playing a central role in shaping our interactions with and memories of the environment. Visual perception begins with the detection of light, where photons, focused by the cornea and lens, reach the retina. Within this multi-layered neural structure, light is converted into electrochemical signals, visual information is processed and relayed to the brain via the optic nerve. The retina is made up of various neuronal and glial cells, including rod and cone photoreceptors (PRs), which capture photons using photopigments. Through an elaborate cascade, PRs convert light into electrical signals that are transmitted to bipolar cells and then to retinal ganglion cells (RGCs). Axons from RGCs project along the inner retinal layer to form the optic nerve which transmits the visual information to the higher brain areas, particularly the visual cortex in the occipital lobe [1]. The inhibitory horizontal and amacrine cells refine these signals within the retinal circuitry, for example, enhancing contrast and motion detection (Fig. 8.1).

The mammalian retina consists of about 100 cell types, including PRs (one rod and two to three cone types, depending on the species), >50 types of interneurons (horizontal, amacrine and bipolar cells), and >20 types of RGCs [2, 3]. The key

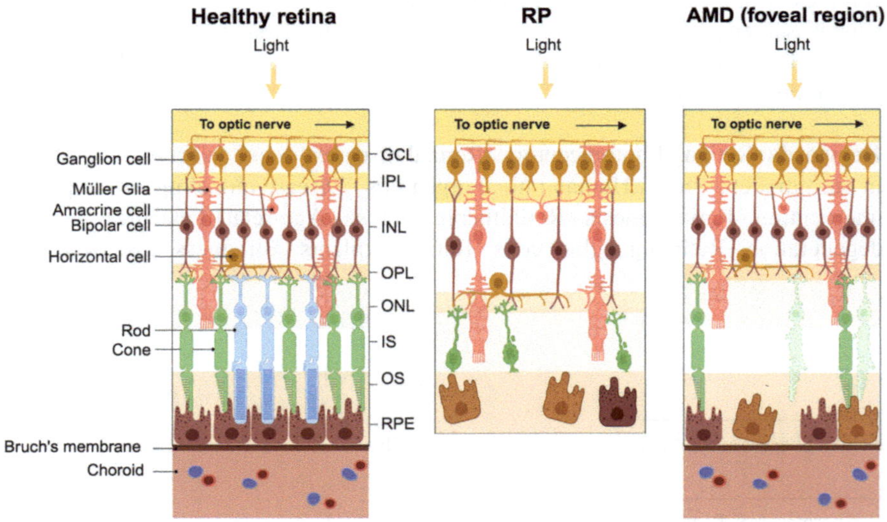

Fig. 8.1 Illustration of the multi-layered human retina. Healthy retina (left), morphology of a RP affected retina (middle), morphology of the foveal region of an AMD affected retina (right). *GCL* Ganglion cell layer, *IPL* inner plexiform layer, *INL* inner nuclear layer, *OPL* outer plexiform layer, *ONL* outer nuclear layer, *IS* inner segments, *OS* outer segments, *RPE* retinal pigment epithelium. (Created with BioRender.com)

aspects of functional connectivity in the retina are highlighted by the interactions between different cell types, such as PRs, interneurons, and RGCs, in visual processing. This includes quantifying input strength, understanding the mechanisms of color perception and measuring visual information at single cell resolution. In addition, these findings highlight the importance of retinal circuitry in both basic neuroscience and potential clinical applications [4]. Disruptions in these processes can lead to disorders, ranging from mild to severe vision loss, with significant impact on the quality of life [5]. Another cell type in the retina are the retinal pigment epithelial (RPE) cells. These are monolayered cells with their apical side facing the outer segments of the PR and their basal side facing Bruch's membrane, which separates the RPE from the choroidal capillaries. The RPE also forms the outer blood–retinal barrier. Tight junctions between the RPE and the adjacent endothelium therefore provide the retina with an immune privilege [6, 7].

There are many retinal diseases that cause blindness. For example, retinitis pigmentosa (RP) is a group of inherited retinal degenerative disorders (IRDs) that primarily affect rod PRs and lead to progressive vision loss. Patients typically present with night blindness and a gradual loss of peripheral vision, progressing to central vision loss due to degeneration of cone PRs (Fig. 8.1). RP affects 1 in 4000 individuals and is associated with more than 70 genetic mutations [8]. RP has been extensively studied using well-established mouse models [9–11]. The retinal degeneration (rd) mouse model is one of the most extensively studied models of RP and exhibits biochemical and morphological changes similar to those observed in human RP [12]. The rd1 mouse model harbors a nonsense mutation in exon 7 of the phosphodiesterase 6B (*Pde6b*) gene [13]. The mutation is associated with one of the earliest onset and most severe forms of human RP, accounting for 5% of all RP cases [14]. These models provide a valuable tool for investigating disease mechanisms, in particular the pathways leading to PR apoptosis, a hallmark of RP pathology. The knowledge gained from rd mice is contributing to a better understanding of retinal degeneration and potential therapeutic strategies [15]. Research has identified specific genetic mutations associated with RP, including those affecting the rhodopsin (*RHO*) gene. These mutations result in mislocalized or dysfunctional rhodopsin protein in PRs, ultimately triggering their degeneration [16]. Other mutations in genes responsible for rod PR function such as *RPGR* [17] and *USH2A* [18] result in different forms of RP. Elucidation of these genetic mechanisms is essential to advance targeted therapeutic strategies, such as gene editing, pharmacological interventions, and cell-based therapies, to preserve retinal function and slow disease progression.

Another retinal disease is age-related macular degeneration (AMD), the leading cause of vision loss in the elderly. AMD is characterized by the progressive degeneration of the macula, the central region of the retina responsible for sharp, detailed vision (Fig. 8.1) [19, 20]. AMD is broadly classified into two main types: Dry AMD, the more common form, is characterized by the gradual thinning of the macula and the accumulation of drusen, which are lipid and protein deposits. These deposits also cause gradual atrophy of the RPE and the overlying retina [21]. The second form of AMD is wet AMD or neovascular AMD which involves the abnormal growth of blood vessels from the choroid into the RPE or subretinal tissue, also known as choroidal neovascularization [22–24]. These vessels are dysfunctional and leaky, leading to accumulation of fluid and blood in the macula or under the retina resulting in RPE detachment [25]. Understanding these mechanisms is critical for the development of targeted therapies aimed at slowing disease progression and preserving vision. Transgenic mouse models expressing human genes associated with AMD, such as complement factor H (CFH) and apolipoprotein E (APOE), have been developed to study the genetic and molecular mechanisms underlying the disease [26]. These mouse models exhibit some key features of AMD, such as drusen-like deposits, that resemble those found in human cases of dry AMD.These deposits lead to RPE dysfunction which in turn causes PR degeneration. Inflammatory and oxidative stress pathways are critical contributors to AMD pathogenesis [27].

While the animal retina provides valuable insights, important anatomical differences, such as the absence of a macula and variations in color vision, limit the direct translation of findings to humans. These challenges, combined with growing ethical concerns about animal experimentation and adherence to the 3Rs (replacement, reduction, refinement) principles [28, 29], have driven the development of alternative in vitro model systems for preclinical research.

8.1.2 Studying the Retina In Vitro

Understanding the morphologic, physiologic, and functional aspects of retinal pathologies is essential for advancing therapies. Traditionally, the study of these diseases has relied on histologic analysis of surgically removed animal and *post-mortem* human retinal tissue [30, 31]. Retinal explants help to bridge the gap between in vitro cell culture and in vivo studies. Maintaining these retinal explants from different species in culture for different applications also offers several advantages as an experimental model, such as preserving the intricate laminar structure of the retina and ensuring prolonged PR cell viability, which is essential for studying retinal function and disease mechanisms in vitro [32, 33]. This simplifies the experimental setup compared to whole animal models, making them easier to manipulate and analyze. Porcine eyes are readily available as a by-product of the food industry, making them a valuable resource for research. In particular, porcine retinal explants are of great interest due to their close resemblance to the human retina in terms of relative size and scleral thickness [128]. Since a single pig can provide a large number of samples, this approach minimizes animal waste while also reducing data variability. Non-human primates have retinal anatomy and physiology that closely resembles that of humans and have been used in ex vivo experiments to study and develop treatments for retinal degeneration [34].

Among these, human retinal explants have emerged as a clinically relevant model, particularly for the evaluation of gene delivery methods such as adeno-associated virus (AAV) vectors. These explants are critical for optimizing treatment dosages and minimizing toxicity [35]. In addition, organotypic retinal explants are increasingly used to model retinal neurodegenerative diseases, providing a platform to study disease mechanisms and test potential therapies. By maintaining interactions between retinal cell types under controlled conditions, these systems provide valuable insights into complex retinal processes [36]. In this context, researchers have successfully observed activated vessel growth in human retinal explants and investigated the effects of ischemia and altered glucose metabolism on neovascularization [37]. They concluded that this model is an excellent system to study retinal angiogenesis.

However, maintaining the viability of retinal explants in vitro requires rapid isolation and continuous oxygenation to maintain its electrical activity [38]. A major drawback is their short viability in culture, which limits long-term studies of disease progression or treatment effects. In addition, maintaining retinal neurons in culture is technically challenging, as preserving interactions between neurons, glial cells, and vascular cells is critical for studying neurovascular diseases [30]. Explants can also exhibit immunoreactivity, especially in co-culture systems, leading to inflammation and reduced viability. Isolation from systemic factors further alters cellular responses compared to in vivo conditions. The limited availability of donor retinal tissue, particularly for certain pathologies and at various stages of disease, further complicates this approach. Despite these challenges, retinal explants remain an essential resource for vision research.

To complement the knowledge gained from retinal explants, pluripotent stem cells have emerged as an unlimited source of diverse cell types for retinal research. Embryonic stem cells (ESCs) were first discovered in mice in 1981 [39] and later in humans in 1998 [40], providing valuable models for disease research and drug testing. However, ethical concerns about the destruction of embryos led to the development of induced pluripotent stem cells (iPSCs) in 2006 (mice) [41] and 2007 (humans) [42]. This breakthrough addressed ethical concerns while advancing stem cell research by enabling iPSCs to differentiate into retinal cells through specific fate determining factors [43].

While 2D PSC-based cultures have facilitated advances in developmental biology, drug discovery, disease modeling, and cell replacement therapies [44, 45], they are unable to recapitulate the intricate cell–cell interactions and structural complexity of the retina. To better recapitulate in vivo retinogenesis, including essential interactions and signaling pathways, 3D hROs have been developed as a more robust and effective platform for studying retinal pathologies.

8.1.3 3D In Vitro Human Organoid Model System

Organoids are self-organizing or scaffolded tissues derived from PSCs that simulate the structural and functional characteristics of native organs. Since the 1980s, organoid research has progressed significantly, with successful models of liver [46], intestine [47, 48], prostate [49], brain [50], and retina [51]. In vitro models such as

3D ROs have contributed to the understanding of retinal cell types and mechanisms in both healthy and diseased states and have provided further insight into potential therapeutic approaches [52]. In this context, numerous protocols for generating hROs have been developed and refined over the past decades. These advances have demonstrated that hROs have the remarkable ability to mimic the five-layered structure of the retina, providing a promising platform for studying retinal development, disease mechanisms, and testing potential therapies [38, 53–57].

Early efforts to create 3D retinal tissues in vitro aimed to mimic the in vivo environment by enhancing cell–cell interactions, leading to the development of ROs. ROs advanced with the differentiation of mouse ESCs [58] and human ESCs [59] into eye-field progenitors expressing CRX and opsins. A major breakthrough in 2011 was the generation of stratified retinal layers from mouse ESCs using embryoid body (EB) aggregates and Matrigel, forming optic cup-like vesicles that mimic in vivo retinogenesis. This method was adapted to human ESCs by incorporating Wnt pathway and sonic hedgehog signaling antagonists to generate stratified ROs capable of forming nascent synapses [60].

Subsequent protocol refinements improved RO production, increasing the yield and quality of PR-rich ROs without the need for optic vesicle evagination [61]. In 2014, human stem cell-derived ROs exhibited functional PRs with light-responsive outer segments, with the first recorded light responses observed in 27-week-old ROs [57]. However, functional responses were limited, occurring in only 2 of 13 PRs examined, highlighting the need for further improvements.

To improve RO survival, culture conditions were optimized using bioreactors, which significantly improved laminar stratification and increased the production of mature PRs with nascent outer segment-like structures [62, 63]. One million PRs were harvested from 120 ROs, sufficient for transplantation into mice. Advances such as hydrogel-based cultures and insulin growth factor-1 (IGF1) supplementation further accelerated PR maturation and improved interactions with co-cultured RPE [64, 65]. These strategies enabled the formation of PR outer segments that resembled human retinogenisis in vivo [66]. ROs replicated human retinal developmental milestones, such as cilia and outer segment formation, enhanced by retinoic acid and taurine supplementation. Although proper retinal layer stratification was achieved with photosensitive PRs, variability in iPSC lines and protocols highlighted the need for refinement [67].

The Agarose Microwell Array Seeding and Scraping (AMASS) protocol addressed efficiency and scalability issues and allowed for rapid generation of large batches of ROs. This technique uses agarose microwell arrays for uniform seeding of single cells, resulting in EB formation by day 7. EBs are transferred to Matrigel-coated plates, where they develop into optic cup-like structures under defined growth conditions. At day 28, the optic cups are scraped in a checkerboard pattern and cultured as floating 3D ROs. These ROs mature into five-layered structures that reflect the morphology and stratification of the adult human retina. Transcriptomic analyses confirm that these ROs accurately recapitulate the cellular diversity and temporal gene expression of in vivo retinogenesis, with mature synapses demonstrating transient functional viability [38]. Unlike previous protocols requiring manual EB formation and optic cup dissection, the AMASS protocol standardizes EB size and simplifies RO production, reducing labor and media consumption while improving reproducibility.

8.1.4 Disease Modeling in hROs

RP is a complex form of IRD for which there are currently no effective treatments due to the lack of an efficient human in vitro model system, making RP a significant clinical challenge. Approximately, 15% of RP cases follow an autosomal dominant inheritance pattern, often associated with mutations in pre-mRNA processing factors (PRPFs). These PRPFs are essential components of the U4/U6.U5 tri-snRNP subunit of the spliceosome, a key complex in pre-mRNA splicing [68]. hROs have been developed as in vitro model systems to identify the pathological mechanisms of IRDs. To investigate the pathomechanisms of RP, they characterized the cellular phenotypes and splicing programs in hROs derived from RP11 patients with *PRPF31* mutations. These hROs showed aberrant splicing in retinal cells and dysregulated ciliogenesis in PRs. In situ gene editing using CRISPR/Cas9 successfully rescued these abnormalities at the protein level, highlighting the potential of hROs for disease modeling and targeted therapeutic development [69].

Another notable feature of this IRD is observed in cases of autosomal dominant retinitis pigmentosa (adRP) linked to the RP17 locus. In this subtype, complex structural variants (SVs) located on chromosome 17q22 have been implicated, highlighting the diverse genetic mechanisms underlying adRP. The impact of the RP17 SVs on 3D chromatin organization was examined on genome compartmentalization into topologically associating domains (TADs), and the epigenetic landscape of the region. TADs facilitate enhancer–promoter interactions within the 3D nuclear space, and their disruption can result in pathogenic gene expression changes due to ectopic regulatory interactions. To investigate these disease mechanisms, fibroblasts from affected individuals and controls were reprogrammed into iPSCs and differentiated into PR progenitor cells and ROs. This study demonstrated these altered TAD structures at the RP17 locus in ROs lead to disrupted 3D chromatin architecture. These results underscore the pathogenic potential of SVs and highlight the need to reexamine rare Mendelian diseases where causal genes and variants remain unconfirmed in larger cohorts [70].

RP pathology was also demonstrated in hROs with *RPGR* gene frameshift mutation. The results suggest that dysregulation causes ciliopathies. The study also showed that gene correction effectively rescued ciliopathy, a disorder caused by ciliary defects in retinal cells essential for PR function. Transcriptome analysis confirmed this restoration, showing gene expression levels comparable to those of healthy controls [71]. Mutations in *USH2A* are the leading genetic cause of non-syndromic RP. hROs were generated from iPSCs derived from *USH2A* mutant RP patients and their healthy controls. hROs with the respective mutation resulted in defective organization of neuronal retinal layers during retinogenesis. *USH2A*-ROs also exhibited defective retinal progenitor development with reduced laminin levels. Furthermore, RPE cells in *USH2A*-ROs show lower levels of RPE65, abnormal morphology and higher neuronal apoptotic tendencies. These results in hROs recapitulate the pathogenesis of RP in in vitro 3D models and provide an opportunity for more efficient molecular disease screening [72].

Loss of PR is a major cause of blindness, yet the underlying mechanisms of PR degeneration remain poorly understood. Understanding the gene expression that regulates PR development is critical for advancing targeted therapeutic strategies. Neuronal leucine zipper (*NRL*) plays a key role in rod PR development, and mutations in *NRL* have been associated with S-opsin syndrome and RP in mice. Using hROs, a similar pathogenic mechanism has been recapitulated in a human 3D model system, providing valuable insights into disease progression and potential therapeutic approaches [73].

Defects in the biogenesis or function of primary cilia lead to a spectrum of syndromic disorders collectively termed ciliopathies. Mutations in the *CEP290* gene, a key regulator of ciliary function, are implicated in Leber congenital amaurosis (LCA) and Joubert syndrome and related disorders (JSRD). To investigate its role, patient-derived hROs were analyzed and confirmed the function of CEP290 protein in regulating ciliary protein gating. These findings highlight its impact on cellular biogenesis and trafficking, particularly in PRs [74].

X-linked juvenile retinoschisis (XLRS) is an early onset degenerative disorder of the central retina, characterized by a distinctive retinal splitting phenotype. Patient-specific XLRS-hROs carrying the *RS1* mutation have been generated that exhibit structural defects similar to those observed in patient retinas. These phenotypes include reduced rod and cone cell frequencies, connecting cilium abnormalities, and accelerated paxillin dynamics, along with progressive loss of expression of *IQCB1* and *OPA1*, two genes associated with inherited eye diseases [75]. One study investigated the role of PR cilium actin regulator (PCARE) in F-actin assembly within PR cilia, a process essential for outer segment disc formation. These discs are critical for phototransduction, the conversion of light into electrical signals in the retina. The results demonstrate that regulation of actin assembly is essential for the structural integrity and functionality of PR outer segments. Collectively, these findings underscore the critical role of actin dynamics in the early stages of PR outer segment disk formation and highlight the complex molecular interactions that are essential for maintaining overall retinal health. These findings provide valuable insights into the pathogenesis of diseases such as RP and may inform therapeutic strategies aimed at preserving PR function and preventing vision loss [76].

Novel hypomorphic mutations in the *TRNT1* gene, which are essential for tRNA processing, have been associated with early onset RP, highlighting the genetic basis of the disease. To investigate the underlying mechanisms, iPSCs were generated from dermal fibroblasts of patients with confirmed *TRNT1* mutations, and these iPSCs were differentiated into ROs. By using patient-specific hROs, the research provides valuable insights into the cellular and molecular mechanisms driving *TRNT1*-associated RP. The study revealed that *TRNT1* mutations disrupt autophagy processes in retinal cells, which is critical, because autophagy is essential for cellular maintenance and function, especially in the highly specialized and metabolically active ROs. This approach not only advances the understanding of the disease but also opens avenues for potential therapeutic strategies aimed at restoring function or preventing degeneration in affected individuals [77]. RP2, a GTPase-activating protein, regulates the trafficking of lipidated proteins to PR outer segments. Mutations in *RP2* account for ~15% of X-linked retinitis pigmentosa (XLRP) cases [78]. Human iPSC-derived ROs effectively model retinal disease and reveal a human-specific RP2-deficient phenotype not observed in animal models. PR cell death in RP2-deficient ROs correlates with rod maturation and rhodopsin expression. Notably, AAV-mediated RP2 restoration successfully rescued this phenotype, demonstrating its potential as a therapeutic strategy for RP2-driven XLRP [79]. The absence of this phenotype in existing RP2 XLRP animal models underscores the unique insights provided by patient-derived ROs [80].

The retina shares many structural similarities with the brain, given it is part of the central nervous system. Previous studies have identified Alzheimer's disease (AD)-related phenotypes in the retina, including those observed in RGCs [81]. Another study aimed to develop human iPSC-derived ROs as a model for early AD phenotypes. ROs closely mimicked retinal development and structure, making them a promising tool for studying AD-related biomarkers before the onset of clinical symptoms. Using patient-derived iPSCs with familial AD mutations in *PSEN1* (A246E) and *PSEN2* (N141I), along with unaffected controls, the researchers successfully differentiated ROs and identified hallmarks of AD phenotypes, including increased Aβ42:Aβ40 ratio and elevated hyperphosphorylated tau (pTau). Further transcriptional profiling revealed differentially expressed genes (DEGs) associated with synaptic dysfunction in AD ROs. These findings highlight their potential as a powerful tool for early AD detection and intervention in AD, paving the way for future biomarker development and therapeutic strategies [82].

Age-related neurodegenerative diseases are associated with a dynamic, complex pathogenesis involving PR degeneration, Müller glial (MG) pathogenesis, and RPE atrophy [83, 84]. Tumor necrosis factor (TNF) is a pro-inflammatory factor that regulates various types of cell death and has been implicated in PR degeneration [85]. Epidermal growth factor (*EGF*) and heparin-binding EGF (*HBEGF*) have also been identified as AMD risk genes in glial pathologies [86, 87] and associated with disease progression [88].

These factors have been investigated in hROs, which demonstrated their impact on significant pathological changes in the outer retina leading to PR cell extrusion, a hallmark of severe retinal damage. These factors also helped to recapitulate a dynamic phenotype of retinal degeneration in a human model system. Overall, this model provides valuable insights into the mechanisms of vision loss in AMD and related diseases [89]. Mimicking age-related diseases such as AMD in hRO systems relies on the induction of disease-causing factors and remains highly complex. This is also due to the fact that late-stage hROs mimic the fetal developmental stage of the human retina, as shown by transcriptomic analysis [38]. Furthermore, the lack of functional vasculature and poorly developed RPE-choroid interaction, which are central to AMD pathogenesis, precludes the use of hROs to study the neovascular form of AMD (wet AMD). External factors such as UV exposure and dietary components remain to be investigated to understand their impact on AMD. For future experiments with hROs, their maturation can be accelerated by using co-culture techniques, as demonstrated in [90], or by prolonging culture duration, by using bioreactors [62].

8.1.5 Therapeutic Testing in hROs/Cell Replacement

Transcriptomic analyses of hROs show a remarkable molecular similarity to the human retina, thereby establishing these in vitro models as a highly advanced and reliable platform for investigating therapeutic improvements [38]. Optogenetic tools have great potential as a possible therapeutic strategy for the treatment of late-stage retinal degeneration [91]. Optogenetic approaches involve the expression of transgenes encoding photosensitive proteins in residual retinal cells, thereby rendering these cells photosensitive [92–95].

In one study, the trafficking efficiency and safety of several opsins, including Jaws, ChrimsonR, and humanized CatCh (hCatCh), were evaluated in hiPSC-derived ROs. While all opsins elicited light responses, they exhibited different membrane trafficking efficiencies. Some opsins colocalized with the endoplasmic reticulum, while others were predominantly cytosolic and lacked membrane insertion. These findings suggest that additional signal sequences may improve membrane trafficking and increase efficiency [96]. hROs also serve as promising preclinical models in the field of gene therapy, aiding in the optimization of AAV vector variants [97–99].

hROs offer robust and clinically relevant alternatives for the in vitro generation of PRs for the application of cell replacement therapies. Cell replacement therapies have significant potential for the treatment of retinal diseases due to the unique characteristics of the eye, including its isolated structure and partial immune privilege, which help to limit cell migration and immune rejection [100]. Since hROs allow for the production of quantifiable numbers of PRs, recent studies have shown that PRs isolated using the CD73 marker successfully integrate into mouse retinas following transplantation, representing a promising step forward for cell-based replacement therapies. Although this integration shows potential, further research is needed to assess whether these results can be translated into meaningful improvements in visual function [101, 102]. Similarly, a similar approach using the CD90 marker for RGCs demonstrated successful integration after transplantation, further highlighting the potential of hROs in retinal regeneration [103].

Cell replacement therapies performed in the early 2000s using fetal mouse retinas showed that PRs isolated from neonatal mouse retinal tissue could integrate into host retinas [104]. However, subsequent studies showed that there was a material transfer between transplanted and host PR cells rather than physical cellular integration [105–107]. Recent studies using hRO-derived PRs suggest that material transfer may occur more readily in same-species transplantation than in cross-species transplantation, making it challenging to fully investigate this phenomenon in cross-species models [108]. These findings underscore the importance of the transition from mouse tissue-derived PRs to hRO-derived cells as a model for retinal cell replacement, offering the potential for more precise, effective, and translational therapeutic strategies for retinal diseases. Optimization of hRO protocols to generate cone-rich ROs for transplantation then later resulted in survival of transplanted cones for up to 6 months [109]. In this approach, cone cells were isolated from the ROs and later transplanted, which resulted in the integration of the cells into the host PR layer, which also interacted with Müller glia and other retinal neurons, and restored cone-mediated light responses in cone-deficient mice, with donor-host interactions enhancing maturation. However, a major drawback of isolating cells from hROs for use in cell therapy is that it still requires fluorescent proteins such as green fluorescent protein (GFP), as effective cell sorting has been a major problem to date due to the lack of cell surface markers that facilitate effective sorting of donor cells [109].

In recent years, hROs have also been used as a source of retinal sheets or PRs isolated from hROs to improve and advance the field of regenerative therapy [110–113]. Transplantation of embryonic stem cell(ESC)/iPSC-derived retinas is thought to be beneficial because neighboring retinal cells, such as horizontal cells or Müller glia cells, can be incorporated to support graft cell survival, synapse formation, and functional maturation [111, 114, 115]. A preclinical study investigated the long-term survival and functional integration of human iPSC-derived retinas in models of retinal degeneration, specifically in rats and non-human primates. It showed that transplanted retinal grafts not only survived for extended periods—up to 5 months in rats and over 2 years in monkeys—but also matured into functional PRs within the host retina. Functional assessments, including visual behavior tests, optical coherence tomography, and electroretinography, confirmed that the grafts contributed to visual processing. These findings underscore the therapeutic potential of human iPSC-derived retinal tissues for the treatment of retinal degenerative diseases such as RP and AMD, and lay the groundwork for future clinical applications [116]. A recent clinical study demonstrated that the allogeneic iPSC-derived RO sheets transplanted into two patients with advanced stage RP stably integrated and resulted in graft survival for 2 years, accompanied by increased retinal thickness at the graft side in both patients [117]. Furthermore, the changes in visual function were slower than in the untreated eyes, highlighting allogeneic iPSC-derived RO sheets as a promising therapeutic tool to overcome blindness. However, it has been observed that in some cases the use of retinal sheets led to complications such as rosette formation leading to damage of the subretinal space [118] as well as difficulties in host-graft cell interactions within the retina [119].

8.1.6 Current Drawbacks of 3D-ROs

ROs have greatly advanced our understanding of retinal development and interactions at the cellular level. However, the field still faces certain limitations. For example, the RPE is an important cell type that interacts with PRs and supports their outer segments by recycling 11-trans retinal to 11-cis retinal in the visual cycle and promoting PR maturation. However, ROs lack a complete RPE layer or have limited patches, exposing most PRs to the neural medium and limiting their maturation and function [120]. This absence of the essential retinal niche between the RPE and PRs is critical for understanding retinal diseases, limiting their use in modeling, drug screening, and transplantation, and prompting active research in this area.

Another limitation of current RO protocols is the inability to prolong RGC survival. There are several reasons for this. It has been observed that limited extension of RGC axons into long-range connections, unlike in vivo, results in temporary neuronal networks that die prematurely. RGCs require brain connectivity for neurotrophin support; studies using RGC axotomy models and anencephalic brains show that disruption of this connection leads to rapid RGC degeneration in the human fetal retina [121]. Early degeneration of RGCs is also due to a lack of nutrition and media exchange within the inner core, resulting in limited access to inner cell types such as RGCs. Time-consuming protocols for generating ROs have resulted in inconsistent organoid quality [129]. In addition, ROs exhibit batch-to-batch variability, which poses a challenge to reproducibility [61, 122]. In addition to the lack of information on 3D models, this limits drug screening applications, especially their knowledge on drug efficacy and toxicity accuracy [123].

Assembloids, which are self-organizing 3D structures created by fusing different types of organoids or organoid-like tissues, offer a significant advance in human retinal disease modeling by providing a more complex and biologically relevant system. These hybrid models allow the study of interactions between different retinal cell types and other tissue types, better mimicking the cellular diversity and the microenvironment [124]. The successful generation of an assembly of dorsal and ventral forebrain organoids with migrating interneurons within integrated cortical circuits has been established [125]. Recently, liver, bile, and pancreas system assemblies have also shown successful integration of these cell types [126]. This system remains to be established and elucidated in the retina field. Assembloids also present inherent technical challenges such as high cost and reduced reproducibility. Assembloids are far from achieving cellular complexity in long-term culture with sustained functional stability. The lack of a functional vasculature is one of the limiting factors as well as the limited availability of nutrients in growing organoids [127]. This results in unstable organoids/assembloids with limited growth and longevity. Limited cell types, such as those of vascular or immune origin, are essential for cell–cell interaction and thus help to regulate stabilized function.

ROs represent a groundbreaking advance in ophthalmic research, providing unprecedented insight into retinal development, cellular interactions, and disease modeling. They provide a valuable platform for studying retinal diseases, testing therapeutics, and exploring regenerative medicine. Continued innovation in this area holds promise for the development of more physiologically accurate models and potential clinical applications.

8.2 Conclusion

Vision is a complex sensory system that is critical for visualizing and interacting with the world around us. Disruption of the visual circuitry can lead to visual impairment, with diseases such as RP and AMD having a significant impact on vision. Retinal explants serve as a crucial bridge between in vitro and in vivo models, providing a physiologically relevant platform to study retinal biology, disease mechanisms, and therapeutic interventions. Their ability to preserve the laminar structure of the retina, support PR viability, and facilitate pharmacological and gene therapy research underscores their value in vision research. However, limitations such as species-specific differences, short viability, technical challenges, and lack of systemic factors must be carefully considered when interpreting experimental results. While human retinal explants provide clinically relevant insights, their limited availability and short culture life pose significant challenges.

To address these limitations, pluripotent stem cell-derived retinal models, particularly 3D hROs, have emerged as transformative tools for recapitulating in vivo retinogenesis, cellular interactions, and disease progression. Unlike 2D cultures, hROs better mimic the structural and functional complexity of the human retina, making them invaluable for disease modeling, drug discovery, and regenerative medicine. As advances in retinal research continue to evolve, the integration of retinal explants with stem cell-based models holds great promise for the development of innovative therapies for retinal degenerative diseases.

The evolution of hRO technology has significantly advanced our ability to model retinal development, disease mechanisms, and potential therapeutic strategies. Over the past decades, protocol refinements—from early EB aggregation techniques to the incorporation of bioreactors, hydrogel-based cultures, and growth factor supplementation—have improved RO maturation, PR functionality, and structural fidelity. Despite these advances, variability in differentiation results and scalability challenges have remained barriers to widespread application. The development of standardized methods has addressed these limitations by improving efficiency, reproducibility, and scalability, providing a robust platform for high-throughput studies. As research progresses, further optimization of culture conditions and functional integration strategies will be essential to bridge the gap between in vitro models and clinical applications, ultimately advancing regenerative therapies for retinal degenerative diseases.

Various forms of RP, including those associated with mutations in pre-mRNA processing factors (PRPFs) and other genes such as *RPGR*, *USH2A*, and RP2, have been modeled using hROs. These models have proven essential for understanding disease mechanisms, including disrupted splicing, ciliogenesis defects, and epigenetic alterations in chromatin architecture. hROs derived from patient-specific iPSCs have shown promise in therapeutic approaches, such as gene correction for ciliopathies and structural reorganization of retinal layers in RP. In addition, studies of the *TRNT1* gene and other mutations, including those associated with X-linked RP (RP2), have revealed cellular and molecular dysfunctions, such as impaired autophagy and defective protein trafficking, further highlighting the potential of hROs in personalized medicine. In addition, hROs have been used to model complex retinal diseases such as AMD, providing valuable insights into pathogenesis and potential therapeutic interventions.

hROs have become essential for studying PR biology and modeling retinal diseases using patient-specific iPSCs. Their molecular similarity to the human retina has solidified their role in disease modeling and facilitated the identification of new therapeutic agents and strategies, such as the improvement of gene therapy approaches and the optimization of AAV vectors. hROs also serve as valuable sources for cell transplantation studies, aiding in the isolation of PRs and RGCs for regenerative therapies. While fetal mouse retinal transplantation has shown promise, ethical concerns have led to the use of iPSC-derived retinal tissues, which have demonstrated long-term survival and functional integration in animal models and even clinical trials. Despite these advances, challenges remain, including complications such as rosette formation and difficulties with host-graft integration. In addition, the lack of effective cell sorting methods and the need for fluorescent protein labeling complicate cell isolation from hROs for therapeutic purposes. Nevertheless, recent advances in optimizing hRO protocols, particularly for generating cone-rich ROs, hold promise for improving cell-based therapies and advancing clinical applications for retinal degenerative diseases. These findings underscore the growing utility of hROs as powerful in vitro platforms for disease modeling and therapeutic discovery, advancing our understanding of RP and related retinal degenerative diseases.

Despite these advantages, several limitations continue to challenge their full potential. One critical issue is the lack of a complete RPE layer in ROs, which hinders proper PR maturation and function. The RPE is essential for promoting PR maturation, but without it, most PRs are directly exposed to the neural medium, limiting their ability to fully develop. Another major challenge is the inability to prolong RGC survival, due to the lack of neural connectivity and support from brain-derived neurotrophins, as well as insufficient vascularization for nutrient and oxygen exchange. In addition, the time-consuming protocols used to generate ROs often result in inconsistent organoid quality and batch-to-batch variability, which affects reproducibility and complicates their use in drug screening. Assembloids, which integrate different organoid types to mimic more complex cellular systems, hold promise for overcoming some of these challenges in the future. While recent advances have demonstrated successful integration of different organoid types, including forebrain, liver, and pancreatic organoids, they still fall short in terms of long-term functional stability and the inclusion of critical cell types such as vascular or immune cells. The lack of a functional vasculature in both organoids and assembloids further limits their growth and longevity. Despite these challenges, hROs represent a transformative platform for studying retinal disease, advancing therapeutic development, and exploring regenerative medicine. Ongoing innovations aimed at overcoming these limitations hold great promise for improving physiological accuracy and opening new clinical applications in the future.

Acknowledgments V.B. acknowledges support from the Deutsche Forschungsgemeinschaft (DFG—project number 531985111), the Pro Retina Foundation, and the TRA Life and Health (University of Bonn) as part of the Excellence Strategy of the German Federal and State Governments. V.B. is also supported by the Cluster of Excellence—ImmunoSensation2 EXC-2151-390873048 at the University of Bonn.

References

1. Yu DY, Cringle SJ, Balaratnasingam C, Morgan WH, Yu PK, Su EN (2013) Retinal ganglion cells: energetics, compartmentation, axonal transport, cytoskeletons and vulnerability. Prog Retin Eye Res 36:217–246. https://doi.org/10.1016/J.PRETEYERES.2013.07.001
2. Euler T, Haverkamp S, Schubert T, Baden T (2014) Retinal bipolar cells: elementary building blocks of vision. Nat Rev Neurosci 15(8):507–519. https://doi.org/10.1038/NRN3783
3. Masland RH (2012) The neuronal organization of the retina. Neuron 76(2):266–280. https://doi.org/10.1016/J.NEURON.2012.10.002
4. Field GD, Gauthier JL, Sher A, Greschner M, MacHado TA, Jepson LH, Shlens J, Gunning DE, Mathieson K, Dabrowski W, Paninski L, Litke AM, Chichilnisky EJ (2010) Functional connectivity in the retina at the resolution of photoreceptors. Nature 467(7316):673–677. https://doi.org/10.1038/NATURE09424
5. Cahill MT, Banks AD, Stinnett SS, Toth CA (2005) Vision-related quality of life in patients with bilateral severe age-related macular degeneration. Ophthalmology 112(1):152–158. https://doi.org/10.1016/J.OPHTHA.2004.06.036
6. Simó R, Villarroel M, Corraliza L, Hernández C, Garcia-Ramírez M (2010) The retinal pigment epithelium: something more than a constituent of the blood-retinal barrier—implications for the pathogenesis of diabetic retinopathy. J Biomed Biotechnol 2010:190724. https://doi.org/10.1155/2010/190724
7. Strauss O (2005) The retinal pigment epithelium in visual function. Physiol Rev 85(3):845–881. https://doi.org/10.1152/PHYSREV.00021.2004
8. Zhou Y, Li S, Huang L, Yang Y, Zhang L, Yang M, Liu W, Ramasamy K, Jiang Z, Sundaresan P, Zhu X, Yang Z (2018) A splicing mutation in aryl hydrocarbon receptor associated with retinitis pigmentosa. Hum Mol Genet 27(14):2563–2572. https://doi.org/10.1093/HMG/DDY165
9. Dräger UC, Hubel DH (1978) Studies of visual function and its decay in mice with hereditary retinal degeneration. J Comp Neurol 180(1):85–114. https://doi.org/10.1002/CNE.901800107
10. Moshiri A (2021) Animals models of inherited retinal disease. Int Ophthalmol Clin 61(3):113–130. https://doi.org/10.1097/IIO.0000000000000368
11. Zhang L, Liu W, Wang HY, Qiang W, Wang R, Cui ZL, Zhang ZM (2024) The temporal progression of retinal degeneration and early-stage idebenone treatment in the Pde6brd1/rd1 mouse model of retinal dystrophy. Sci Rep 14(1):2019. https://doi.org/10.1038/S41598-024-52391-Y
12. Chang B, Hawes NL, Pardue MT, German AM, Hurd RE, Davisson MT, Nusinowitz S, Rengarajan K, Boyd AP, Sidney SS, Phillips MJ, Stewart RE, Chaudhury R, Nickerson JM, Heckenlively JR, Boatright JH (2007) Two mouse retinal degenerations caused by missense

mutations in the beta-subunit of rod cGMP phosphodiesterase gene. Vis Res 47(5):624–633. https://doi.org/10.1016/J.VISRES.2006.11.020

13. Bowes C, Li T, Danciger M, Baxter LC, Applebury ML, Farber DB (1990) Retinal degeneration in the rd mouse is caused by a defect in the beta subunit of rod cGMP-phosphodiesterase. Nature 347(6294):677–680. https://doi.org/10.1038/347677A0
14. Cheng LL, Han RY, Yang FY, Yu XP, Xu JL, Min QJ, Tian J, Ge XL, Zheng SS, Lin YW, Zheng YH, Qu J, Gu F (2016) Novel mutations in PDE6B causing human retinitis pigmentosa. Int J Ophthalmol 9(8):1094–1099. https://doi.org/10.18240/IJO.2016.08.02
15. Matuk Y (1984) Retinitis pigmentosa and retinal degeneration in animals: a review. Can J Biochem Cell Biol 62(6):535–546. https://doi.org/10.1139/O84-071
16. Sancho-Pelluz J, Tosi J, Hsu CW, Lee F, Wolpert K, Refidaff C, Greenberg JP, Tsang SH, Lin CS (2012) Mice with a D190N mutation in the gene encoding rhodopsin: a model for human autosomal-dominant retinitis pigmentosa. Mol Med 18(1):549–555. https://doi.org/10.2119/MOLMED.2011.00475
17. Talib M, Van Schooneveld MJ, Thiadens AA, Fiocco M, Wijnholds J, Florijn RJ, Schalij-Delfos NE, Van Genderen MM, Putter H, Cremers FPM, Dagnelie G, Ten Brink JB, Klaver CCW, Van Den Born LI, Hoyng CB, Bergen AA, Boon CJF (2019) Clinical and genetic characteristics of male patients with RPGR-associated retinal dystrophies: a long-term follow-up study. Retina 39(6):1186–1189. https://doi.org/10.1097/IAE.0000000000002125
18. Toualbi L, Toms M, Moosajee M (2020) USH2A-retinopathy: from genetics to therapeutics. Exp Eye Res 201:108330. https://doi.org/10.1016/J.EXER.2020.108330
19. Flaxman SR, Bourne RRA, Resnikoff S, Ackland P, Braithwaite T, Cicinelli MV, Das A, Jonas JB, Keeffe J, Kempen J, Leasher J, Limburg H, Naidoo K, Pesudovs K, Silvester A, Stevens GA, Tahhan N, Wong T, Taylor H et al (2017) Global causes of blindness and distance vision impairment 1990–2020: a systematic review and meta-analysis. Lancet Glob Health 5(12):e1221–e1234. https://doi.org/10.1016/S2214-109X(17)30393-5
20. Slijkerman RWN, Song F, Astuti GDN, Huynen MA, van Wijk E, Stieger K, Collin RWJ (2015) The pros and cons of vertebrate animal models for functional and therapeutic research on inherited retinal dystrophies. Prog Retin Eye Res 48:137–159. https://doi.org/10.1016/J.PRETEYERES.2015.04.004
21. Somasundaran S, Constable IJ, Mellough CB, Carvalho LS (2020) Retinal pigment epithelium and age-related macular degeneration: a review of major disease mechanisms. Clin Experiment Ophthalmol 48(8):1043–1056. https://doi.org/10.1111/CEO.13834
22. Ferris FL, Wilkinson CP, Bird A, Chakravarthy U, Chew E, Csaky K, Sadda SR (2013) Clinical classification of age-related macular degeneration. Ophthalmology 120(4):844–851. https://doi.org/10.1016/J.OPHTHA.2012.10.036
23. Ruan Y, Jiang S, Gericke A (2021) Age-related macular degeneration: role of oxidative stress and blood vessels. Int J Mol Sci 22(3):1–22. https://doi.org/10.3390/IJMS22031296
24. Yeo NJY, Chan EJJ, Cheung C (2019) Choroidal neovascularization: mechanisms of endothelial dysfunction. Front Pharmacol 10:1363. https://doi.org/10.3389/FPHAR.2019.01363
25. Chirco KR, Sohn EH, Stone EM, Tucker BA, Mullins RF (2017) Structural and molecular changes in the aging choroid: implications for age-related macular degeneration. Eye 31(1):10–25. https://doi.org/10.1038/EYE.2016.216
26. Toomey CB, Johnson LV, Bowes Rickman C (2018) Complement factor H in AMD: bridging genetic associations and pathobiology. Prog Retin Eye Res 62:38–57. https://doi.org/10.1016/J.PRETEYERES.2017.09.001
27. Fleckenstein M, Keenan TDL, Guymer RH, Chakravarthy U, Schmitz-Valckenberg S, Klaver CC, Wong WT, Chew EY (2021) Age-related macular degeneration. Nat Rev Dis Primers 7(1):31. https://doi.org/10.1038/S41572-021-00265-2
28. Kiani AK, Pheby D, Henehan G, Brown R, Sieving P, Sykora P, Marks R, Falsini B, Capodicasa N, Miertus S, Lorusso L, Dondossola D, Tartaglia GM, Ergoren MC, Dundar M, Michelini S, Malacarne D, Bonetti G, Dautaj A et al (2022) Ethical considerations

regarding animal experimentation. J Prev Med Hyg 63(2 Suppl 3):E255–E266. https://doi.org/10.15167/2421-4248/JPMH2022.63.2S3.2768

29. Tannenbaum J, Bennett BT (2015) Russell and Burch's 3Rs then and now: the need for clarity in definition and purpose. J Am Assoc Lab Anim Sci 54(2):120
30. Elmasry K, Moustafa M, Al-Shabrawey M (2022) Retinal explant of the adult mouse retina as an ex vivo model for studying retinal neurovascular diseases. J Vis Exp (190). https://doi.org/10.3791/63966
31. Malani M, Nirmal J (2022) Retinal pathophysiological evaluation in a rat model. J Vis Exp (183). https://doi.org/10.3791/63111
32. Murali A, Ramlogan-Steel CA, Andrzejewski S, Steel JC, Layton CJ (2019) Retinal explant culture: a platform to investigate human neuro-retina. Clin Experiment Ophthalmol 47(2):274–285. https://doi.org/10.1111/CEO.13434
33. Vats A, Xi Z, Byrne LC, Chen Y (2025) Retinal explant culture from mouse, human, and nonhuman primates and its applications in vision research. Methods Mol Biol 2848:169–186. https://doi.org/10.1007/978-1-0716-4087-6_11
34. Xu W, Dong Y, Li Y, Hu Z, Paquet-Durand F, Jiao K (2022) Organotypic retinal explant cultures from macaque monkey. J Vis Exp (186). https://doi.org/10.3791/64178
35. Wu WH, Tso A, Breazzano MP, Jenny LA, Levi SR, Tsang SH, Quinn PMJ (2023) Culture of human retinal explants for ex vivo assessment of AAV gene delivery. Methods Mol Biol 2560:303–311. https://doi.org/10.1007/978-1-0716-2651-1_28
36. Ahmed Z (2023) Preparation of retinal explant cultures. Methods Mol Biol 2708:25–31. https://doi.org/10.1007/978-1-0716-3409-7_3
37. Knott RM, Robertson M, Muckersie E, Folefac VA, Fairhurst FE, Wileman SM, Forrester JV (1999) A model system for the study of human retinal angiogenesis: activation of monocytes and endothelial cells and the association with the expression of the monocarboxylate transporter type 1 (MCT-1). Diabetologia 42(7):870–877. https://doi.org/10.1007/S001250051240
38. Cowan CS, Renner M, De Gennaro M, Gross-Scherf B, Goldblum D, Hou Y, Munz M, Rodrigues TM, Krol J, Szikra T, Cuttat R, Waldt A, Papasaikas P, Diggelmann R, Patino-Alvarez CP, Galliker P, Spirig SE, Pavlinic D, Gerber-Hollbach N et al (2020) Cell types of the human retina and its organoids at single-cell resolution. Cell 182(6):1623–1640.e34. https://doi.org/10.1016/J.CELL.2020.08.013
39. Evans MJ, Kaufman MH (1981) Establishment in culture of pluripotential cells from mouse embryos. Nature 292(5819):154–156. https://doi.org/10.1038/292154A0
40. Thomson JA, Itskovitz-Eldor J, Shapiro SS, Waknitz MA, Swiergiel JJ, Marshall VS, Jones JM (1998) Embryonic stem cell lines derived from human blastocysts. Science 282(5391):1145–1147. https://doi.org/10.1126/science.282.5391.1145
41. Takahashi K, Yamanaka S (2006) Induction of pluripotent stem cells from mouse embryonic and adult fibroblast cultures by defined factors. Cell 126(4):663–676. https://doi.org/10.1016/J.CELL.2006.07.024
42. Takahashi K, Tanabe K, Ohnuki M, Narita M, Ichisaka T, Tomoda K, Yamanaka S (2007) Induction of pluripotent stem cells from adult human fibroblasts by defined factors. Cell 131(5):861–872. https://doi.org/10.1016/J.CELL.2007.11.019
43. Krohne TU, Westenskow PD, Kurihara T, Friedlander DF, Lehmann M, Dorsey AL, Li W, Zhu S, Schultz A, Wang J, Siuzdak G, Ding S, Friedlander M (2012) Generation of retinal pigment epithelial cells from small molecules and OCT4 reprogrammed human induced pluripotent stem cells. Stem Cells Transl Med 1(2):96–109. https://doi.org/10.5966/SCTM.2011-0057
44. Barber AC, Hippert C, Duran Y, West EL, Bainbridge JWB, Warre-Cornish K, Luhmann UFO, Lakowski J, Sowden JC, Ali RR, Pearson RA (2013) Repair of the degenerate retina by photoreceptor transplantation. Proc Natl Acad Sci U S A 110(1):354–359. https://doi.org/10.1073/PNAS.1212677110
45. Tucker BA, Mullins RF, Stone EM (2014) Stem cells for investigation and treatment of inherited retinal disease. Hum Mol Genet 23(R1):R9–R16. https://doi.org/10.1093/HMG/DDU124

46. Huch M, Dorrell C, Boj SF, Van Es JH, Li VSW, Van De Wetering M, Sato T, Hamer K, Sasaki N, Finegold MJ, Haft A, Vries RG, Grompe M, Clevers H (2013) In vitro expansion of single Lgr5+ liver stem cells induced by Wnt-driven regeneration. Nature 494(7436):247–250. https://doi.org/10.1038/NATURE11826
47. Jung P, Sato T, Merlos-Suárez A, Barriga FM, Iglesias M, Rossell D, Auer H, Gallardo M, Blasco MA, Sancho E, Clevers H, Batlle E (2011) Isolation and in vitro expansion of human colonic stem cells. Nat Med 17(10):1225–1227. https://doi.org/10.1038/NM.2470
48. Sato T, Vries RG, Snippert HJ, Van De Wetering M, Barker N, Stange DE, Van Es JH, Abo A, Kujala P, Peters PJ, Clevers H (2009) Single Lgr5 stem cells build crypt-villus structures in vitro without a mesenchymal niche. Nature 459(7244):262–265. https://doi.org/10.1038/NATURE07935
49. Gao D, Vela I, Sboner A, Iaquinta PJ, Karthaus WR, Gopalan A, Dowling C, Wanjala JN, Undvall EA, Arora VK, Wongvipat J, Kossai M, Ramazanoglu S, Barboza LP, Di W, Cao Z, Zhang QF, Sirota I, Ran L et al (2014) Organoid cultures derived from patients with advanced prostate cancer. Cell 159(1):176–187. https://doi.org/10.1016/J.CELL.2014.08.016
50. Lancaster MA, Renner M, Martin CA, Wenzel D, Bicknell LS, Hurles ME, Homfray T, Penninger JM, Jackson AP, Knoblich JA (2013) Cerebral organoids model human brain development and microcephaly. Nature 501(7467):373–379. https://doi.org/10.1038/NATURE12517
51. Eiraku M, Takata N, Ishibashi H, Kawada M, Sakakura E, Okuda S, Sekiguchi K, Adachi T, Sasai Y (2011) Self-organizing optic-cup morphogenesis in three-dimensional culture. Nature 472(7341):51–58. https://doi.org/10.1038/NATURE09941
52. Artero Castro A, Rodríguez Jimenez FJ, Jendelova P, Erceg S (2019) Deciphering retinal diseases through the generation of three dimensional stem cell-derived organoids: concise review. Stem Cells 37(12):1496–1504. https://doi.org/10.1002/STEM.3089
53. Harkin J, Peña KH, Gomes C, Hernandez M, Lavekar SS, So K, Lentsch K, Feder EM, Morrow S, Huang KC, Tutrow KD, Morris A, Zhang C, Meyer JS (2024) A highly reproducible and efficient method for retinal organoid differentiation from human pluripotent stem cells. Proc Natl Acad Sci U S A 121(25):e2317285121. https://doi.org/10.1073/PNAS.2317285121
54. Kelley RA, Chen HY, Swaroop A, Li T (2020) Accelerated development of rod photoreceptors in retinal organoids derived from human pluripotent stem cells by supplementation with 9- cis retinal. STAR Protoc 1(1):100033. https://doi.org/10.1016/J.XPRO.2020.100033
55. Mayerl SJ, Bajgai S, Ludwig AL, Jager LD, Williams BN, Bacig C, Stoddard C, Sinha D, Philpot BD, Gamm DM (2022) Human retinal organoids harboring IMPG2 mutations exhibit a photoreceptor outer segment phenotype that models advanced retinitis pigmentosa. Stem Cell Reports 17(11):2409–2420. https://doi.org/10.1016/J.STEMCR.2022.09.004
56. Rashidi H, Leong YC, Venner K, Pramod H, Fei QZ, Jones OJR, Moulding D, Sowden JC (2022) Generation of 3D retinal tissue from human pluripotent stem cells using a directed small molecule-based serum-free microwell platform. Sci Rep 12(1):6646. https://doi.org/10.1038/S41598-022-10540-1
57. Zhong X, Gutierrez C, Xue T, Hampton C, Vergara MN, Cao LH, Peters A, Park TS, Zambidis ET, Meyer JS, Gamm DM, Yau KW, Canto-Soler MV (2014) Generation of three-dimensional retinal tissue with functional photoreceptors from human iPSCs. Nat Commun 5:4047. https://doi.org/10.1038/NCOMMS5047
58. Nakano T, Ando S, Takata N, Kawada M, Muguruma K, Sekiguchi K, Saito K, Yonemura S, Eiraku M, Sasai Y (2012) Self-formation of optic cups and storable stratified neural retina from human ESCs. Cell Stem Cell 10(6):771–785. https://doi.org/10.1016/J.STEM.2012.05.009
59. Meyer JS, Shearer RL, Capowski EE, Wright LS, Wallace KA, McMillan EL, Zhang SC, Gamm DM (2009) Modeling early retinal development with human embryonic and induced pluripotent stem cells. Proc Natl Acad Sci U S A 106(39):16698–16703. https://doi.org/10.1073/PNAS.0905245106

60. Eguizabal C, Montserrat N, Vassena R, Barragan M, Garreta E, Garcia-Quevedo L, Vidal F, Giorgetti A, Veiga A, Izpisua Belmonte JC (2011) Complete meiosis from human induced pluripotent stem cells. Stem Cells 29(8):1186–1195. https://doi.org/10.1002/STEM.672
61. Völkner M, Zschätzsch M, Rostovskaya M, Overall RW, Busskamp V, Anastassiadis K, Karl MO (2016) Retinal organoids from pluripotent stem cells efficiently recapitulate Retinogenesis. Stem Cell Reports 6(4):525–538. https://doi.org/10.1016/J.STEMCR.2016.03.001
62. DiStefano T, Chen HY, Panebianco C, Kaya KD, Brooks MJ, Gieser L, Morgan NY, Pohida T, Swaroop A (2018) Accelerated and improved differentiation of retinal organoids from pluripotent stem cells in rotating-wall vessel bioreactors. Stem Cell Reports 10(1):300–313. https://doi.org/10.1016/J.STEMCR.2017.11.001
63. Ovando-Roche P, West EL, Branch MJ, Sampson RD, Fernando M, Munro P, Georgiadis A, Rizzi M, Kloc M, Naeem A, Ribeiro J, Smith AJ, Gonzalez-Cordero A, Ali RR (2018) Use of bioreactors for culturing human retinal organoids improves photoreceptor yields. Stem Cell Res Ther 9(1):156. https://doi.org/10.1186/S13287-018-0907-0
64. Hunt NC, Hallam D, Chichagova V, Steel DH, Lako M (2018) The application of biomaterials to tissue engineering neural retina and retinal pigment epithelium. Adv Healthc Mater 7(23):e1800226. https://doi.org/10.1002/ADHM.201800226
65. Mellough CB, Collin J, Khazim M, White K, Sernagor E, Steel DHW, Lako M (2015) IGF-1 signaling plays an important role in the formation of three-dimensional laminated neural retina and other ocular structures from human embryonic stem cells. Stem Cells 33(8):2416–2430. https://doi.org/10.1002/STEM.2023
66. Hallam D, Hilgen G, Dorgau B, Zhu L, Yu M, Bojic S, Hewitt P, Schmitt M, Uteng M, Kustermann S, Steel D, Nicholds M, Thomas R, Treumann A, Porter A, Sernagor E, Armstrong L, Lako M (2018) Human-induced pluripotent stem cells generate light responsive retinal organoids with variable and nutrient-dependent efficiency. Stem Cells 36(10):1535–1551. https://doi.org/10.1002/STEM.2883
67. Lowe A, Harris R, Bhansali P, Cvekl A, Liu W (2016) Intercellular adhesion-dependent cell survival and ROCK-regulated Actomyosin-driven forces mediate self-formation of a retinal organoid. Stem Cell Reports 6(5):743–756. https://doi.org/10.1016/J.STEMCR.2016.03.011
68. Nguyen THD, Galej WP, Bai XC, Savva CG, Newman AJ, Scheres SHW, Nagai K (2015) The architecture of the spliceosomal U4/U6.U5 tri-snRNP. Nature 523(7558):47–52. https://doi.org/10.1038/NATURE14548
69. Buskin A, Zhu L, Chichagova V, Basu B, Mozaffari-Jovin S, Dolan D, Droop A, Collin J, Bronstein R, Mehrotra S, Farkas M, Hilgen G, White K, Pan KT, Treumann A, Hallam D, Bialas K, Chung G, Mellough C et al (2018) Disrupted alternative splicing for genes implicated in splicing and ciliogenesis causes PRPF31 retinitis pigmentosa. Nat Commun 9(1):4234. https://doi.org/10.1038/S41467-018-06448-Y
70. de Bruijn SE, Fiorentino A, Ottaviani D, Fanucchi S, Melo US, Corral-Serrano JC, Mulders T, Georgiou M, Rivolta C, Pontikos N, Arno G, Roberts L, Greenberg J, Albert S, Gilissen C, Aben M, Rebello G, Mead S, Raymond FL et al (2020) Structural variants create new topological-associated domains and ectopic retinal enhancer-gene contact in dominant retinitis Pigmentosa. Am J Hum Genet 107(5):802–814. https://doi.org/10.1016/J.AJHG.2020.09.002
71. Deng WL, Gao ML, Lei XL, Lv JN, Zhao H, He KW, Xia XX, Li LY, Chen YC, Li YP, Pan D, Xue T, Jin ZB (2018) Gene correction reverses ciliopathy and photoreceptor loss in iPSC-derived retinal organoids from retinitis Pigmentosa patients. Stem Cell Reports 10(4):1267–1281. https://doi.org/10.1016/J.STEMCR.2018.02.003
72. Guo Y, Wang P, Ma JH, Cui Z, Yu Q, Liu S, Xue Y, Zhu D, Cao J, Li Z, Tang S, Chen J (2019) Modeling retinitis pigmentosa: retinal organoids generated from the iPSCs of a patient with the USH2A mutation show early developmental abnormalities. Front Cell Neurosci 13:361. https://doi.org/10.3389/FNCEL.2019.00361
73. Kallman A, Capowski EE, Wang J, Kaushik AM, Jansen AD, Edwards KL, Chen L, Berlinicke CA, Joseph Phillips M, Pierce EA, Qian J, Wang TH, Gamm DM, Zack DJ (2020)

Investigating cone photoreceptor development using patient-derived NRL null retinal organoids. Commun Biol 3(1):82. https://doi.org/10.1038/S42003-020-0808-5
74. Shimada H, Lu Q, Insinna-Kettenhofen C, Nagashima K, English MA, Semler EM, Mahgerefteh J, Cideciyan AV, Li T, Brooks BP, Gunay-Aygun M, Jacobson SG, Cogliati T, Westlake CJ, Swaroop A (2017) In vitro modeling using ciliopathy-patient-derived cells reveals distinct cilia dysfunctions caused by CEP290 mutations. Cell Rep 20(2):384–396. https://doi.org/10.1016/J.CELREP.2017.06.045
75. Huang KC, Wang ML, Chen SJ, Kuo JC, Wang WJ, Nhi Nguyen PN, Wahlin KJ, Lu JF, Tran AA, Shi M, Chien Y, Yarmishyn AA, Tsai PH, Yang TC, Jane WN, Chang CC, Peng CH, Schlaeger TM, Chiou SH (2019) Morphological and molecular defects in human three-dimensional retinal organoid model of X-linked juvenile Retinoschisis. Stem Cell Reports 13(5):906–923. https://doi.org/10.1016/J.STEMCR.2019.09.010
76. Corral-Serrano JC, Lamers IJC, Van Reeuwijk J, Duijkers L, Hoogendoorn ADM, Yildirim A, Argyrou N, Ruigrok RAA, Letteboer SJF, Butcher R, Van Essen MD, Sakami S, Van Beersum SEC, Palczewski K, Cheetham ME, Liu Q, Boldt K, Wolfrum U, Ueffing M et al (2020) PCARE and WASF3 regulate ciliary F-actin assembly that is required for the initiation of photoreceptor outer segment disk formation. Proc Natl Acad Sci U S A 117(18):9922–9931. https://doi.org/10.1073/PNAS.1903125117/SUPPL_FILE/PNAS.1903125117.SD01.XLSX
77. Sharma TP, Wiley LA, Whitmore SS, Anfinson KR, Cranston CM, Oppedal DJ, Daggett HT, Mullins RF, Tucker BA, Stone EM (2017) Patient-specific induced pluripotent stem cells to evaluate the pathophysiology of TRNT1-associated retinitis pigmentosa. Stem Cell Res 21:58–70. https://doi.org/10.1016/J.SCR.2017.03.005
78. Breuer DK, Yashar BM, Filippova E, Hiriyanna S, Lyons RH, Mears AJ, Asaye B, Acar C, Vervoort R, Wright AF, Musarella MA, Wheeler P, MacDonald I, Iannaccone A, Birch D, Hoffman DR, Fishman GA, Heckenlively JR, Jacobson SG et al (2002) A comprehensive mutation analysis of RP2 and RPGR in a north American cohort of families with X-linked retinitis pigmentosa. Am J Hum Genet 70(6):1545–1554. https://doi.org/10.1086/340848
79. Lane A, Jovanovic K, Shortall C, Ottaviani D, Panes AB, Schwarz N, Guarascio R, Hayes MJ, Palfi A, Chadderton N, Farrar GJ, Hardcastle AJ, Cheetham ME (2020) Modeling and rescue of RP2 retinitis Pigmentosa using iPSC-derived retinal organoids. Stem Cell Reports 15(1):67–79. https://doi.org/10.1016/J.STEMCR.2020.05.007
80. Zhang H, Hanke-Gogokhia C, Jiang L, Li X, Wang P, Gerstner CD, Frederick JM, Yang Z, Baehr W (2015) Mistrafficking of prenylated proteins causes retinitis pigmentosa 2. FASEB J 29(3):932–942. https://doi.org/10.1096/FJ.14-257915
81. Blanks JC, Hinton DR, Sadun AA, Miller CA (1989) Retinal ganglion cell degeneration in Alzheimer's disease. Brain Res 501(2):364–372. https://doi.org/10.1016/0006-8993(89)90653-7
82. Lavekar SS, Harkin J, Hernandez M, Gomes C, Patil S, Huang KC, Puntambekar SS, Lamb BT, Meyer JS (2023) Development of a three-dimensional organoid model to explore early retinal phenotypes associated with Alzheimer's disease. Sci Rep 13(1):13827. https://doi.org/10.1038/S41598-023-40382-4
83. Sarks JP, Sarks SH, Killingsworth MC (1988) Evolution of geographic atrophy of the retinal pigment epithelium. Eye 2(Pt 5):552–577. https://doi.org/10.1038/EYE.1988.106
84. Wu KHC, Madigan MC, Billson FA, Penfold PL (2003) Differential expression of GFAP in early v late AMD: a quantitative analysis. Br J Ophthalmol 87(9):1159–1166. https://doi.org/10.1136/BJO.87.9.1159
85. Murakami Y, Notomi S, Hisatomi T, Nakazawa T, Ishibashi T, Miller JW, Vavvas DG (2013) Photoreceptor cell death and rescue in retinal detachment and degenerations. Prog Retin Eye Res 37:114–140. https://doi.org/10.1016/J.PRETEYERES.2013.08.001
86. Han X, Gharahkhani P, Mitchell P, Liew G, Hewitt AW, MacGregor S (2020) Genome-wide meta-analysis identifies novel loci associated with age-related macular degeneration. J Hum Genet 65(8):657–665. https://doi.org/10.1038/S10038-020-0750-X

87. Logue MW, Schu M, Vardarajan BN, Farrell J, Lunetta KL, Jun G, Baldwin CT, DeAngelis MM, Farrer LA (2014) Search for age-related macular degeneration risk variants in Alzheimer disease genes and pathways. Neurobiol Aging 35(6):1510.e7–1510.e18. https://doi.org/10.1016/J.NEUROBIOLAGING.2013.12.007
88. Wagner BD, Patnaik JL, Palestine AG, Frazer-Abel AA, Baldermann R, Holers VM, Mathias MT, Mandava N, Lynch AM (2022) Association of Systemic Inflammatory Factors with progression to advanced age-related macular degeneration. Ophthalmic Epidemiol 29(2):139–148. https://doi.org/10.1080/09286586.2021.1910314
89. Völkner M, Wagner F, Steinheuer LM, Carido M, Kurth T, Yazbeck A, Schor J, Wieneke S, Ebner LJA, Del Toro Runzer C, Taborsky D, Zoschke K, Vogt M, Canzler S, Hermann A, Khattak S, Hackermüller J, Karl MO (2022) HBEGF-TNF induce a complex outer retinal pathology with photoreceptor cell extrusion in human organoids. Nat Commun 13(1):6183. https://doi.org/10.1038/S41467-022-33848-Y
90. Akhtar T, Xie H, Khan MI, Zhao H, Bao J, Zhang M, Xue T (2019) Accelerated photoreceptor differentiation of hiPSC-derived retinal organoids by contact co-culture with retinal pigment epithelium. Stem Cell Res 39:101491. https://doi.org/10.1016/J.SCR.2019.101491
91. Lindner M, Gilhooley MJ, Hughes S, Hankins MW (2022) Optogenetics for visual restoration: from proof of principle to translational challenges. Prog Retin Eye Res 91:101089. https://doi.org/10.1016/J.PRETEYERES.2022.101089
92. Berry MH, Holt A, Salari A, Veit J, Visel M, Levitz J, Aghi K, Gaub BM, Sivyer B, Flannery JG, Isacoff EY (2019) Restoration of high-sensitivity and adapting vision with a cone opsin. Nat Commun 10(1):1221. https://doi.org/10.1038/S41467-019-09124-X
93. Bi A, Cui J, Ma YP, Olshevskaya E, Pu M, Dizhoor AM, Pan ZH (2006) Ectopic expression of a microbial-type rhodopsin restores visual responses in mice with photoreceptor degeneration. Neuron 50(1):23–33. https://doi.org/10.1016/J.NEURON.2006.02.026
94. Busskamp V, Duebel J, Balya D, Fradot M, Viney TJ, Siegert S, Groner AC, Cabuy E, Forster V, Seeliger M, Biel M, Humphries P, Paques M, Mohand-Said S, Trono D, Deisseroth K, Sahel JA, Picaud S, Roska B (2010) Genetic reactivation of cone photoreceptors restores visual responses in retinitis pigmentosa. Science 329(5990):413–417. https://doi.org/10.1126/SCIENCE.1190897
95. Lagali PS, Balya D, Awatramani GB, Münch TA, Kim DS, Busskamp V, Cepko CL, Roska B (2008) Light-activated channels targeted to on bipolar cells restore visual function in retinal degeneration. Nat Neurosci 11(6):667–675. https://doi.org/10.1038/NN.2117
96. Garita-Hernandez M, Guibbal L, Toualbi L, Routet F, Chaffiol A, Winckler C, Harinquet M, Robert C, Fouquet S, Bellow S, Sahel JA, Goureau O, Duebel J, Dalkara D (2018) Optogenetic light sensors in human retinal organoids. Front Neurosci 12:789. https://doi.org/10.3389/FNINS.2018.00789
97. Garita-Hernandez M, Routet F, Guibbal L, Khabou H, Toualbi L, Riancho L, Reichman S, Duebel J, Sahel JA, Goureau O, Dalkara D (2020) AAV-mediated gene delivery to 3D retinal organoids derived from human induced pluripotent stem cells. Int J Mol Sci 21(3):994. https://doi.org/10.3390/IJMS21030994
98. Gonzalez-Cordero A, Goh D, Kruczek K, Naeem A, Fernando M, Kleine Holthaus SM, Takaaki M, Blackford SJI, Kloc M, Agundez L, Sampson RD, Borooah S, Ovando-Roche P, Mehat MS, West EL, Smith AJ, Pearson RA, Ali RR (2018) Assessment of AAV vector tropisms for mouse and human pluripotent stem cell-derived RPE and photoreceptor cells. Hum Gene Ther 29(10):1124–1139. https://doi.org/10.1089/HUM.2018.027
99. Völkner M, Pavlou M, Büning H, Michalakis S, Karl MO (2021) Optimized adeno-associated virus vectors for efficient transduction of human retinal organoids. Hum Gene Ther 32(13–14):694–706. https://doi.org/10.1089/HUM.2020.321
100. Mead B, Berry M, Logan A, Scott RAH, Leadbeater W, Scheven BA (2015) Stem cell treatment of degenerative eye disease. Stem Cell Res 14(3):243–257. https://doi.org/10.1016/J.SCR.2015.02.003

101. Gagliardi G, Ben M'Barek K, Chaffiol A, Slembrouck-Brec A, Conart JB, Nanteau C, Rabesandratana O, Sahel JA, Duebel J, Orieux G, Reichman S, Goureau O (2018) Characterization and transplantation of CD73-positive photoreceptors isolated from human iPSC-derived retinal organoids. Stem Cell Reports 11(3):665–680. https://doi.org/10.1016/j.stemcr.2018.07.005
102. Gasparini SJ, Llonch S, Borsch O, Ader M (2019) Transplantation of photoreceptors into the degenerative retina: current state and future perspectives. Prog Retin Eye Res 69:1–37. https://doi.org/10.1016/J.PRETEYERES.2018.11.001
103. Rabesandratana O, Chaffiol A, Mialot A, Slembrouck-Brec A, Joffrois C, Nanteau C, Rodrigues A, Gagliardi G, Reichman S, Sahel JA, Chédotal A, Duebel J, Goureau O, Orieux G (2020) Generation of a transplantable population of human iPSC-derived retinal ganglion cells. Front Cell Dev Biol 8:585675. https://doi.org/10.3389/FCELL.2020.585675
104. MacLaren RE, Pearson RA, MacNeil A, Douglas RH, Salt TE, Akimoto M, Swaroop A, Sowden JC, Ali RR (2006) Retinal repair by transplantation of photoreceptor precursors. Nature 444(7116):203–207. https://doi.org/10.1038/NATURE05161
105. Pearson RA, Gonzalez-Cordero A, West EL, Ribeiro JR, Aghaizu N, Goh D, Sampson RD, Georgiadis A, Waldron PV, Duran Y, Naeem A, Kloc M, Cristante E, Kruczek K, Warre-Cornish K, Sowden JC, Smith AJ, Ali RR (2016) Donor and host photoreceptors engage in material transfer following transplantation of post-mitotic photoreceptor precursors. Nat Commun 7:13029. https://doi.org/10.1038/NCOMMS13029
106. Santos-Ferreira T, Llonch S, Borsch O, Postel K, Haas J, Ader M (2016) Retinal transplantation of photoreceptors results in donor-host cytoplasmic exchange. Nat Commun 7:13028. https://doi.org/10.1038/NCOMMS13028
107. Singh MS, Balmer J, Barnard AR, Aslam SA, Moralli D, Green CM, Barnea-Cramer A, Duncan I, MacLaren RE (2016) Transplanted photoreceptor precursors transfer proteins to host photoreceptors by a mechanism of cytoplasmic fusion. Nat Commun 7:13537. https://doi.org/10.1038/NCOMMS13537
108. Ho MT, Kawai K, Abdo D, Comanita L, Ortin-Martinez A, Ueno Y, Tsao E, Rastgar-Moghadam A, Xue C, Cui H, Wallace VA, Shoichet MS (2024) Transplanted human photoreceptors transfer cytoplasmic material but not to the recipient mouse retina. Stem Cell Res Ther 15(1):79. https://doi.org/10.1186/S13287-024-03679-3
109. Gasparini SJ, Tessmer K, Reh M, Wieneke S, Carido M, Völkner M, Borsch O, Swiersy A, Zuzic M, Goureau O, Kurth T, Busskamp V, Zeck G, Karl MO, Ader M (2022) Transplanted human cones incorporate into the retina and function in a murine cone degeneration model. J Clin Invest 132(12):e154619. https://doi.org/10.1172/JCI154619
110. Assawachananont J, Mandai M, Okamoto S, Yamada C, Eiraku M, Yonemura S, Sasai Y, Takahashi M (2014) Transplantation of embryonic and induced pluripotent stem cell-derived 3D retinal sheets into retinal degenerative mice. Stem Cell Reports 2(5):662–674. https://doi.org/10.1016/J.STEMCR.2014.03.011
111. Mandai M (2023) Pluripotent stem cell-derived retinal organoid/cells for retinal regeneration therapies: a review. Regen Ther 22:59–67. https://doi.org/10.1016/J.RETH.2022.12.005
112. Radtke ND, Seiler MJ, Aramant RB, Petry HM, Pidwell DJ (2002) Transplantation of intact sheets of fetal neural retina with its retinal pigment epithelium in retinitis pigmentosa patients. Am J Ophthalmol 133(4):544–550. https://doi.org/10.1016/S0002-9394(02)01322-3
113. Shirai H, Mandai M, Matsushita K, Kuwahara A, Yonemura S, Nakano T, Assawachananont J, Kimura T, Saito K, Terasaki H, Eirakug M, Sasai Y, Takahashi M (2016) Transplantation of human embryonic stem cell-derived retinal tissue in two primate models of retinal degeneration. Proc Natl Acad Sci U S A 113(1):E81–E90. https://doi.org/10.1073/PNAS.1512590113/-/DCSUPPLEMENTAL
114. Bringmann A, Pannicke T, Biedermann B, Francke M, Iandiev I, Grosche J, Wiedemann P, Albrecht J, Reichenbach A (2009) Role of retinal glial cells in neurotransmitter uptake and metabolism. Neurochem Int 54(3–4):143–160. https://doi.org/10.1016/J.NEUINT.2008.10.014

115. Wang JS, Kefalov VJ (2011) The cone-specific visual cycle. Prog Retin Eye Res 30(2):115–128. https://doi.org/10.1016/J.PRETEYERES.2010.11.001
116. Tu HY, Watanabe T, Shirai H, Yamasaki S, Kinoshita M, Matsushita K, Hashiguchi T, Onoe H, Matsuyama T, Kuwahara A, Kishino A, Kimura T, Eiraku M, Suzuma K, Kitaoka T, Takahashi M, Mandai M (2019) Medium- to long-term survival and functional examination of human iPSC-derived retinas in rat and primate models of retinal degeneration. EBioMedicine 39:562–574. https://doi.org/10.1016/J.EBIOM.2018.11.028
117. Hirami Y, Mandai M, Sugita S, Maeda A, Maeda T, Yamamoto M, Uyama H, Yokota S, Fujihara M, Igeta M, Daimon T, Fujita K, Ito T, Shibatani N, Morinaga C, Hayama T, Nakamura A, Ueyama K, Ono K et al (2023) Safety and stable survival of stem-cell-derived retinal organoid for 2 years in patients with retinitis pigmentosa. Cell Stem Cell 30(12):1585–1596.e6. https://doi.org/10.1016/J.STEM.2023.11.004
118. Lin B, McLelland BT, Mathur A, Aramant RB, Seiler MJ (2018) Sheets of human retinal progenitor transplants improve vision in rats with severe retinal degeneration. Exp Eye Res 174:13–28. https://doi.org/10.1016/J.EXER.2018.05.017
119. Yamasaki S, Tu HY, Matsuyama T, Horiuchi M, Hashiguchi T, Sho J, Kuwahara A, Kishino A, Kimura T, Takahashi M, Mandai M (2021) A genetic modification that reduces ON-bipolar cells in hESC-derived retinas enhances functional integration after transplantation. iScience 25(1):103657. https://doi.org/10.1016/J.ISCI.2021.103657
120. Singh RK, Nasonkin IO (2020) Limitations and promise of retinal tissue from human pluripotent stem cells for developing therapies of blindness. Front Cell Neurosci 14:179. https://doi.org/10.3389/FNCEL.2020.00179
121. Hendrickson A, Djajadi H, Erickson A, Possin D (2006) Development of the human retina in the absence of ganglion cells. Exp Eye Res 83(4):920–931. https://doi.org/10.1016/J.EXER.2006.04.017
122. Capowski EE, Samimi K, Mayerl SJ, Phillips MJ, Pinilla I, Howden SE, Saha J, Jansen AD, Edwards KL, Jager LD, Barlow K, Valiauga R, Erlichman Z, Hagstrom A, Sinha D, Sluch VM, Chamling X, Zack DJ, Skala MC, Gamm DM (2019) Reproducibility and staging of 3D human retinal organoids across multiple pluripotent stem cell lines. Development 146(1):dev171686. https://doi.org/10.1242/DEV.171686/VIDEO-4
123. Aasen DM, Vergara MN (2020) New drug discovery paradigms for retinal diseases: a focus on retinal organoids. J Ocul Pharmacol Ther 36(1):18–24. https://doi.org/10.1089/JOP.2018.0140
124. Paşca SP, Arlotta P, Bateup HS, Camp JG, Cappello S, Gage FH, Knoblich JA, Kriegstein AR, Lancaster MA, Ming GL, Muotri AR, Park IH, Reiner O, Song H, Studer L, Temple S, Testa G, Treutlein B, Vaccarino FM (2022) A nomenclature consensus for nervous system organoids and assembloids. Nature 609(7929):907–910. https://doi.org/10.1038/S41586-022-05219-6
125. Birey F, Andersen J, Makinson CD, Islam S, Wei W, Huber N, Fan HC, Metzler KRC, Panagiotakos G, Thom N, O'Rourke NA, Steinmetz LM, Bernstein JA, Hallmayer J, Huguenard JR, Pasca SP (2017) Assembly of functionally integrated human forebrain spheroids. Nature 545(7652):54–59. https://doi.org/10.1038/NATURE22330
126. Koike H, Iwasawa K, Ouchi R, Maezawa M, Giesbrecht K, Saiki N, Ferguson A, Kimura M, Thompson WL, Wells JM, Zorn AM, Takebe T (2019) Modelling human hepato-biliary-pancreatic organogenesis from the foregut-midgut boundary. Nature 574(7776):112–116. https://doi.org/10.1038/S41586-019-1598-0
127. Huch M, Knoblich JA, Lutolf MP, Martinez-Arias A (2017) The hope and the hype of organoid research. Development 144(6):938–941. https://doi.org/10.1242/DEV.150201
128. Olsen TW, Sanderson S, Feng X, Hubbard WC (2002) Porcine sclera: Thickness and surface area. Invest Ophthalmol Vis. Sci. 43 (8):2529–2532.
129. Regent F, Chen HY, Kelley RA, Qu Z, Swaroop A, Li T (2020) A simple and efficient method for generating human retinal organoids. Mol Vis 26:97–105.

Chapter 9
Retinal Organoid Differentiation Strategies and Their Application in Retinal Degenerative Disease Modeling

Chia Fei Liu, Nonthaphat Kent Wong, Shea Ping Yip, and Chien-Ling Huang

Abstract The retina is a specialized, light-sensitive tissue comprising multiple cell types that convert light into neural signals for vision. Retinal degenerative diseases remain a major cause of irreversible vision loss worldwide, underscoring the need for suitable in vitro models to elucidate disease mechanisms and investigate therapeutic strategies. Three-dimensional (3D) retinal organoids derived from human pluripotent stem cells (hPSCs), including human embryonic and induced pluripotent stem cells, have emerged as a powerful tool for modeling these conditions. Several protocols have been established, notably fully 3D floating cultures, a hybrid 3D/2D/3D approach initiated with embryoid bodies, and a combined 2D/3D system starting with confluent hPSC cultures. Despite their promise, challenges persist in scalability and reproducibility, fully recapitulating the complexity of the native retina. This chapter reviews the current state of retinal organoid research and highlights the application of gene editing and reprogramming to create patient-specific models of diseases such as age-related macular degeneration, glaucoma, and retinitis pigmentosa. By more accurately reflecting in vivo architecture compared to conventional cell lines and animal models, hPSC-derived retinal organoids offer valuable platforms for disease modeling, drug screening, and regenerative medicine, with ongoing innovations poised to address remaining hurdles.

C. F. Liu
Department of Health Technology and Informatics, The Hong Kong Polytechnic University, Hong Kong, China

N. K. Wong · S. P. Yip (✉) · C.-L. Huang (✉)
Department of Health Technology and Informatics, The Hong Kong Polytechnic University, Hong Kong, China

Centre for Eye and Vision Research, Hong Kong Science Park, Hong Kong, China
e-mail: shea.ping.yip@polyu.edu.hk; cl.huang@polyu.edu.hk

R. C. B. Wong and Y. Hayashi (eds.), *Advances in Using Stem Cells to Treat Neural Diseases*, Stem Cell Biology and Regenerative Medicine 77,
https://doi.org/10.1007/978-3-031-94101-6_9

Keywords Retinal organoid · Human pluripotent stem cell · Embryonic stem cell · Induced pluripotent stem cell · Differentiation strategy · 3D floating culture · 2D/3D culture · 3D/2D/3D culture · Retinal degenerative disease · Disease modeling · Age-related macular degeneration · Glaucoma · Retinitis pigmentosa

9.1 Introduction

Eyes are remarkably complex sensory organs with specialized anatomical structures and physiology that enable precise visual function. The retina, situated at the innermost layer of the eye, is pivotal in converting incoming light photons into neural signals, which the brain interprets as three-dimensional (3D) images.

Retinal degenerative diseases refer to a class of disorders that primarily affect the neural retina, resulting in the progressive loss of cells such as the retinal pigment epithelium (RPE), retinal ganglion cells (RGCs), and photoreceptors. These conditions are among the most prevalent causes of irreversible visual impairment worldwide. Recent epidemiological data estimate that more than 243 million individuals are affected by age-related macular degeneration (AMD), around 95 million by glaucoma, and over one million by retinitis pigmentosa (RP) [1–3]. Such widespread occurrence imposes substantial health burdens, diminishing patients' quality of life, and imposing significant socioeconomic costs on families and healthcare systems. Despite advanced clinical interventions, most forms of retinal degeneration still lack definitive treatment, underscoring the need for innovative models to elucidate disease mechanisms and develop targeted therapies.

In response to these challenges, major progress in stem cell research and developmental biology has introduced new avenues for producing in vitro models based on human pluripotent stem cells (hPSCs), including human embryonic stem cells (hESCs) and human induced pluripotent stem cells (hiPSCs). These hPSC-derived models more accurately mirror the complexity of in vivo systems when compared with traditional cell lines and animal models. Over the last decade, significant strides have been made in differentiating hPSCs into all principal retinal cell lineages. One of the most notable breakthroughs is the generation of 3D retinal organoids, which faithfully recapitulate the development, structure, and function of the native human retina.

Multiple laboratories have demonstrated that both hESCs and hiPSCs can be differentiated into 3D retinal organoids via stepwise protocols. These methods yield organoids containing key retinal cell types, including RPE, RGCs, horizontal cells, amacrine cells, bipolar cells, photoreceptors, and Müller glia [4–11]. This platform provides invaluable opportunities for studying retinal development and advancing cell therapy research. Furthermore, refined reprogramming techniques have enabled patient-derived hiPSCs to be converted into 3D retinal organoids that retain the individual's genetic background, offering powerful models for diseases such as AMD, glaucoma, and RP. As a result, these patient-specific organoids show great promise for discovery of effective therapies.

Despite these encouraging developments, challenges persist, particularly regarding scalability, variability in differentiation efficiency, and difficulties in fully reproducing the functional intricacies of the native retina. These limitations pose hurdles not only for advancing retinal organoid technology, but also for broader applications in the stem cell field. In this chapter, we provide an in-depth review of retinal organoid research, focusing on current methodologies for organoid generation and their application in modeling retinal degenerative diseases.

9.2 Retinal Organoids Generated from Human Pluripotent Stem Cells

hPSCs exhibit an extraordinary capacity for self-renewal, reorganization, and differentiation into numerous specialized cell types. They have transformed modern biological and medical research by shedding light on both developmental processes and disease progression. Among hPSCs, hESCs and hiPSCs are commonly utilized to derive functional retinal cell lineages and 3D retinal organoids for modeling retinal degenerative diseases (Fig. 9.1).

Embryonic stem cells (ESCs) arise during the early embryonic stage, only a few days post-fertilization. These pluripotent cells feature the ability to self-renew indefinitely and differentiate into any somatic cell type, including cells of the human retina. Their extensive proliferative capacity and plasticity make them ideal candidates for retinogenesis research, regenerative therapy, and disease modeling. Retinal lineages and organoids can be generated from hESCs through carefully orchestrated differentiation protocols, often involving Wnt inhibition, modulation of BMP and Hedgehog pathways, and supplementation of extracellular matrix (ECM) to simulate early developmental events [5, 12]. Despite the substantial advancements hESC-derived retinal organoids have brought to the fields of retinal research, regenerative medicine, and disease modeling, their widespread application is tempered by the ethical and regulatory considerations inherent in the use of human embryos [13, 14]. Consequently, over the past decade, hiPSCs have gained prominence as a very attractive alternative, particularly for patient-specific retinal organoids that offer personalized models for investigating disease pathogenesis and potential therapies.

Induced pluripotent stem cell (iPSC) technology was first described by Shinya Yamanaka and Kazutoshi Takahashi in 2006 [15]. They successfully reprogrammed mouse fibroblasts into pluripotent cells using the transcription factors Oct3/4, Sox2, Klf4, and c-Myc. In 2007, similar methods were applied to human somatic cells, confirming that iPSCs could be generated from different types of human somatic cells [16, 17]. Since then, iPSC technology has evolved rapidly and gained extensive traction. Aside from circumventing many ethical concerns linked to hESCs, iPSCs are valued for their accessibility, reproducibility, and potential in personalized medicine. Because iPSCs can be derived from both healthy donors and patients,

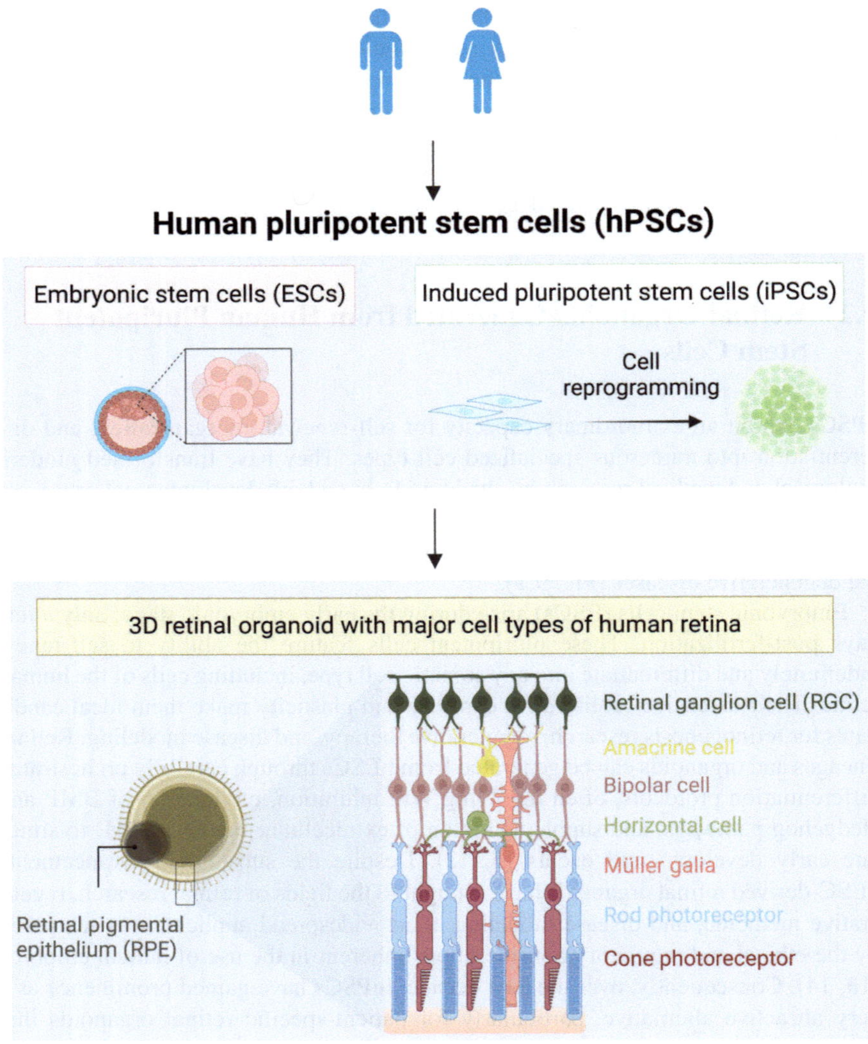

Fig. 9.1 Retinal organoids derived from human pluripotent stem cells (hPSCs). hPSCs such as human embryonic stem cells (hESCs) and human induced pluripotent stem cells (hiPSCs) can be used for retinal organoid generation. The three-dimensional (3D) retinal organoids derived from hPSCs contain major retina cell types such as RPE, RGC, amacrine cell, bipolar cell, horizontal cell, Müller glia, and rod and cone photoreceptors. Note that the retinal organoids mimic the genuine retina to a large extent with some distinctive differences such as lack of blood vessels and mature status of different cell types. (Created in ©BioRender—https://BioRender.com)

they provide an unparalleled resource for generating various retinal cell types [18, 19] and complex 3D organoids [20, 21]. These in vitro models serve as powerful tools for elucidating disease mechanisms, assessing cell transplantation strategies, and developing innovative therapeutic approaches.

9.3 Current Methods for Generating Retinal Organoids

Building on advances in stem cell biology and innovative cell reprogramming techniques, numerous research teams have established protocols to generate 3D retinal organoids that closely mimic human retinal development and disease pathology.

In 2011, Sasai's group introduced a self-organizing 3D culture system that enabled mouse embryonic stem cells (mESCs) to undergo optic cup morphogenesis. These mESC-derived organoids exhibited the major neural retinal cell types—retinal ganglion cells, horizontal cells, amacrine cells, bipolar cells, photoreceptors, and Müller glia—arranged in a multilayered architecture akin to that of the postnatal eye [4, 22]. Around the same time, Meyer and colleagues reported optic vesicle-like structures derived from hPSCs, featuring multipotent retinal progenitor cells (RPCs) and the capacity to differentiate into photoreceptor-like and RPE cells [20]. Building on this work, Nakano et al. refined the protocol for hESCs, opening avenues for 3D retinal organoid-based disease modeling [5].

Currently, three major strategies have emerged for producing 3D retinal organoids (Fig. 9.2). The first employs a fully 3D "floating" culture system (Fig. 9.2a) [4–6, 23–35]. Eiraku et al. first reported this method, using mESC aggregates in a floating culture with serum-free, growth factor-reduced medium (the so-called SFEBq culture, which stands for serum-free floating culture of embryoid body-like aggregates with quick aggregation) to generate 3D retinal structures [4, 22]. Nakano et al. later adapted it to hESCs, providing a foundational protocol involving embryoid body (EB) formation, selective inhibitors (ROCK and Wnt), and timed supplementation of small molecules such as SAG (Smoothened agonist) and CHIR99021 to encourage morphogenesis of optic vesicle and optic cup [5]. This strategy has since been further optimized by multiple groups to reduce the culture period, improve yield, and advance the maturation state of 3D organoids [6, 23]. For instance, Mellough et al. introduced insulin-like growth factor (IGF-1) to accelerate retinal differentiation and generate ocular structures such as corneas and lenses [23]. Kuwahara et al. employed a transient early treatment with BMP4 and a novel induction-reversal protocol using GSK3 and FGF inhibitors to enhance differentiation efficiency to over 80% [6]. These protocols have been successfully applied to mouse and human stem cells, enabling diverse investigations of RGCs and the generation of light-responsive retinal organoids [25, 29].

The second and third strategies employ a combination of two-dimensional (2D) and 3D culture steps. The second approach generally forms EBs in suspension culture with neural induction medium before plating them onto adherent surfaces to promote eye field specification, followed by manual isolation of neural retinal

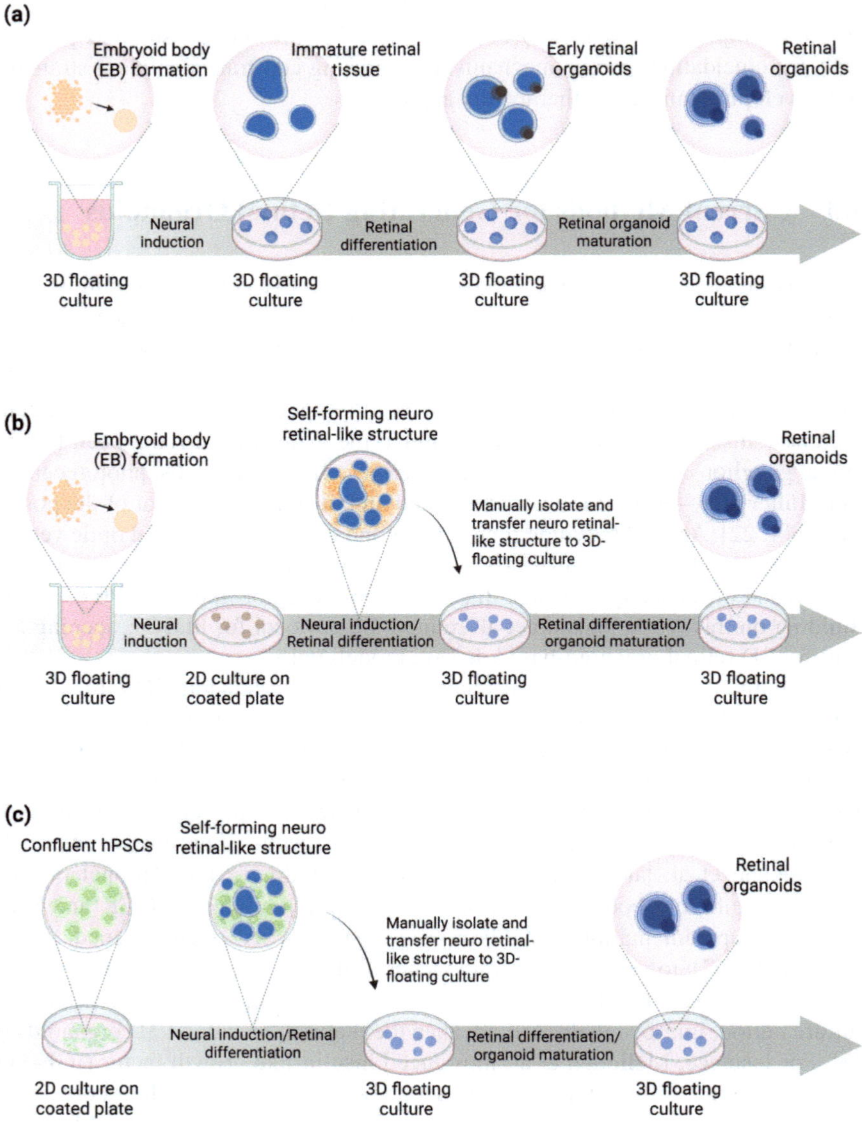

Fig. 9.2 Schematic overview of three major approaches used for 3D retinal organoid generation. 3D retinal organoids have been successfully derived from hPSCs through three major approaches: (**a**) an entirely 3D floating culture system [4–6], (**b**) a hybrid 3D/2D/3D approach initiating with EB formation [7–9], and (**c**) a combined 2D/3D system starting with confluent hPSC culture [10, 11, 39]. (Created in ©BioRender—https://BioRender.com)

regions and transfer into floating cultures to form mature 3D organoids (Fig. 9.2b) [7–9, 36–38]. Zhong et al. described this "forced aggregate" protocol to produce fully laminated 3D retinal organoids from hiPSCs, featuring well-developed, light-responsive outer segments [7]. Capowski and colleagues refined the approach by introducing BMP4 during early differentiation and adapting the method for multiple hiPSC lines, enhancing yield and structural integrity of optic vesicles [8]. More recently, Harkin et al. reported a further streamlined protocol to produce consistently uniform retinal organoids across various cell lines [9].

Both the 3D-only and the combined 3D/2D/3D approaches generally involve multiple steps and rely on exogenous factors to drive retinal differentiation. To simplify the process and reduce the use of external growth factors, some groups have begun differentiating hPSCs in confluent 2D cultures, enabling spontaneous neural retina formation (Fig. 9.2c) [10, 11, 39–42]. Reichman et al. pioneered this method under xenogeneic conditions in 2014 [10] and later refined it into a xeno-free and feeder-free protocol, meeting good manufacturing practice standards [11]. By removing fibroblast growth factor 2 and transforming growth factor beta from the culture medium, these methods trigger endogenous BMP and Wnt signaling at early developmental stages. The resulting self-formed retinal organoids contain multiple cell populations relevant to cell-based therapies, including RGCs and photoreceptors [11]. Building on this, Gozlan et al. generated a "bankable" population of RPCs, and Gonzalez-Cordero's group described a modified 2D/3D protocol that supports the formation of a robust outer nuclear layer (ONL) with cone and rod photoreceptors [39, 41].

Taken together, these diverse protocols have greatly advanced our ability to model human retinal development and function in vitro. Moreover, recent progress in cell reprogramming and gene editing has expanded the utility of hPSC-derived retinal organoids, establishing them as an indispensable platform for studying retinal degenerative diseases, probing disease mechanisms, and testing prospective therapeutic interventions.

9.4 Retinal Organoids as Models of Retinal Degenerative Disease

The advent of cell reprogramming, allowing somatic cells from patient biopsies to be transformed into iPSCs, has opened new frontiers in the in vitro modeling of retinal degenerative conditions such as AMD, glaucoma, and RP.

AMD is a leading cause of visual impairment worldwide, characterized by the progressive degeneration of RPE cells and photoreceptors, culminating in central vision loss and potential blindness [43]. Mesch et al. established long-term cultured human iPSC-derived RPE organoids (up to 360 days) to explore AMD pathology. These "aged" organoids exhibited classic hallmarks of AMD, including enhanced autofluorescence, lipid accumulation, and drusen-associated protein deposits (e.g., apolipoprotein E and tissue inhibitor of metalloproteinases 3), underscoring their suitability for studying the molecular mechanisms of RPE deterioration and drusen formation [44]. Seeking therapeutic strategies, Thomas et al. applied a co-grafting approach and transplanted hESC-derived retinal organoids along with RPE cells into a rat model of retinal degeneration, aiming to reconstruct damaged retinal tissue and restore visual function [45]. More recently, Völkner et al. treated hiPSC-derived retinal organoids with heparin-binding EGF-like growth factor and tumor necrosis factor to induce a pathology resembling AMD, characterized by photoreceptor loss, glial activation, and retinal layer disorganization [46]. They identified potential therapeutic targets in the mechanosensory PIEZO1 receptor, MAPK signaling, and actomyosin pathways, offering new directions for drug development.

Glaucoma, the second most frequent cause of irreversible blindness globally, is marked by the gradual degeneration of RGCs and damage of the optic nerve head and nerve fiber layer, often manifesting as peripheral vision deficits [1, 47, 48]. Although the molecular basis of RGC degeneration remains elusive, stem cell-based retinal organoids now provide cutting-edge tools to probe disease mechanisms. VanderWall et al. investigated the effects of the OPTN(E50K) mutation, a known risk factor for glaucoma, by introducing it into hPSCs via CRISPR/Cas9 gene editing [49]. In their 3D retinal organoid model, mutant RGCs displayed neurodegenerative phenotypes—such as neurite retraction, autophagy dysfunction, and increased excitability, along with reduced expression of BRN3B and ISL1. Treating these mutant organoids with rapamycin partially rescued some defects, linking autophagy pathways to RGC survival. In a related study, Daniszewski and colleagues generated retinal organoids from both healthy and patient-derived hiPSCs to investigate primary open-angle glaucoma (POAG) [50]. Using single-cell RNA sequencing, they detected 97 expression quantitative trait loci specific to RGC populations. By integrating genome-wide association studies and transcriptome-wide association studies, seven candidate genes (*PPP1R17*, *RASD1*, *NXPH1*, *IGFBPL1*, *SAPCD2*, *KRTAP5-AS1*, and *TK1*) emerged as potential contributors to POAG pathogenesis.

RP, a group of inherited retinal degenerative diseases, is marked by the progressive loss of rod and cone photoreceptors, typically presenting as night blindness during childhood, followed by peripheral vision deficits in early adulthood, and ultimately total blindness [3, 51]. Over 70 genes have been linked to RP, reflecting its highly heterogeneous genetic backdrop. Many RP cases involve a single-gene defect, such as *USH2A*, whose mutations are the most frequent cause of autosomal recessive RP. Guo's group modeled USH2A-associated RP by using retinal organoids derived from patient-specific iPSCs carrying the c.8559-2A > G/c.9127_9129delTCC mutation [52]. These organoids displayed improper laminar organization together with compromised basement membrane structures and dysregulated genes associated with apoptosis (*HIF1A*, *ADARB1*, and *CASP3*) and RPC maintenance (*PAX6* and *RAX*). Mutations in other genes, including *IMPG2*, *PDE6B*, *CRB1*, *RP2*, and *PRPF31*, have also been examined in patient-derived iPSC organoids to understand disease onset and progression [53–57]. For instance, Lene et al. have demonstrated that *RP2* mutations cause significant rod photoreceptor cell loss and thinning of the outer nuclear layer in organoids [54]. Similarly, iPSCs with a *PDE6B* mutation form organoids that appear structurally normal at early stages but develop pronounced photoreceptor abnormalities at later time points, mirroring the late-onset nature of RP [53]. Beyond elucidating disease mechanisms, these patient-derived retinal organoids also serve as valuable platforms for gene therapy, notably using adeno-associated virus vectors. Restoration of RP2 and CRB1 expression has shown promising rescue effects in organoids bearing respective mutations [54, 57], and CRISPR/Cas9-based approaches have also been tested for *PRPF31*-mutated organoids [56]. Such developments signal exciting prospects for individualized therapies targeting the root causes of RP.

9.5 Current Limitations and Future Perspectives

Although hPSC-derived retinal organoids have significantly advanced our ability to model retinal degenerative diseases, several challenges must be overcome to fully unlock their potential in both research and clinical settings.

One main obstacle is the considerable time, financial cost, and labor required for large-scale production of retinal organoids. As retinal organoids often require months of differentiation, similar to the timeline of human embryonic development, this protracted maturation process hampers both data collection and high-throughput screening.

Additionally, no matter whether researchers employ exclusively 3D cultures or 2D/3D hybrid approaches, extensive manual handling is frequently necessary. This includes isolating optic cup-like structures from 3D aggregates or dissecting neural retina-like domains from 2D monolayers, which can be laborious, time-consuming, and susceptible to operator-dependent variability. Consequently, these procedural complexities reduce yield, complicate reproducibility, and hinder standardization in high-throughput applications [58]. While Regent et al. have proposed novel methods to streamline the isolation of neural retina-like structures [59], variability in generation efficiency, cellular composition, and retinal organoid maturation remains a pressing issue.

Another significant concern is variation among different hiPSC lines, especially patient-derived cell lines, because cell line-specific traits can influence early differentiation efficiency [34]. For instance, BMP4 or IGF1 signaling, commonly used to direct retinal organoid formation, can produce different outcomes depending on the hiPSC line and the specifics of each protocol [33]. Likewise, in a study using five separate fibroblast-derived hiPSC lines, all produced light-responsive organoids but with markedly varying efficiencies [29]. Epigenetic memory from the donor cells likely contributes to this variability, emphasizing the need to carefully select and characterize starting cells, reprogramming strategies, and differentiation protocols.

Current retinal organoid systems also fail to replicate the full structural and functional complexity of the mature retina. A key limitation is the lack of vascularization, which deprives inner retinal layers of essential oxygen and nutrient supplies

[60]. This issue is especially critical for RGCs, which degenerate over extended culture due in part to limited metabolic support [61, 62]. Furthermore, the absence of synaptic connections between RGCs and their postsynaptic targets in the brain may exacerbate this vulnerability since proper trophic support from downstream neurons is essential for long-term RGC survival [7, 37].

In addition, while photoreceptor-bearing outer segments have been observed in some organoid models, they lack fully elaborated, disk-rich structures. This deficit is at least partially attributable to insufficient contact with RPE cells, which are vital for outer segment maintenance, renewal, and phagocytosis. RPE cells also synthesize essential growth factors and support the inter-photoreceptor matrix, underscoring their pivotal role in sustaining photoreceptor health.

Moving forward, researchers should focus on several key areas. First, establishing standardized protocols for retinal organoid generation is critical for improved reproducibility across laboratories [63]. Techniques such as bioreactors may also enhance organoid yield and reduce labor, enabling broader use in high-throughput applications [28, 40]. Second, the integration of vascular endothelial cells [61, 62], RPE [64], or forebrain organoids [41] could help recapitulate more physiologically relevant microenvironment. Advances in microfluidics and organ-on-a-chip technologies offer promising strategies to introduce perfusion, facilitate nutrient exchange, and potentially promote vascularization [65]. Finally, biomaterials, bioprinting, and tissue engineering approaches may help create more sophisticated organoid architectures with precise spatial control over cell organization and ECM composition, thereby improving cell function and survival [66, 67]. Continued progress in these areas could further establish hPSC-derived retinal organoids as a pivotal platform for studying retinal degenerative diseases, accelerating drug discovery, and advancing regenerative medicine and personalized treatment strategies.

9.6 Conclusions

Human pluripotent stem cell-derived retinal organoids have emerged as a powerful platform for modeling retinal diseases, closing the gap between traditional cell lines and animal models by more accurately replicating the intricate structure and function of the human retina in vitro. Such organoids offer invaluable opportunities in disease modeling, drug screening, and regenerative medicine. In this chapter, we have outlined key protocols for generating retinal organoids and explored their application in studying major degenerative diseases, including age-related macular degeneration, glaucoma, and retinitis pigmentosa. We also highlighted persistent technical hurdles that limit current organoid utility, namely the challenges of scalability, reproducibility, vascularization, and functional interaction with adjacent tissues. Nevertheless, ongoing advances in stem cell biology, tissue engineering, and biofabrication are expected to further refine retinal organoid technology, paving the way for meaningful clinical and translational breakthroughs.

References

1. Tham YC et al (2014) Global prevalence of glaucoma and projections of glaucoma burden through 2040 a systematic review and meta-analysis. Ophthalmology 121(11):2081–2090
2. Wong WL et al (2014) Global prevalence of age-related macular degeneration and disease burden projection for 2020 and 2040: a systematic review and meta-analysis. Lancet Global Health 2(2):E106–E116
3. Hartong DT, Berson EL, Dryja TP (2006) Retinitis pigmentosa. Lancet 368(9549):1795–1809
4. Eiraku M, Sasai Y (2012) Mouse embryonic stem cell culture for generation of three-dimensional retinal and cortical tissues. Nat Protoc 7(1):69–79
5. Nakano T et al (2012) Self-formation of optic cups and storable stratified neural retina from human ESCs. Cell Stem Cell 10(6):771–785
6. Kuwahara A et al (2015) Generation of a ciliary margin-like stem cell niche from self-organizing human retinal tissue. Nat Commun 6:15
7. Zhong XF et al (2014) Generation of three-dimensional retinal tissue with functional photoreceptors from human iPSCs. Nat Commun 5:14
8. Capowski EE et al (2019) Reproducibility and staging of 3D human retinal organoids across multiple pluripotent stem cell lines. Development 146(1):13
9. Harkin J et al (2024) A highly reproducible and efficient method for retinal organoid differentiation from human pluripotent stem cells. Proc Natl Acad Sci USA 121(25):12
10. Reichman S et al (2014) From confluent human iPS cells to self-forming neural retina and retinal pigmented epithelium. Proc Natl Acad Sci USA 111(23):8518–8523
11. Reichman S et al (2017) Generation of storable retinal organoids and retinal pigmented epithelium from adherent human iPS cells in Xeno-free and feeder-free conditions. Stem Cells 35(5):1176–1188
12. Meyer JS et al (2009) Modeling early retinal development with human embryonic and induced pluripotent stem cells. Proc Natl Acad Sci USA 106(39):16698–16703
13. Park SJ et al (2024) Advancements in human embryonic stem cell research: clinical applications and ethical issues. Tissue Eng Regen Med 21(3):379–394
14. Pera ME (2001) Scientific considerations relating to the ethics of the use of human embryonic stem cells in research and medicine. Reprod Fertil Dev 13(1):23–29
15. Takahashi K, Yamanaka S (2006) Induction of pluripotent stem cells from mouse embryonic and adult fibroblast cultures by defined factors. Cell 126(4):663–676
16. Yu JY et al (2007) Induced pluripotent stem cell lines derived from human somatic cells. Science 318(5858):1917–1920
17. Park IH et al (2008) Reprogramming of human somatic cells to pluripotency with defined factors. Nature 451(7175):141–1U1
18. Parameswaran S et al (2010) Induced pluripotent stem cells generate both retinal ganglion cells and photoreceptors: therapeutic implications in degenerative changes in glaucoma and age-related macular degeneration. Stem Cells 28(4):695–703
19. Riazifar H et al (2014) Chemically induced specification of retinal ganglion cells from human embryonic and induced pluripotent stem cells. Stem Cells Transl Med 3(4):424–432
20. Meyer JS et al (2011) Optic vesicle-like structures derived from human pluripotent stem cells facilitate a customized approach to retinal disease treatment. Stem Cells 29(8):1206–1218
21. Phillips MJ et al (2014) Modeling human retinal development with patient-specific induced pluripotent stem cells reveals multiple roles for visual system Homeobox 2. Stem Cells 32(6):1480–1492
22. Eiraku M et al (2011) Self-organizing optic-cup morphogenesis in three-dimensional culture. Nature 472(7341):51–U73
23. Mellough CB et al (2015) IGF-1 signaling plays an important role in the formation of three-dimensional laminated neural retina and other ocular structures from human embryonic stem cells. Stem Cells 33(8):2416–2430

24. Chen HY et al (2016) Three-dimensional retinal organoids from mouse pluripotent stem cells mimic in vivo development with enhanced stratification and rod photoreceptor differentiation. Mol Vis 22:1077–1094
25. Maekawa Y et al (2016) Optimized culture system to induce neurite outgrowth from retinal ganglion cells in three-dimensional retinal aggregates differentiated from mouse and human embryonic stem cells. Curr Eye Res 41(4):558–568
26. Völkner M et al (2016) Retinal organoids from pluripotent stem cells efficiently recapitulate retinogenesis. Stem Cell Reports 6(4):525–538
27. Browne AW et al (2017) Structural and functional characterization of human stem-cell-derived retinal organoids by live imaging. Invest Ophthalmol Vis Sci 58(9):3311–3318
28. DiStefano T et al (2018) Accelerated and improved differentiation of retinal organoids from pluripotent stem cells in rotating-wall vessel bioreactors. Stem Cell Reports 10(1):300–313
29. Hallam D et al (2018) Human-induced pluripotent stem cells generate light responsive retinal organoids with variable and nutrient-dependent efficiency. Stem Cells 36(10):1535–1551
30. Ueda K et al (2018) Generation of three-dimensional retinal organoids expressing rhodopsin and S- and M-cone opsins from mouse stem cells. Biochem Biophys Res Commun 495(4):2595–2601
31. Brooks MJ et al (2019) Improved retinal organoid differentiation by modulating Signaling pathways revealed by comparative transcriptome analyses with development in vivo. Stem Cell Reports 13(5):891–905
32. Chichagova V et al (2023) Incorporating microglia-like cells in human induced pluripotent stem cell-derived retinal organoids. J Cell Mol Med 27(3):435–445
33. Chichagova V et al (2020) Human iPSC differentiation to retinal organoids in response to IGF1 and BMP4 activation is line- and method-dependent. Stem Cells 38(2):195–201
34. Mellough CB et al (2019) Systematic comparison of retinal organoid differentiation from human pluripotent stem cells reveals stage specific, cell line, and methodological differences. Stem Cells Transl Med 8(7):694–706
35. Wahlin KJ et al (2017) Photoreceptor outer segment-like structures in long-term 3D retinas from human pluripotent stem cells. Sci Rep 7:15
36. Ohlemacher SK et al (2015) Generation of highly enriched populations of optic vesicle−like retinal cells from human pluripotent stem cells. Curr Protoc Stem Cell Biol 32(1):1H.8.1–1H.8.20
37. Fligor CM et al (2018) Three-dimensional retinal organoids facilitate the investigation of retinal ganglion cell development, organization and neurite outgrowth from human pluripotent stem cells. Sci Rep 8:14
38. Kim S et al (2019) Generation, transcriptome profiling, and functional validation of cone-rich human retinal organoids. Proc Natl Acad Sci USA 116(22):10824–10833
39. Gonzalez-Cordero A et al (2017) Recapitulation of human retinal development from human pluripotent stem cells generates transplantable populations of cone photoreceptors. Stem Cell Reports 9(3):820–837
40. Ovando-Roche P et al (2018) Use of bioreactors for culturing human retinal organoids improves photoreceptor yields. Stem Cell Res Ther 9:14
41. Fernando M et al (2022) Differentiation of brain and retinal organoids from confluent cultures of pluripotent stem cells connected by nerve-like axonal projections of optic origin. Stem Cell Reports 17(6):1476–1492
42. Gozlan S et al (2023) Bankable human iPSC-derived retinal progenitors represent a valuable source of multipotent cells. Commun Biol 6(1):12
43. Flores R et al (2021) Age-related macular degeneration: pathophysiology, management, and future perspectives. Ophthalmologica 244(6):495–511
44. Mesch L et al (2021) Aged human iPSC-RPE organoid cultures display hallmarks of drusen formation. bioRxiv
45. Thomas BB et al (2021) Co-grafts of human embryonic stem cell derived retina organoids and retinal pigment epithelium for retinal reconstruction in immunodeficient retinal degenerate Royal College of surgeons rats. Front Neurosci 15:17

46. Völkner M et al (2022) HBEGF-TNF induce a complex outer retinal pathology with photoreceptor cell extrusion in human organoids. Nat Commun 13(1):22
47. Steinmetz JD et al (2021) Causes of blindness and vision impairment in 2020 and trends over 30 years, and prevalence of avoidable blindness in relation to VISION 2020: the right to sight: an analysis for the global burden of disease study. Lancet Global Health 9(2):E144–E160
48. Quigley HA (2011) Glaucoma. Lancet 377(9774):1367–1377
49. VanderWall KB et al (2020) Retinal ganglion cells with a glaucoma OPTN(E50K) mutation exhibit neurodegenerative phenotypes when derived from three-dimensional retinal organoids. Stem Cell Reports 15(1):52–66
50. Daniszewski M et al (2022) Retinal ganglion cell-specific genetic regulation in primary open-angle glaucoma. Cell Genomics 2(6):20
51. Wang Q et al (2001) Update on the molecular genetics of retinitis pigmentosa. Ophthalmic Genet 22(3):133–154
52. Guo YL et al (2019) Modeling retinitis pigmentosa: retinal organoids generated from the iPSCs of a patient with the USH2A mutation show early developmental abnormalities. Front Cell Neurosci 13:17
53. Gao ML et al (2020) Patient-specific retinal organoids recapitulate disease features of late-onset retinitis pigmentosa. Front Cell Dev Biol 8:14
54. Lane A et al (2020) Modeling and rescue of RP2 retinitis Pigmentosa using iPSC-derived retinal organoids. Stem Cell Reports 15(1):67–79
55. Mayerl SJ et al (2022) Human retinal organoids harboring IMPG2 mutations exhibit a photoreceptor outer segment phenotype that models advanced retinitis pigmentosa. Stem Cell Reports 17(11):2409–2420
56. Rodrigues A et al (2022) Modeling PRPF31 retinitis pigmentosa using retinal pigment epithelium and organoids combined with gene augmentation rescue. NPJ Regen Med 7(1):22
57. Boon N et al (2023) AAV-mediated gene augmentation therapy of CRB1 patient-derived retinal organoids restores the histological and transcriptional retinal phenotype. Stem Cell Reports 18(5):1123–1137
58. Huch M et al (2017) The hope and the hype of organoid research. Development 144(6):938–941
59. Regent F et al (2020) A simple and efficient method for generating human retinal organoids. Mol Vis 26:97–105
60. Lancaster MA, Knoblich JA (2014) Organogenesis in a dish: modeling development and disease using organoid technologies. Science 345(6194):1247125
61. Busskamp V, Sharma K (2023) Advanced vascularized human stem cell-derived retinal organoids. Invest Ophthalmol Vis Sci 64(8):1894–1894
62. Sharma K, Busskamp V, de Almodóvar CR (2024) Vascularization of human iPSC-derived 3D retinal organoids. Invest Ophthalmol Vis Sci 65(7):4505–4505
63. Kim J, Koo BK, Knoblich JA (2020) Human organoids: model systems for human biology and medicine. Nat Rev Mol Cell Biol 21(10):571–584
64. Akhtar T et al (2019) Accelerated photoreceptor differentiation of hiPSC-derived retinal organoids by contact co-culture with retinal pigment epithelium. Stem Cell Res 39:101491
65. Wimmer RA et al (2019) Human blood vessel organoids as a model of diabetic vasculopathy. Nature 565(7740):505–510
66. Skylar-Scott MA et al (2019) Biomanufacturing of organ-specific tissues with high cellular density and embedded vascular channels. Sci Adv 5(9):eaaw2459
67. Skardal A et al (2017) Multi-tissue interactions in an integrated three-tissue organ-on-a-chip platform. Sci Rep 7(1):8837

Chapter 10
Cell Therapy for Parkinson's Disease with Pluripotent Stem Cells

Asuka Morizane

Abstract One of the anticipated applications of stem cell research in clinics is Parkinson's disease (PD). Open-label clinical trials of cell therapies for PD have been conducted using aborted fetal tissues since the 1980s. Some patients showed favorable outcomes that lasted more than 10 years. These therapies, however, suffer from several problems, such as a short supply of donor materials, unstable quality of the donor cells, and ethical restrictions. As an alternative donor source to fetal tissues, researchers have developed methods to differentiate dopaminergic neural progenitors from stem cells, especially pluripotent stem cells (PSCs). PSC-derived donor cells have been prepared at clinical grade, and their efficacies and safeties were confirmed in non-clinical studies, leading to clinical trials. While promising, there are several practical points to improve in stem cell-based therapies for PD: the quality of donor cells, immune control to avoid graft rejection, surgical procedures for grafting, and the indications for the cell therapy.

Keywords Pluripotent stem cells · Parkinson's disease · Transplantation · Graft · Cell therapy · Central nervous system · Dopamine neuron · Clinical trial · Fetal ventral mesencephalon (fVM) · Positron emission tomography (PET) · Donor · Host · Immune control · Surgery

---

A. Morizane (✉)
Department of Regenerative Medicine, Center for Clinical Research and Innovation, Kobe City Medical Center General Hospital, Kobe, Hyogo, Japan
e-mail: asuka_morizane@kcho.jp

R. C. B. Wong and Y. Hayashi (eds.), *Advances in Using Stem Cells to Treat Neural Diseases*, Stem Cell Biology and Regenerative Medicine 77,
https://doi.org/10.1007/978-3-031-94101-6_10

10.1 Introduction

Parkinson's disease (PD) is defined as a neurodegenerative disorder that has clinically evident Parkinsonism with pathological neuronal cell loss in substantia nigra pars compacta and alpha-synuclein deposition [1]. Parkinsonism is characterized by bradykinesia with resting tremor or muscle rigidity due to an impaired nigrostriatal pathway. The biggest risk factor for PD is aging. Therefore, with populations aging, the number of patients is expected to increase from six million worldwide to over 12 million by 2040, leading to the so-called Parkinson Pandemic [2]. Besides age-related changes, several factors, such as autophagy, mitochondria, metabolism, and oxidative stress, are implicated in the PD pathophysiology. Intestinal microbiota may also affect general pathological conditions in PD.

While current treatments for PD exist, they target the symptoms. Levodopa is still the main drug for PD and is effective in the early phase. However, levodopa has a high risk of motor complications, including wearing-off phenomenon and levodopa-induced dyskinesia (LID) in a long clinical course. In addition to levodopa, dopamine agonists and MAO-B inhibitors are used. However, 5–10 years after the start of these treatments, it becomes difficult to control PD symptoms by medication alone. Device-aided therapy (DAT) can improve the activity of daily living (ADL) and quality of life (QOL) in some patients with advanced PD. Deep brain stimulation (DBS) in the subthalamic nucleus (STN) and globus pallidus internus (GPi) was the first DAT introduced three decades ago. It can manage motor complications such as wearing off and dyskinesia in advanced PD by reducing medication doses. Levodopa-carbidopa intestinal gel (LCIG) is another DAT, in which a tube is set in the jejunum via a gastrostomy, and a gel formulation of levodopa/carbidopa is administered continuously. Yet no current treatment can halt the progression of the disease. Thus, despite many therapeutic options, there is an urgent need to provide a treatment that will restore the lost neural functions or reverse the disease progression.

10.2 Historical Background of Cell Therapy in Parkinson's Disease

Early efforts to replace striatal dopamine neurons have depended on various cell types. The first approaches used autologous dopamine-producing cells such as adrenal medulla cells [3], carotid body cells [4], and sympathetic ganglionic tissues [5]. Because the effects of those approaches in early trials were limited due to the poor survival of the cells, those approaches were given up. In another trial for PD, human retinal pigment epithelial (RPE) cells were attached to gelatin microcarriers (Spheramine) and injected stereotaxically into the putamen. Although RPE cells lack axons or synaptic connections with host tissue, they can produce some dopamine, levodopa, and other growth factors. An open-label clinical trial demonstrated improvements in the Unified Parkinson's Disease Rating Scale (UPDRS) motor score in the OFF state at months 6, 12, 24, and 48 by 33%, 48%, 41%, and 48%, respectively [6–8]. This approach reached a phase II, double-blind, randomized, controlled trial, but the trial failed to replicate the clinical benefit [9]. These results indicate that cell replacement needs the donor dopamine neurons to survive for a long time and make functional connections in the host neural circuit to release dopamine in a regulated manner.

The first clinical trial of cell transplantation in PD was performed in 1987 using human ventral mesencephalon (VM) of aborted embryos [10]. Up to now, about 450 PD patients worldwide have received this treatment in open-label trials and two double-blind, placebo-controlled trials. The open-labeled trials reported substantial and long-lasting functional effect of the graft. However, the subsequent double-blind placebo-controlled trials did not prove clinical benefit [11, 12]. Some cases developed serious side effects, including graft-induced dyskinesia, which showed different clinical features from LID [13]. The failure of the trials has been attributed to the clinical study design, such as the short period of the study, the way of evaluation, tissue dosing, and inadequate immunosuppression. On the other hand, a meta-analysis of trials using VM tissue showed that 30% of patients have a significant improvement in motor scores in the OFF state [14]. Some objective analyzes using positron emission tomography (PET) showed a long-term effect on the dopamine synthesis to the point that pharmacological treatment was unnecessary [15]. Several post-mortem reports on patients with VM grafts showed that the grafted dopaminergic (DA) neurons survived for decades, extending fiber outgrowth and innervation of the host striatum [16–18]. Interestingly, some autopsy studies of VM grafts noted that alpha-synuclein pathology known as Lewy bodies were transferred to the grafted cells from the pathological host brains [16, 17], leading to the research concept of alpha-synuclein propagation in PD pathology [19].

The above led to a new trial of fetal VM cell transplantation, TRANSEURO (NCT01898390), an open-label trial in Europe [20]. This multicentral trial had two major arms: (1) an observational study charting the natural history of younger onset, early stage PD ($n > 100$); and (2) a transplant arm that included patients randomly selected from the observational cohort. In the transplant arm, 11 patients received a graft during 2015–2018, and the results will be reported soon. This clinical trial aims to replace donor VM tissues with ones derived from stem cells in the future.

10.3 Pluripotent Stem Cells for Parkinson's Disease

Human pluripotent stem cells (hPSCs) have the potential to differentiate into all cells of the human body. Human embryonic stem cells (hESCs) are hPSCs derived from the inner cell mass from the early blastocyst and normally retrieved from surplus embryos in artificial insemination. The establishment of hESCs was first reported in 1998 [21]. An alternative route to generating hPSCs was developed by Yamanaka and colleagues in 2007 by reprogramming somatic cells with a cocktail of four transcription factors: Oct3/4, Nanog, KLF4, and c-Myc [22]. These cells were named induced pluripotent stem cells (iPSCs). The methods to establish hiPSCs have been further improved to increase their efficiency, stability, and safety. Initially, skin fibroblasts were commonly used as the original material of somatic cells, but now peripheral blood cells are normally used. c-Myc, one of the original reprogramming genes, has been replaced for reducing the risk of tumorigenicity [23]. The technology of gene delivery has also changed from original retroviral vectors to mRNA [24], episomal vectors [25] or Sendai virus [26]. Additionally, the original protocol used mouse stromal cell feeders and animal-derived components in the culture medium. Now feeder-free and xeno-free culture protocols have been established [23]. There are several advantages of hiPSCs over hESCs. First, the original cells are easier to access and have lower ethical hurdles compared with blastocysts. Second, they present the possibility of autologous transplantation. Lastly, cells can be derived from pre-screened donors with medical information, including the donor's health state and family history, making iPSCs suitable for establishing cell banks.

According to basic research and clinical results on human fetal VM transplantation, donor cells for PD transplantation need to be DA neural progenitors that can mature into authentic midbrain A9-type DA neurons. Knowledge in developmental biology has provided many clues to the induction of DA neurons from hPSCs. Most protocols currently use a dual SMAD inhibition strategy that inhibits both bone morphogenic protein (BMP) and transforming growth factor beta (TGF-β)/Activin/Nodal pathways. Chambers et al. first described dual SMAD inhibition using Noggin and SB431542 [27]. Noggin, a protein inhibiting BMP, is often replaced by small molecules such as Dorsomorphin and LDN193189. Physiological midbrain DA neurons are derived from the mesencephalic floorplate, which is located in the ventral midline part of the developing midbrain. To induce neural progenitors of the midbrain floor plate from hPSCs, sonic hedgehog (SHH) and canonical WNT signaling need to be finely tuned in the differentiation protocol. The gradient of the WNT signal decides the anterior-posterior fate of the neural tube in development. Usually, a small molecule, such as CHIR99021, a GSK3 inhibitor that accelerates WNT signaling, is added to the differentiation culture medium to specify the anterior-posterior axis. A lower CHIR99021 concentration induces the forebrain phenotype, and a higher concentration induces the hindbrain phenotype. Authentic midbrain DA neural progenitors (mDANs) derived from hPSCs and expressing the transcriptional markers FOXA2 and LMX1A efficiently survive and function in animal models of PD [28]. Several groups have established clinical grade protocols [29–32] for the induction of mDANs from hESC and hiPSCs, demonstrating the efficacy and safety of these cells in non-clinical studies. As well as proper patterning to authentic A9 DA neurons, it is crucial to use the right differentiation stage of the donor cells. If the cells are harvested too early, the graft matures and differentiates into off-target cells. If the cells are grafted too late, they lose their survivability and fail to integrate into the host brain. The optimal time window for transplantation for current protocols is between differentiation days 16–30.

10.4 Clinical Trials of Human PSC-Based Cell Therapy for PD (Table 10.1)

The first hPSC-based cell therapy for PD was reported by Dr. Kwang-Soo-Kim's group in 2020 [33]. It was a case report of autologous grafts with the patient's own iPSCs. The graft survived over 24 months without any immune suppression, which was confirmed by fluorine-18-L-dihydroxyphenylalanine (^{18}F-DOPA) PET. Although the symptomatic recovery was limited, there was no serious adverse event.

A group in Kyoto University, Japan started a clinical trial on allogeneic hiPSC-based cell therapy for PD in 2018 (jRCT2090220384) [34, 35]. This trial was a single-arm, non-randomized and open phase I/II study with seven subjects. In this trial, hiPSCs were differentiated for 12–14 days, and then CORIN+ cells were isolated by cell sorting. The sorted cells were cultured in 96-well plates as cell aggregates until day 30 and then used for the transplantation [32]. At 2 years after the transplantation, no serious adverse events were reported. Based on these results, Sumitomo Pharma is planning new trials in the USA.

A group in the USA sponsored by BlueRock Therapeutics started a clinical trial in 2021 using hESCs (Clinicaltrials.gov; NCT04802733). This was an international, open-label, non-randomized study involving 12 subjects with a 24-month observation period. They used the hESC-derived cell product Bemdaneprocel. Bemdaneprocel is a cryopreserved product at day 16 of the differentiation [29]. At 2 years after surgery, the average MDS-UPDRS score part III change was 8.3 for the low-dose group and 21.9 for the high-dose group (https://www.bayer.com/en/us/cell-therapy-bemdaneprocel). Based on these results, a Phase II study is planned.

In Europe, STEM-PD trial (Clinicaltrials.gov; NCT05635409) has been conducted since 2022 as an academic international (Skanes University Hospital, Lund, Sweden and Cambridge University Hospital, UK), phase I/IIa, single-arm trial. Like Bemdaneprocel, this trial used a cryopreserved cell product at day 16 of the differentiation from hESCs [30]. The study results are expected to be reported in 2028.

A Korean team at Yonsei University Health System, Severance Hospital started the TED-A9 study (Clinicaltrials.gov; NCT05887466) in 2023, which is expected to be completed in 2026. They used hESC-derived A9 dopamine progenitor cells differentiated for 25 days [31].

All these clinical trials represent the first-generation of hPSC-based cell products for PD as allogeneic transplantation. On the other hand, following the case study by Dr. Kwang-Soo-Kim's group, an autologous trial involving eight sporadic PD was approved by the same group. Also, an autologous hiPSC-graft trial, ASPIRO, has been designed for nine subjects by Aspen Neuroscience, a private biotechnology company in the USA [36]. This approach, with autologous iPSCs, is advantageous from an immunological perspective. However, challenges related to cost and cell production instability still remain to be overcome. Technological breakthroughs are needed for autologous cell therapy to be practical.

Table 10.1 Donor cells in representative trials

	STEM-PD	BlueRock therapeutics	Kyoto trial	MGH trial	Korea	Aspen neuroscience
Cell source	ESCs (RC17)	ESCs (WA09)	iPSCs (QHJI01s04)	iPSCs (auto)	ESCs (SNU-hES32)	iPSCs (auto)
Cryopreservation	Yes	Yes	No	No	Yes (at day 19)	Yes
Days of differentiation	16	16	28	28	25	18
Format of cell product	Single cells	Single cells	Cell aggregates	Single cells	Single cells	Single cells
Factor for neural induction	Noggin, SB	LDN, SB	LDN, A83–01	LDN, SB	Dorsomorphin, SB	LDN, SB
Factor for midbrain dopamine neural patterning	SHH, FGF8b, CHIR	SHH, CHIR (boost)	Purmorphamine, FGF8, CHIR	SHH, purmorphamine, FGF8, CHIR	SAG, CHIR	SHH, purmorphamine, CHIR
Trial ID	NCT05635409	NCT04802733	UMIN. 000033564	N/A	NCT05887466	NCT06344026
Dose (millions of cells/side)	3.5/7.1	0.9/2.7	2.4/4.8	4	1.58/3.15	No dose available
Immune suppression	Tacrolimus, basiliximab, azathioprine, steroids	Tacrolimus, basiliximab, steroids	Tacrolimus	N/A	Yes (not described in detail)	N/A
Reference	[30]	[29, 44]	[32, 34, 35]	[33]	[31]	[36]

ESCs embryonic stem cells, *iPSCs* induced pluripotent stem cells, *LDN* LDN193189, *SB* SB431542, *SHH* sonic hedgehog, *FGF8* fibroblast growth factor 8, *CHIR* CHIR-99021, *SAG* smoothened agonist

10.5 Key Elements for Optimal Outcomes

Despite the above clinical trials, there is still room for improvement in this treatment modality. One of the remaining challenges is to improve the survivability of the cells. Less than 10% of transplanted donor cells survive and mature into dopaminergic neurons in the host brain [37–39]. Overall, the key elements for cell therapy can be divided into four components called "DISH": donor cell (D), immune control (I), surgery (S), and host (H) (Fig. 10.1).

First, the quality of the donor (D) cells is essential for cell survival and needs to be optimized in the context of purity, maturation, and viability. Notably, the percentage of target cells, namely mDANs, in the donor cell population can never reach 100%. Furthermore, these progenitor cells represent heterogeneous maturation stages, but an optimal maturity stage is desired for engraftment. If the cells are too mature, the viability declines after transplantation. At the same time, if they are patterned insufficiently, some might differentiate into undesired cells. Common undesired cells, such as choroid plexus epithelial cells (CPECs) [32, 40] and vascular leptomeningeal cells (VLMCs) [41], have been reported to be included in grafts, although in small percentages. Two possibilities should be considered for undesired cells in the graft: the contamination of off-target cells and the differentiation of common progenitors into undesired cells after the transplantation. Single-cell RNA sequencing reveals the gene expression and cell lineages during the differentiation of hPSCs to mDANs at the single-cell level [40–43]. Research to eliminate off-target cells and generate donor populations with higher purity is ongoing and generally involves either refining the differentiation protocol [44] or sorting out off-target cells by cell surface markers [40, 43].

Second, immune control (I) is important for avoiding graft rejection, although the impact of immune reactions is not as great in the central nervous system as in

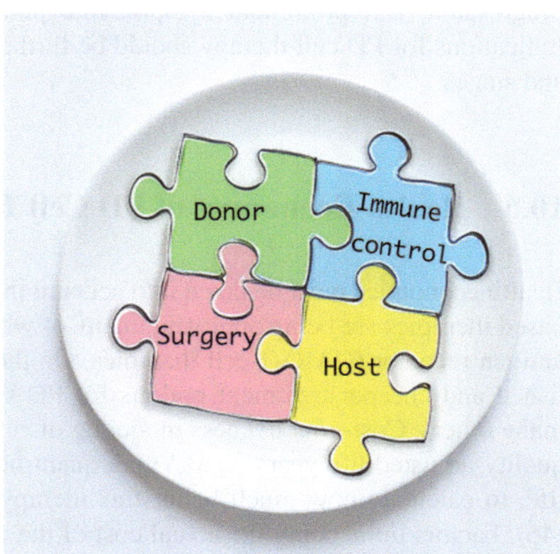

Fig. 10.1 Key elements for cell therapy "DISH." All pieces are indispensable for graft survival and function

other organs. Two possible approaches are mostly being explored: the establishment of an appropriate regimen for post-operative immunosuppression in recipients and modifying the donor cells. Allogeneic donor cells from hPSCs are commonly used under standard immunosuppression. More personalized immune control will be possible by HLA-matched haplotype grafts assuming an HLA-homo haplotype iPSC bank, the development of low-immunogenic iPS cells by gene editing technologies, or cost-effective autologous iPSCs.

Third, proper surgery (S) is required to deliver the donor cells to the target of the host's brain by minimizing implantation trauma [45]. The decision of the target and the needle trajectory by preoperative scanned brain imaging is an important step. The right choice of a delivery instrumental system, including the needle, syringe, and stereotactic device, is crucial. The surgery also needs to avoid implantation in the wrong site due to brain shifts caused by the loss of cerebrospinal fluid during the operation. When loading donor cells into a needle, extreme care should be taken; for example: the cells should not be damaged mechanically, and air should not be aspirated through the surgical needle. When injecting the cells into the brain, there are several important parameters, such as the flow rate of injection, the volume per point, and the number of points and trajectories.

Fourth, the factors in the host (H) are also important for survival and efficacy. Because PD is a progressive disease, there might be an optimal time window for the cell therapy to be effective. As in many cases of PD, chronic inflammation exists, and it might affect graft survival. Furthermore, because older patients are likely to have more degenerated post-synaptic DA neurons, they are unlike to benefit at the end stage of the disease even if the desired cells are grafted. Indeed, one double-blind trial of fetal VM showed improvement only in patients below 60 years old [11]. Furthermore, the timing of the indication for cell therapy needs to be carefully considered, as medication for PD is also effective in the early stage. There are several subtypes of PD, such as with or without genetical background, fast or slow progression, and prominent symptomatic patterns other than motor symptoms. Indications for PD cell therapy should be further explored in terms of PD subtypes and stages.

10.6 Health Economics of PD Cell Therapy

Health economics must be taken into account in an aging society. A number of cell-based therapies are being approved, many of which are expensive, such as chimeric antigen receptor (CAR)-T-cell therapies. On the other hand, there are many established and cheaper treatment options for PD such as medicine, DBS, LCIG, and many others. Cost-effectiveness modeling of cell therapy for PD is underway using quality-adjusted life years (QALYs), a quantified measure of a patient's quality of life, to calculate how much better this therapy is compared to existing therapies [46]. Factors influencing the actual cost of the cell preparation include the scale of production in a single batch, strategies for immune control, and the efficiency and stability of the differentiation of the cells. In the future, economies of scale are expected to be achieved through mass cultures using automated equipment and robotics.

10.7 Conclusion

First-generation stem cell products for PD have reached clinical trials and are moving toward market authorization. Based on the results of clinical trials, further refinement of this therapy is desirable. Nevertheless, it is expected that cell therapy will become a standard treatment option for PD in the future.

References

1. Berg D, Postuma RB, Bloem B, Chan P, Dubois B, Gasser T, Goetz CG, Halliday GM, Hardy J, Lang AE, Litvan I, Marek K, Obeso J, Oertel W, Olanow CW, Poewe W, Stern M, Deuschl G (2014) Time to redefine PD? Introductory statement of the MDS task force on the definition of Parkinson's disease. Mov Disord 29:454–462. https://doi.org/10.1002/mds.25844
2. Dorsey ER, Sherer T, Okun MS, Bloem BR (2018) The emerging evidence of the Parkinson pandemic. J Parkinsons Dis 8:S3–S8. https://doi.org/10.3233/JPD-181474
3. Peterson DI, Price ML, Small CS (1989) Autopsy findings in a patient who had an adrenal-to-brain transplant for Parkinson's disease. Neurology 39:235–238. https://doi.org/10.1212/wnl.39.2.235
4. Arjona V, Mínguez-Castellanos A, Montoro RJ, Ortega A, Escamilla F, Toledo-Aral JJ, Pardal R, Méndez-Ferrer S, Martín JM, Pérez M, Katati MJ, Valencia E, García T, López-Barneo J (2003) Autotransplantation of human carotid body cell aggregates for treatment of Parkinson's disease. Neurosurgery 53:321–328.; ; discussion 328. https://doi.org/10.1227/01.neu.0000073315.88827.72
5. Itakura T, Komai N, Ryujin Y, Ooiwa Y, Nakai M, Yasui M (1994) Autologous transplantation of the cervical sympathetic ganglion into the parkinsonian brain: case report. Neurosurgery 35:155–157; discussion 157. https://doi.org/10.1227/00006123-199407000-00026
6. Stover NP, Bakay RA, Subramanian T, Raiser CD, Cornfeldt ML, Schweikert AW, Allen RC, Watts RL (2005) Intrastriatal implantation of human retinal pigment epithelial cells attached to microcarriers in advanced Parkinson disease. Arch Neurol 62:1833–1837. https://doi.org/10.1001/archneur.62.12.1833
7. Bakay RA, Raiser CD, Stover NP, Subramanian T, Cornfeldt ML, Schweikert AW, Allen RC, Watts R (2004) Implantation of Spheramine in advanced Parkinson's disease (PD). Front Biosci 9:592–602. https://doi.org/10.2741/1217
8. Stover NP, Watts RL (2008) Spheramine for treatment of Parkinson's disease. Neurotherapeutics 5:252–259. https://doi.org/10.1016/j.nurt.2008.02.006
9. Gross RE, Watts RL, Hauser RA, Bakay RA, Reichmann H, von Kummer R, Ondo WG, Reissig E, Eisner W, Steiner-Schulze H, Siedentop H, Fichte K, Hong W, Cornfeldt M, Beebe K, Sandbrink R, Spheramine Investigational Group (2011) Intrastriatal transplantation of microcarrier-bound human retinal pigment epithelial cells versus sham surgery in patients with advanced Parkinson's disease: a double-blind, randomised, controlled trial. Lancet Neurol 10:509–519. https://doi.org/10.1016/S1474-4422(11)70097-7

10. Brundin P, Strecker RE, Lindvall O, Isacson O, Nilsson OG, Barbin G, Prochiantz A, Forni C, Nieoullon A, Widner H (1987) Intracerebral grafting of dopamine neurons. Experimental basis for clinical trials in patients with Parkinson's disease. Ann N Y Acad Sci 495:473–496. https://doi.org/10.1111/j.1749-6632.1987.tb23695.x
11. Freed CR, Greene PE, Breeze RE, Tsai WY, DuMouchel W, Kao R, Dillon S, Winfield H, Culver S, Trojanowski JQ (2001) Transplantation of embryonic dopamine neurons for severe Parkinson's disease. N Engl J Med 344:710–719. https://doi.org/10.1056/NEJM200103083441002
12. Olanow CW, Goetz CG, Kordower JH, Stoessl AJ, Sossi V, Brin MF, Shannon KM, Nauert GM, Perl DP, Godbold J, Freeman TB (2003) A double-blind controlled trial of bilateral fetal nigral transplantation in Parkinson's disease. Ann Neurol 54:403–414. https://doi.org/10.1002/ana.10720
13. Greene PE, Fahn S, Eidelberg D, Bjugstad KB, Breeze RE, Freed CR (2021) Persistent dyskinesias in patients with fetal tissue transplantation for Parkinson disease. NPJ Parkinsons Dis 7:38. https://doi.org/10.1038/s41531-021-00183-w
14. Barker RA, Barrett J, Mason SL, Björklund A (2013) Fetal dopaminergic transplantation trials and the future of neural grafting in Parkinson's disease. Lancet Neurol 12:84–91. https://doi.org/10.1016/S1474-4422(12)70295-8
15. Kefalopoulou Z, Politis M, Piccini P, Mencacci N, Bhatia K, Jahanshahi M, Widner H, Rehncrona S, Brundin P, Björklund A, Lindvall O, Limousin P, Quinn N, Foltynie T (2014) Long-term clinical outcome of Fetal cell transplantation for Parkinson disease: two case reports. JAMA Neurol 71:83–87. https://doi.org/10.1001/jamaneurol.2013.4749
16. Li J-Y, Englund E, Holton JL, Soulet D, Hagell P, Lees AJ, Lashley T, Quinn NP, Rehncrona S, Björklund A, Widner H, Revesz T, Lindvall O, Brundin P (2008) Lewy bodies in grafted neurons in subjects with Parkinson's disease suggest host-to-graft disease propagation. Nat Med 14:501–503. https://doi.org/10.1038/nm1746
17. Kordower JH, Chu Y, Hauser RA, Freeman TB, Olanow CW (2008) Lewy body-like pathology in long-term embryonic nigral transplants in Parkinson's disease. Nat Med 14:504–506. https://doi.org/10.1038/nm1747
18. Li W, Englund E, Widner H, Mattsson B, van Westen D, Lätt J, Rehncrona S, Brundin P, Björklund A, Lindvall O, Li J-Y (2016) Extensive graft-derived dopaminergic innervation is maintained 24 years after transplantation in the degenerating parkinsonian brain. Proc Natl Acad Sci USA 113:6544–6549. https://doi.org/10.1073/pnas.1605245113
19. Dehay B, Vila M, Bezard E, Brundin P, Kordower JH (2016) Alpha-synuclein propagation: new insights from animal models. Mov Disord 31:161–168. https://doi.org/10.1002/mds.26370
20. Barker RA, TRANSEURO Consortium (2019) Designing stem-cell-based dopamine cell replacement trials for Parkinson's disease. Nat Med 25:1045–1053. https://doi.org/10.1038/s41591-019-0507-2
21. Thomson JA, Itskovitz-Eldor J, Shapiro SS, Waknitz MA, Swiergiel JJ, Marshall VS, Jones JM (1998) Embryonic stem cell lines derived from human blastocysts. Science 282:1145–1147. https://doi.org/10.1126/science.282.5391.1145
22. Takahashi K, Tanabe K, Ohnuki M, Narita M, Ichisaka T, Tomoda K, Yamanaka S (2007) Induction of pluripotent stem cells from adult human fibroblasts by defined factors. Cell 131:861–872. https://doi.org/10.1016/j.cell.2007.11.019
23. Nakagawa M, Taniguchi Y, Senda S, Takizawa N, Ichisaka T, Asano K, Morizane A, Doi D, Takahashi J, Nishizawa M, Yoshida Y, Toyoda T, Osafune K, Sekiguchi K, Yamanaka S (2014) A novel efficient feeder-free culture system for the derivation of human induced pluripotent stem cells. Sci Rep 4:3594. https://doi.org/10.1038/srep03594
24. Mandal PK, Rossi DJ (2013) Reprogramming human fibroblasts to pluripotency using modified mRNA. Nat Protoc 8:568–582. https://doi.org/10.1038/nprot.2013.019
25. Okita K, Matsumura Y, Sato Y, Okada A, Morizane A, Okamoto S, Hong H, Nakagawa M, Tanabe K, Tezuka K-I, Shibata T, Kunisada T, Takahashi M, Takahashi J, Saji H, Yamanaka S (2011) A more efficient method to generate integration-free human iPS cells. Nat Methods 8:409–412. https://doi.org/10.1038/nmeth.1591

26. Fusaki N, Ban H, Nishiyama A, Saeki K, Hasegawa M (2009) Efficient induction of transgene-free human pluripotent stem cells using a vector based on Sendai virus, an RNA virus that does not integrate into the host genome. Proc Jpn Acad Ser B Phys Biol Sci 85:348–362. https://doi.org/10.2183/pjab.85.348
27. Chambers SM, Fasano CA, Papapetrou EP, Tomishima M, Sadelain M, Studer L (2009) Highly efficient neural conversion of human ES and iPS cells by dual inhibition of SMAD signaling. Nat Biotechnol 27:275–280. https://doi.org/10.1038/nbt.1529
28. Kriks S, Shim J-W, Piao J, Ganat YM, Wakeman DR, Xie Z, Carrillo-Reid L, Auyeung G, Antonacci C, Buch A, Yang L, Beal MF, Surmeier DJ, Kordower JH, Tabar V, Studer L (2011) Dopamine neurons derived from human ES cells efficiently engraft in animal models of Parkinson's disease. Nature 480:547–551. https://doi.org/10.1038/nature10648
29. Piao J, Zabierowski S, Dubose BN, Hill EJ, Navare M, Claros N, Rosen S, Ramnarine K, Horn C, Fredrickson C, Wong K, Safford B, Kriks S, El Maarouf A, Rutishauser U, Henchcliffe C, Wang Y, Riviere I, Mann S, Bermudez V, Irion S, Studer L, Tomishima M, Tabar V (2021) Preclinical efficacy and safety of a human embryonic stem cell-derived midbrain dopamine progenitor product, MSK-DA01. Cell Stem Cell 28:217–229.e7. https://doi.org/10.1016/j.stem.2021.01.004
30. Kirkeby A, Nelander J, Hoban DB, Rogelius N, Bjartmarz H, Storm P, Fiorenzano A, Adler AF, Vale S, Mudannayake J, Zhang Y, Cardoso T, Mattsson B, Landau AM, Glud AN, Sørensen JC, Lillethorup TP, Lowdell M, Carvalho C, Bain O, van Vliet T, Lindvall O, Björklund A, Harry B, Cutting E, Widner H, Paul G, Barker RA, Parmar M, Novo Nordisk Cell Therapy R&D (2023) Preclinical quality, safety, and efficacy of a human embryonic stem cell-derived product for the treatment of Parkinson's disease, STEM-PD. Cell Stem Cell 30:1299–1314.e9. https://doi.org/10.1016/j.stem.2023.08.014
31. Park S, Park CW, Eom JH, Jo MY, Hur HJ, Choi SK, Lee JS, Nam ST, Jo KS, Oh YW, Lee J, Kim S, Kim DH, Park CY, Kim SJ, Lee HY, Cho MS, Kim DS, Kim DW (2024) Preclinical and dose-ranging assessment of hESC-derived dopaminergic progenitors for a clinical trial on Parkinson's disease. Cell Stem Cell 31:25–38.e8. https://doi.org/10.1016/j.stem.2023.11.009
32. Doi D, Magotani H, Kikuchi T, Ikeda M, Hiramatsu S, Yoshida K, Amano N, Nomura M, Umekage M, Morizane A, Takahashi J (2020) Pre-clinical study of induced pluripotent stem cell-derived dopaminergic progenitor cells for Parkinson's disease. Nat Commun 11:3369. https://doi.org/10.1038/s41467-020-17165-w
33. Schweitzer JS, Song B, Herrington TM, Park TY, Lee N, Ko S, Jeon J, Cha Y, Kim K, Li Q, Henchcliffe C, Kaplitt M, Neff C, Rapalino O, Seo H, Lee IH, Kim J, Kim T, Petsko GA, Ritz J, Cohen BM, Kong SW, Leblanc P, Carter BS, Kim KS (2020) Personalized iPSC-derived dopamine progenitor cells for Parkinson's disease. N Engl J Med 382:1926–1932. https://doi.org/10.1056/NEJMoa1915872
34. Takahashi J (2020) iPS cell-based therapy for Parkinson's disease: a Kyoto trial. Regen Ther 13:18–22. https://doi.org/10.1016/j.reth.2020.06.002
35. Sawamoto N, Doi D, Nakanishi E, Sawamura M, Kikuchi T, Yamakado H, Taruno Y, Shima A, Fushimi Y, Okada T, Kikuchi T, Morizane A, Hiramatsu S, Anazawa T, Shindo T, Ueno K, Morita S, Arakawa Y, Nakamoto Y, Miyamoto S, Takahashi R, Takahashi J (2025) Phase I/II trial of iPS-cell-derived dopaminergic cells for Parkinson's disease. Nature 641:971–977. https://doi.org/10.1038/s41586-025-08700-0
36. Hills R, Mossman JA, Bratt-Leal AM, Tran H, Williams RM, Stouffer DG, Sokolova IV, Sanna PP, Loring JF, Lelos MJ (2023) Neurite outgrowth and gene expression profile correlate with efficacy of human induced pluripotent stem cell-derived dopamine neuron grafts. Stem Cells Dev 32:387–397. https://doi.org/10.1089/scd.2023.0043
37. Kikuchi T, Morizane A, Doi D, Magotani H, Onoe H, Hayashi T, Mizuma H, Takara S, Takahashi R, Inoue H, Morita S, Yamamoto M, Okita K, Nakagawa M, Parmar M, Takahashi J (2017) Human iPS cell-derived dopaminergic neurons function in a primate Parkinson's disease model. Nature 548:592–596. https://doi.org/10.1038/nature23664

38. Nolbrant S, Heuer A, Parmar M, Kirkeby A (2017) Generation of high-purity human ventral midbrain dopaminergic progenitors for in vitro maturation and intracerebral transplantation. Nat Protoc 12:1962–1979. https://doi.org/10.1038/nprot.2017.078
39. Tao Y, Vermilyea SC, Zammit M, Lu J, Olsen M, Metzger JM, Yao L, Chen Y, Phillips S, Holden JE, Bondarenko V, Block WF, Barnhart TE, Schultz-Darken N, Brunner K, Simmons H, Christian BT, Emborg ME, Zhang SC (2021) Autologous transplant therapy alleviates motor and depressive behaviors in parkinsonian monkeys. Nat Med 27:632–639. https://doi.org/10.1038/s41591-021-01257-1
40. Liang L, Tian Y, Feng L, Wang C, Feng G, Stacey GN, Shyh-Chang N, Wu J, Hu B, Li W, Hao J, Wang L, Wang Y (2022) Single-cell transcriptomics reveals the cell fate transitions of human dopaminergic progenitors derived from hESCs. Stem Cell Res Ther 13:412. https://doi.org/10.1186/s13287-022-03104-7
41. Tiklová K, Nolbrant S, Fiorenzano A, Björklund ÅK, Sharma Y, Heuer A, Gillberg L, Hoban DB, Cardoso T, Adler AF, Birtele M, Lundén-Miguel H, Volakakis N, Kirkeby A, Perlmann T, Parmar M (2020) Single cell transcriptomics identifies stem cell-derived graft composition in a model of Parkinson's disease. Nat Commun 11:2434. https://doi.org/10.1038/s41467-020-16225-5
42. La Manno G, Gyllborg D, Codeluppi S, Nishimura K, Salto C, Zeisel A, Borm L, Stott S, Toledo E, Villaescusa J, Lönnerberg P, Ryge J, Barker R, Arenas E, Linnarsson S (2016) Molecular diversity of midbrain development in mouse, human, and stem cells. Cell 167:566–580.e19. https://doi.org/10.1016/j.cell.2016.09.027
43. Xu P, He H, Gao Q, Zhou Y, Wu Z, Zhang X, Sun L, Hu G, Guan Q, You Z, Zhang X, Zheng W, Xiong M, Chen Y (2022) Human midbrain dopaminergic neuronal differentiation markers predict cell therapy outcomes in a Parkinson's disease model. J Clin Invest 132:e156768. https://doi.org/10.1172/JCI156768
44. Kim TW, Piao J, Koo SY, Kriks S, Chung SY, Betel D, Socci ND, Choi SJ, Zabierowski S, Dubose BN, Hill EJ, Mosharov EV, Irion S, Tomishima MJ, Tabar V, Studer L (2021) Biphasic activation of WNT Signaling facilitates the derivation of midbrain dopamine neurons from hESCs for translational use. Cell Stem Cell 28:343–355.e5. https://doi.org/10.1016/j.stem.2021.01.005
45. Maheshwari S, Akram H, Bulstrode H, Kalia SK, Morizane A, Takahashi J, Natalwala A (2024) Dopaminergic cell replacement for Parkinson's disease: addressing the intracranial delivery hurdle. J Parkinsons Dis 14:415–435. https://doi.org/10.3233/JPD-230328
46. Hjelmgren J, Ghatnekar O, Reimer J, Grabowski M, Lindvall O, Persson U, Hagell P (2006) Estimating the value of novel interventions for Parkinson's disease: an early decision-making model with application to dopamine cell replacement. Parkinsonism Relat Disord 12:443–452. https://doi.org/10.1016/j.parkreldis.2006.04.006

Chapter 11
Drug Discovery and Development for Neurodegenerative Diseases Using Induced Pluripotent Stem Cells

Haruhiko Banno, Taro Okunomiya, and Haruhisa Inoue

Abstract The advent of human-induced pluripotent stem cell (iPSC) technology has enabled the differentiation of iPSCs derived from somatic cells of patients with various diseases into cells of disease target organs. iPSCs are suitable for drug screening using large compound libraries, as they can grow infinitely and be induced to differentiate into various types of cells. Several clinical trials have been conducted as drug repositioning for new indications of already approved drugs based on drug candidates discovered using iPSCs. The use of iPSCs can be applied not only to the discovery of drug candidates through compound screening, but also to the identification of efficacious groups in clinical trials and the prediction of the efficacy of therapeutic drugs in patients. Furthermore, the analysis of iPSCs from clinical trial participants enables comparison of the results of clinical trials with the results of analysis of culture dishes. This paper introduces some examples of recent progress in drug discovery and development for neurodegenerative diseases using iPSCs.

Keywords Induced pluripotent stem cells (iPSCs) · Drug screening · Drug repurposing · Alzheimer's disease · Amyotrophic lateral sclerosis

---

H. Banno
Center for iPS Cell Research and Application (CiRA), Kyoto University, Kyoto, Japan

T. Okunomiya
Center for iPS Cell Research and Application (CiRA), Kyoto University, Kyoto, Japan

RIKEN BioResource Research Center (BRC), Kyoto, Japan

H. Inoue (✉)
Center for iPS Cell Research and Application (CiRA), Kyoto University, Kyoto, Japan

RIKEN BioResource Research Center (BRC), Kyoto, Japan

RIKEN Center for Advanced Intelligence Project (AIP), Kyoto, Japan
e-mail: haruhisa@cira.kyoto-u.ac.jp

R. C. B. Wong and Y. Hayashi (eds.), *Advances in Using Stem Cells to Treat Neural Diseases*, Stem Cell Biology and Regenerative Medicine 77,
https://doi.org/10.1007/978-3-031-94101-6_11

Abbreviations

Aβ	Amyloid beta
AD	Alzheimer's disease
ALS	Amyotrophic lateral sclerosis
APP	Amyloid beta precursor protein
ASO	Antisense oligonucleotide
BH4	Tetrahydrobiopterin
C9orf72	Chromosome 9 open reading frame 72
CCCP	Carbonyl cyanide m-chlorophenyl hydrazone
CMT	Charcot-Marie-Tooth disease
CRISPR-Cas9	Clustered regularly interspaced short palindromic repeats/ CRISPR-associated protein 9
DA	Dopamine
DHA	Docosahexaenoic acid
DMT	Disease modifying therapy
DREADDs	Designer receptors exclusively activated by designer drugs
ER	Endoplasmic reticulum
FD	Familial dysautonomia
FTD	Frontotemporal dementia
FTLD	Frontotemporal lobe degeneration
FUS	Fused in sarcoma
HDE	Heat-diffusion equation
HMSN-P	Hereditary motor and sensory neuropathy with proximal dominant involvement
IKBKAP	Inhibitor of kappa light polypeptide gene enhancer in B-cells, kinase complex-associated protein
iPSC	Induced pluripotent stem cell
iPS-MN	Induced pluripotent stem cell derived-spinal motor neuron
JAK inhibitor	Janus kinase inhibitor
LDH	Lactate dehydrogenase
MEF2C	Myocyte enhancer factor 2C
NPI	Neuropsychiatric inventory
NSAID	Nonsteroidal anti-inflammatory drug
PARK	Parkin RBR E3 ubiquitin protein ligase
PD	Parkinson's disease
PINK1	PTEN-induced kinase 1
PSEN1	Presenilin 1
ROS	Reactive oxygen species
SCA	Spinocerebellar ataxia
SIB-J	Severe impairment battery
SNCA	Synuclein alpha
SOD1	Superoxide dismutase 1
SP	Sporadic

TDP-43	TAR DNA-binding protein
UPS	Ubiquitin proteasome system
WT	Wildtype

11.1 Introduction

Neurodegenerative diseases are a general term for diseases that present with progressive neurological symptoms, resulting in neuronal degeneration and cell death. Alzheimer's disease, which is the most common cause of dementia, Parkinson's disease, which causes tremor and bradykinesia, or amyotrophic lateral sclerosis, which causes muscle weakness and atrophy throughout the body, belong to this group of diseases. As most currently approved treatments for neurodegenerative diseases are replacement therapies that supplement missing neurotransmitters, there is a need to develop disease modifying therapies (DMTs) that can fundamentally halt or inhibit the progression of neurodegenerative diseases.

The first mouse-induced pluripotent stem cells (iPSCs) were reported in 2006, with the first human iPSCs reported in 2007 (Fig. 11.1) [1, 2]. In recent years, it has become possible to induce differentiation of various types of neural cells from iPSCs established from patient-derived somatic cells (disease-specific iPSCs) and to analyse nervous system cells bearing the patient's genetic information in culture

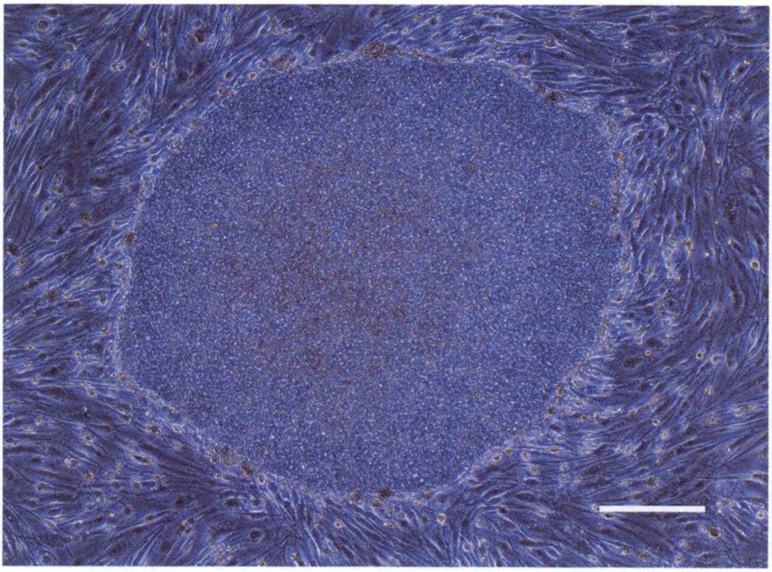

Fig. 11.1 Colonies (aggregates) of human iPSCs established from fibroblasts (The width of the colony is approximately 0.5 mm in actual size. Scale bar = 100 μm). (Source: Centre for iPS Cell Research and Application, Kyoto University)

dishes. As studies using patient-derived neural cells had been almost exclusively limited to pathological studies in autopsy cases, iPSC technology is revolutionising research on neurodegenerative diseases [3]. This article focuses on the status of drug discovery and development for neurodegenerative diseases using disease-specific iPSCs.

11.2 Neurodegenerative Disease Models Using Disease-Specific iPSCs

A common method of drug discovery and development for neurodegenerative diseases is to identify therapeutic agents using models that experimentally reproduce the molecular disease mechanisms. Biopsies of brain tissue to obtain disease-specific neural cells are generally difficult. Therefore, genetically engineered mice and mutant gene transfer into human tumour cell lines have mainly been used as experimental models of diseases. However, it has been noted that even if a candidate drug is effective in animal models of a pathological condition, its efficacy may not be proven in human clinical trials. The reasons for this are that humans and mice have different genetic background and transcription regulation mechanisms are not necessarily identical [4], and in the introduction of mutant genes into disease model mice and human tumour cell lines, the mutant genes are often expressed several to tenfold or more compared to physiological expression levels using strong promoters. The possibility has been raised that the environment for expression of mutant genes may differ from the physiological environment, and that pathological models based on familial diseases may not be appropriate for sporadic diseases, which consist of the majority of neurodegenerative diseases.

Disease models using disease-specific iPSCs could overcome these challenges. They have the same genetic information as the patient and can be analysed under physiological levels of gene expression. It can also be applied to sporadic diseases where the causative gene is unknown. Furthermore, unlike autopsy specimens, the infinite proliferative potential of iPSCs makes it possible to prepare large numbers of various differentiated cells, which can be used as a research resource (Table 11.1).

It has been reported that the induction and analysis of nervous system cells from disease-specific iPSCs can be used to experimentally recapitulate the pathophysiology of neurodegenerative diseases. The pathological conditions to be measured vary according to the characteristics of each disease, and include neuronal cell number reduction, abnormal cell death markers (activation of apoptosis signalling pathways such as caspase-3, increased LDH in the supernatant), morphological abnormalities such as abnormal neurite outgrowth, electrophysiological abnormalities such as overexcitation and inhibition of neuronal activity, reproduction of proteinopathies. Reports are accumulating on the reproduction of pathological conditions such as abnormal protein aggregates and inclusions, release of pathogenic proteins such as amyloid-β (Aβ) and reactivity to DNA damage, endoplasmic reticulum (ER) stress and oxidative stress, which have been previously pointed out.

Table 11.1 Characteristics of study subjects

	Advantages	Disadvantages
Patient-derived specimens (biopsy tissue)	Can analyse the patient's condition directly	Cannot be prepared in large quantities
		Non-diseased tissue may be sampled
		Often not available
Patient-derived specimens (autopsy tissue)	Can examine the pathological changes associated with the patient's symptoms in detail	Difficult to obtain specimens
		Limited analysis methods due to post-mortem changes
Genetically modified animals (e.g., mice)	Can study drug metabolism	Genetically different from humans
	Can reproduce the environment in organs	Artificial control of gene expression
		Cannot reproduce individual human differences
Disease-specific iPSCs	Inherits genetic information from the patient	Differences between the culture environment and the living environment
	Physiological levels of disease gene expression	Lack of ageing information
	Can be applied to sporadic diseases	Differences between differentiated cells and mature living cells
	Can proliferate to study disease mechanisms	

11.3 Drug Discovery and Development for Neurodegenerative Diseases Using Disease-Specific iPSCs

In neurodegenerative diseases, drug discovery using iPSCs is particularly promising. This is because the genetic background of the disease is complex, and it is difficult to construct cellular and animal models that accurately reflect the disease mechanism. Another reason for the expectation of drug discovery using iPSCs is that although therapeutic drugs have been developed, the therapeutic effects of these drugs are limited and there is a great unmet medical need. We have proposed the 'iPSC in vitro trial', which starts with the evaluation of safety and efficacy using disease-specific iPSCs, followed by the identification of target patients (responder) with drug efficacy, and then clinical trials are conducted on the responders [3]. In recent years, knowledge has been accumulating on the evaluation of drug efficacy using iPSCs, drug screening using a library of many existing drugs, and the identification of target population with drug efficacy, and several clinical trials based on the research results have been conducted.

In the evaluation of drug efficacy using iPSCs, there are studies that screen drugs using drug libraries, and studies that analyse neuronal cells differentiated from disease-specific iPSCs, clarify the molecular disease mechanism of the disorder by gene expression analysis, and many others, and then evaluate the drug efficacy of specific inhibitors [5]. Some studies screen a group of drugs and related to the impaired signalling pathway [6].

11.4 Drug Repurposing Using iPSCs

Drug repurposing, as also known as drug repositioning, is a research method that aims to apply existing drugs to new diseases other than their original indications [7–9]. It is not limited to approved drugs, but also includes drugs that have been discontinued from development and whose safety and pharmacokinetics have already been confirmed.

Drug repurposing has advantages for drug discovery. Existing drugs already have accumulated safety, tolerability and pharmacokinetic data, which can reduce the risk of safety problems in humans and significantly shorten the time needed for research and development. Moreover, optimising the design of clinical trials for new indications based on the results of previous clinical trials is possible. Currently, drug development costs are rising due to an increasing number of failures. Drug repositioning can skip steps in drug development, including preclinical studies, and is reported to cost one-tenth of the cost of developing new drugs [9]. In addition, drug repositioning can lead to the discovery of new therapeutic target, treatment pathways and biomarkers, and it may lead to the development of new drugs through structural development of existing drugs. Especially for rare diseases with small markets, drug repositioning in an academic setting has significant advantages.

If drug screening is carried out using patient iPSC-derived disease models, it will be possible to carry out systematic drug repositioning. In the future, further integration of AI and big data analysis will make it possible to predict drug-disease interactions with greater accuracy, which could improve the efficiency of the drug repurposing process and even lead to the discovery of new indications for drugs. With the further accumulation of patient genomic data expected in the future, drug repurposing based on the genetic background of individual patients will become possible, identifying drugs suitable for specific patient groups.

On the other hand, the challenges of drug repurposing include the possibility of patents expiring on existing drugs, the fact that it is not always easy to obtain a patent for a new indication for an existing drug once the patent has expired, and the decline in profitability due to generic competition. In addition, the approval process is not straightforward, as additional clinical trials are required to confirm efficacy and safety for new indications, and in some cases, this does not fit with the international development strategy of pharmaceutical companies.

The process of drug repositioning can be regarded as a strategy for the efficient research and development of new therapeutic drugs. Notwithstanding the challenges involved, the potential benefits are considerable in terms of its capability to address unmet medical needs and its economic impact, and it has the potential to make a sustainable contribution to drug discovery and development.

11.5 Replication of Patient Populations and Clinical Trials Using iPSCs

Drug screening using iPSCs has the advantage of using their proliferative and differentiation potential to perform screening using drug libraries. iPSC-based drug screening in neurological diseases is given in Table 11.2 for representative studies.

When aiming to apply for approval of a drug for a new indication, a large number of preclinical study results are required in addition to the results of clinical trials. At present, it is difficult to evaluate pharmacological effects, data on absorption, distribution, metabolism, and excretion, as well as toxicity and teratogenicity, using only the iPSC evaluation system. However, if safety information on existing drugs can be utilised through drug repositioning, drug development can proceed without the need for new preclinical animal studies, thereby shortening development time and reducing development costs.

The strength of iPSC-based drug repositioning is the ability to utilise iPSCs derived from multiple patients with the disease and compare their efficacy (Fig. 11.2). If drug efficacy can be demonstrated for iPSCs established from multiple patients, the potential of efficacy for the entire target patient population, rather than just one patient, can be taken into account when planning clinical trials. Using sporadic disease-specific iPSCs, the results of compound screening using iPSCs derived from one or a small number of subjects can be generalised to a group of patients with diverse genetic backgrounds. This provides different information than increasing the number of a transgenic mouse. Thus, iPSCs enable not only modelling of diseases, but also modelling of patient populations, that is, of participants in actual clinical trials. Furthermore, if individual iPSCs are established from participants in clinical trials, the clinical therapeutic effects of individual participants can be compared with the results of analyses in iPSC-based culture dishes. This may enable the evaluation of the response to the study drug in cells of the target organ derived from the participant, which has not been possible with conventional blood and imaging biomarkers. The results of drug repositioning using iPSCs reported so far indicate the possibility of reproducing the participant's efficacy of the test drug from the results of analyses in iPSC-based culture dishes. Several clinical trials have been conducted using iPSCs as the starting point (Table 11.3), and their drug developments are detailed below.

11.6 Drug Development for Alzheimer's Disease (AD)

AD is the most common cause of dementia, and the current prevailing hypothesis is that the accumulation of amyloid β (Aβ) protein causes the accumulation of tau protein and neurodegeneration, leading to the development of dementia. Mutations in the genes encoding amyloid precursor protein (*APP*), presenilin 1 (*PSEN1*), and presenilin 2 (*PSEN2*) are causative for familial dominantly inherited AD.

Table 11.2 Drug screening studies for neurological disease using iPSC technology

Disease	Gene	Cell type used	Evaluation of drug efficiency	Number of compounds screened	Candidate compounds	Refs.
ALS	*TDP-43*	Motor neurons	Improvement in neuronal survival, reduction in neurite shortening, improvement in TDP-43 expression	4	Anacardic acid	Egawa et al., 2012 [10]
ALS	SP	Motor neurons	Reduction in the rate of TDP-43 aggregation-positive cells	1,757	Digoxin, Lanatoside C, Proscillaridin A	Burkhardt et al., 2013 [11]
ALS	*SOD1*, *TDP-43*, *C9orf72*, SP	Motor neurons	Improvement in neuronal survival, p62, autophagy, misfolded SOD1	1,416	Bosutinib	Imamura et al., 2017 [12]
ALS	*SOD1*, *TDP-43*, *FUS*, SP	Motor neurons	Composite profile: neurite length, leaked FUS aggregates, stress granules, LDH leakage, cleaved caspase-3, phosphorylated TDP-43 inclusions	1,232	Ropinirole	Fujimori et al., 2018 [13]

ALS	*C9orf72*	Motor neurons	Improvement in neuronal survival under conditions of neurotrophic factor depletion and glutamate excess	800	YM201636, Apilimod	Shi et al., 2018 [14]
ALS	*FUS*	Undifferentiated cells, neurons	Reduction in FUS-eGFP-positive stress granules	~1,600	Rapamycin, Torkinib, Paroxetine, Promethazine, Trimipramine, Chlorpromazine	Marrone et al., 2018 [15]
ALS	*C9orf72*	Cortical neurons	Poly-GA/GP derived from $(G_4C_2)_n$ repeats	1,430	Decitabine, Nelarabine, Entecavir	Czuppa et al., 2022 [16]
ALS	*C9orf72*	Motor neurons, brain organoids	Arsenite-induced TDP-43 cytoplasmic aggregation and loss of TDP-43 splicing activity	3	Rapamycin	Casiraghi et al., 2025 [17]
AD	*APP*, SP	Cortical neurons	Cell survival, ER and oxidative stress	4	DHA	Kondo et al., 2013 [33]
AD	*PSEN1, APP*, SP	Cortical neurons	Reduction in Aβ40 and Aβ42 levels and Aβ42/Aβ40 ratio in culture medium	1,258	Bromocriptine, Cromolyn, Topiramate	Kondo et al., 2017 [18]
AD	*APP*	Neurons	Accumulation of phosphorylated tau	1,684	Statins, Efavirenz	van der Kant et al., 2019 [19]

(continued)

Table 11.2 (continued)

Disease	Gene	Cell type used	Evaluation of drug efficiency	Number of compounds screened	Candidate compounds	Refs.
AD	SP	Brain organoids	Aβ and phosphorylated tau deposition	7	Astaxanthin, Everolimus, Flibanserin, Imipramine, Ibrutinib, Ripasudil, Abemaciclib	Park et al., 2021 [20]
AD, FTD	(Healthy donors), *MAPT*	Neurons	miR-132 transcriptional activity	1,370	Cardiac glycosides	Nguyen et al., 2023 [21]
PD	*SNCA*	Dopaminergic neurons	Activation of MEF2 transcription	~2,000	Isoxazole	Ryan et al., 2013 [6]
PD	*PRKN, PINK1*	Dopaminergic neurons	Improvement in mitochondrial area, ROS accumulation, and cell death under CCCP treatment	320	MRS1220, Tranylcypromine, Flunarizine, Bromocriptine	Yamaguchi et al., 2020 [22]
FD	*IKBKAP*	Neural crest cells	WT-IKBKAP transcript	6,912	SKF-86466	Lee et al., 2012 [23]

Aβ amyloid beta, *AD* Alzheimer's disease, *ALS* amyotrophic lateral sclerosis, *APP* amyloid beta precursor protein, *CCCP* carbonyl cyanide m-chlorophenyl hydrazone, *DHA* docosahexaenoic acid, *FD* familial dysautonomia, *FUS* fused in sarcoma, *IKBKAP* inhibitor of kappa light polypeptide gene enhancer in B-cells, kinase complex-associated protein, *MEF2* myocyte enhancer factor 2, *PRKN* parkin RBR E3 ubiquitin protein ligase, *PD* Parkinson's disease, *PINK1* PTEN induced kinase 1, *PSEN1* presenilin 1, *ROS* reactive oxygen species, *SNCA* synuclein alpha, *SOD1* superoxide dismutase 1, *SP* sporadic, *TDP-43* TAR DNA-binding protein, *WT* wildtype

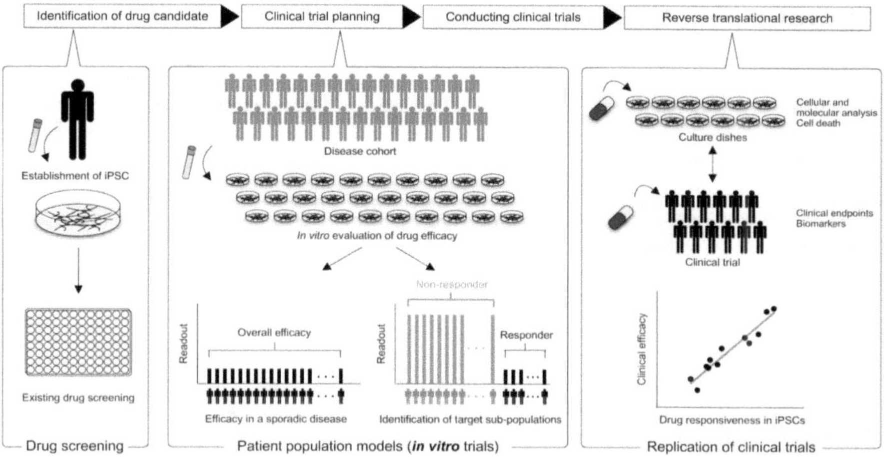

Fig. 11.2 Proposed strategy for iPSC-based drug repositioning. First, patient-derived iPSCs enable identification of drug candidates among existing drugs. Second, iPSCs from multiple patients within a disease cohort allow for the generalization of drug efficacy across sporadic cases or the targeting of specific patient sub-populations with the highest drug efficacy (in vitro trials). Third, iPSCs derived from clinical trial participants represent a promising resource for reverse translational research, enabling the replication of clinical trials in participants' iPSC-derived cells. The association between drug responsiveness of participants' iPSCs-derived cells in culture dishes and clinical efficacy observed in actual clinical trials provides a unique opportunity to elucidate the mechanism of action of trial drugs in human cells

We have shown the efficacy of docosahexaenoic acid (DHA) in neurons derived from patients with familial AD, and that in neurons from patients with sporadic AD, and AD can be stratified by iPSC readouts [33]. On the other hand, van der Kent et al. screened 1684 compounds using neurons from two APP-mutant AD patients, using the accumulation of intracellular phosphorylated tau protein as an indicator, and found statins to be effective [19].

11.6.1 Bromocriptine

We differentiated iPSCs established from familial Alzheimer's disease patients with *PSEN1* mutations (PSEN1-AD) into cortical neurons and screened a library of 1,258 existing drugs, identified six drugs that lower Aβ levels in the culture medium, and found an anti-Aβ cocktail comprising bromocriptine, cromolyn, and topiramate [18]. Furthermore, we identified a group of patients with potential efficacy by analysing neurons from iPSCs derived from familial AD patients, sporadic AD patients and healthy individuals, and by administering this cocktail to them. Of these, bromocriptine markedly reduced the release of Aβ42, which causes Alzheimer's disease, and reduced the Aβ42/40 ratio, an indicator of cytotoxicity, the most.

Table 11.3 List of clinical trials for neurological diseases using iPSC technology in preclinical research

Identifier	Trial name	Disease	Intervention	Trial phase	Start–end
NCT 02450552 [24]	Clinical Trial of Ezogabine (Retigabine) in ALS Subjects	ALS	Ezogabine (Retigabine)	2	2015–2018
NCT 03626012 [25]	A Study to Assess the Safety, Tolerability, and Pharmacokinetics of BIIB078 in Adults with C9ORF72-Associated Amyotrophic Lateral Sclerosis	ALS/FTD (C9orf72)	BIIB078	1	2018–2021
UMIN 000034954 [26]	Phase 1/2a, Double-blind, Placebo-controlled Study with an Open-label Extension of Ropinirole Hydrochloride Extended-Release Tablets—Explorative Assessment of the Safety, Tolerability, and Efficacy after Oral Treatment in Patients with Amyotrophic Lateral Sclerosis (ALS)-	ALS	Ropinirole	1/2a	2018–2021
UMIN 000036295 [27]	Phase 1 Dose Escalation Study of Bosutinib in Patients with Amyotrophic Lateral Sclerosis (ALS)	ALS	Bosutinib	1	2019–2021
NCT 04931862 [28]	Study of WVE-004 in Patients with C9orf72-associated Amyotrophic Lateral Sclerosis (ALS) or Frontotemporal Dementia (FTD) (FOCUS-C9)	ALS/FTD (C9orf72)	WVE-004	1b/2a	2021–2023
jRCT 2051220002 [29]	Phase 1/2 Study of Bosutinib in Patients with Amyotrophic Lateral Sclerosis (ALS)	ALS	Bosutinib	1/2	2022–2024
jRCT 2041200008 [30]/NCT 04413344 [31]	Repurposing Bromocriptine for Abeta Metabolism in Alzheimer's Disease (REBRAnD)	Alzheimer's disease	Bromocriptine	1/2	2020–2022
NCT 05189106 [32]	Neurodegenerative Alzheimer's Disease and Amyotrophic Lateral Sclerosis (NADALS) Basket Trial	Alzheimer's disease and ALS	Baricitinib	1/2	2022–2025

Investigations using compounds with a similar compound structure to bromocriptine showed that the ergoline moiety is important for Aβ metabolism. These results suggest that bromocriptine may be useful as a potential therapeutic candidate for Alzheimer's disease, and we conducted a clinical trial of bromocriptine in Alzheimer's disease patients with *PSEN1* gene mutations (repurposing bromocriptine for Aβ metabolism in Alzheimer's disease study: the REBRAnD study) in patients with *PSEN1* mutations [34]. Based on the iPSC-based drug discovery results, the trial participants were 'enriched', meaning that only certain patients with presumed efficacy are included in the trial, and only the PSEN1-AD patients were included. In the trial, for the first 20 weeks participants received bromocriptine at a daily dose of up to 10 mg or placebo, followed by bromocriptine daily dose of up to 22.5 mg or placebo for 16 weeks in a double-blind design. Quantitative neuropsychological evaluations of AD, such as severe impairment battery (SIB-J) and neuropsychiatric inventory (NPI), were adopted as measures of efficacy.

11.6.2 Baricitinib

Rodriguez et al. developed a system called Drug Repurposing In AD (DRIAD) using machine learning. Their system does not use iPSCs, but learns from post-drug RNA-seq analysis data of cerebral neurons differentiated from neural stem cells and gene expression information from patients with AD at various stages of progression, to verify whether drugs alter molecular markers and to identify candidate. Common trends regarding the target proteins of the drugs were also examined. The group screened 80 phosphatase inhibitors and found JAK inhibitors as candidate drugs [35]. A clinical trial with one of the candidate drugs, baricitinib, is underway in North America [32].

11.7 Drug Development for Amyotrophic Lateral Sclerosis (ALS)

ALS is a disease that causes muscle weakness and atrophy due to degeneration of motor neurons in the brain and spinal cord, and it is still not fully understood how motor neurons degenerate. Riluzole, a glutamate release inhibitor, and edaravone, a free radical scavenger, have been approved for ALS but neither has dramatically slowed the progression of the disease, and the development of a breakthrough treatment is still awaited. We reported that motor neurons derived from ALS patients with *TDP-43* mutations show increased TDP-43 aggregation and insolubility, motor neurite shortening and stress vulnerability [10].

Representative ALS clinical trials originating from drug discoveries using iPSCs are detailed in this section.

11.7.1 Bosutinib

We differentiated iPSCs established from ALS patients into motor neurons and used them to perform a large-scale screening of 1,416 compounds, including existing drugs and development candidates with known pharmacological target molecules [12]. The results showed that 27 compounds improved the survival of motor neurons derived from ALS patient iPSCs. In addition, 14 of the above compounds were inhibitors of the Src/c-Abl pathway. Furthermore, siRNA knockdown of Src and c-Abl inhibited the degeneration of motor neurons derived from ALS patient iPSCs, indicating that inhibition of the Src/c-Abl pathway is a potential therapeutic target for ALS. Bosutinib, an inhibitor of the Src/c-Abl pathway, dose-dependently improved the survival of ALS patient iPSC-derived motor neurons, which prompted further investigation. Bosutinib prolonged the survival of iPSC-derived motor neurons from familial ALS patients with *SOD1* mutations, *TDP-43* mutations, abnormal *C9orf72* repeat elongation and sporadic ALS patients, and mice with *SOD1* mutations. Based on these results, bosutinib is a potential therapeutic candidate for ALS, and we have conducted a clinical trial of bosutinib in ALS patients (iPSC-based drug repurposing for ALS medicine study: iDReAM study) [36]. The phase I study used a combination of a 3 + 3 design, a safety trial in oncology, and a trial design in neurology, in which the study drug is administered after a certain observation period to check for changes in the patient's condition. Thirteen patients who fulfilled the post-observation criteria were allocated to four groups with daily doses of 100, 200, 300, and 400 mg. Of the nine people who took bosutinib for 12 weeks, the decline in the ALS functional rating scale-revised (ALSFRS-R), a measure of motor function in ALS, was halted in five. In those whose ALSFRS-R score did not worsen, their pre-treatment blood neurofilament light chain (neurofilament L: NfL) levels were low. Single-cell analysis also revealed different expression patterns of some genes in the low NfL and high NfL groups. Following the Phase I study, we have conducted a Phase II study.

11.7.2 Ropinirole

Fujimori et al. differentiated iPSCs established from ALS patients with *FUS* or *TDP-43* mutations [10] into motor neurons, conducted a large-scale screening using 1,232 compounds from the US FDA-approved drug library, and found ropinirole as a potential therapeutic agent [13]. Ropinirole ameliorated phenotypes associated with ALS, including accumulation of FUS aggregates, phosphorylated TDP-43, neurite length regression, stress granule formation, neuronal cell death, and mitochondrial dysfunction. It has been speculated that these effects may be due to dopamine-independent actions such as ROS removal [13]. Based on these results, ropinirole was considered to be a potential therapeutic candidate for ALS, and a clinical trial of ropinirole in ALS patients (ropinirole hydrochloride remedy for amyotrophic lateral sclerosis: ROPALS study) was conducted [37]. In the clinical trial, iPSCs from all subjects were established and differentiated into spinal motor neurons and their response to ropinirole was investigated in vitro. After an observation period, 20 subjects who met the criteria were allocated to 13 ropinirole-treated and 7 placebo-treated groups for 24 weeks in a double-blind fashion, followed by an open-labelled trial during which both groups received ropinirole for 4–24 weeks. No serious adverse events were reported in both groups, although minor adverse events were reported in the ropinirole group. In terms of ALSFRS-R, there were no significant differences in the double-blind period, but the worsening was significantly reduced in the ropinirole group during the open-labelled consecutive period. Furthermore, the combined assessment of function and survival (CAFS) score, which adjusted the ALSFRS-R with survival, was significantly higher in the ropinirole group during total treatment periods including the open-label period.

11.7.3 Ezogabine (Retigabine)

Wainger et al. found that in several iPSC-derived familial ALS cell lines, the regulation of the intrinsic membrane potential by voltage-gated potassium channels (Kv) was delayed and the neurons were hyperexcited [38]. Furthermore, they found that ezogabine (retigabine), a Kv7 channel activator, which is used as an antiepileptic drug, suppressed cell death by inhibiting the hyperexcitation of neurons. Based on these results, ezogabine was considered useful as a potential therapeutic candidate for ALS, and a clinical trial of ezogabine in ALS patients was conducted [39]. Of the enrolled patients, 65 participants who met the inclusion criteria were selected for the study after a 6–8 week observation period. The participants were allocated to 23 placebo, 23 to the 600 mg ezogabine group and 19 to the 900 mg ezogabine group and evaluated after a total of 10 weeks of treatment. The results showed that transcranial magnetic stimulation and nerve conduction studies indicated a reduction in hyperexcitability, but there was no clinically significant improvement in ALSFRS-R in the ezogabine group compared to the placebo group [39].

Besides the above, Burkhardt et al. screened 1,757 compounds using motor neurons from one sporadic ALS patient for the presence of TDP-43 aggregates and successfully identified several candidate drugs, including digoxin [11]; Shi et al. conducted 800 compound-drug screening using motor neurons from ALS patients with *C9orf72* repeat sequence elongation and showed that apilimod, a chemical modulator of RAB5 effectors, could ameliorate neuronal cell death [14].

11.8 Drug Development for ALS/FTD

Frontotemporal dementia (FTD) is a clinical form of frontotemporal lobar degeneration (FTLD), a dementia characterised by personality changes and behavioural abnormalities. The genetic and pathological backgrounds of ALS and FTD are diverse, but ALS and FTD share some common factors that induce neurodegeneration. TAR DNA-binding protein of 43 kDa (*TDP-43*), fused in sarcoma (*FUS*), and repetitive abnormal elongation of a six-base repeat sequence (GGGGCC) in intron 1 of the *C9orf72* gene have been reported as factors that induce neurodegeneration common to ALS and FTD.

For frontotemporal lobe degeneration (FTLD), one of the causes of dementia, we have shown that cortical neurons from FTLD patients with *MAPT* mutations produce abnormal excitability and neuronal death. Furthermore, using iPSCs with designer receptors exclusively activated by designer drugs (DREADDs), we generated designer iPSCs and showed that cell death can be suppressed by inhibiting intracellular calcium influx and glutamate receptor antagonists [40]. It has also been reported that administration of antisense oligos (ASOs) to neurons derived from ALS patients with *C9orf72* repeat expansions reduced the number of RNA foci in the nucleus [41, 42].

Liu et al. differentiated iPSCs established from an ALS patient with *C9orf72* repeat expansions into motor neurons and used these cells to investigate the possible protective effect of compound WVE-004 on the motor neurons [43]. WVE-004 is an ASO that recognises the end of a GGGGCC repeat sequence; when WVE-004 is applied to disease-specific iPSC-derived motor neurons, proteins with a structure consisting of repeating glycine, proline dipeptides specified by the GGGGCC repeat sequence (Poly-GP) and RNA aggregates called RNA foci, which are toxic to cells, were inhibited in a dose-dependent manner. In addition, WVE-004 showed no cytotoxic or inflammation-inducing effects. Furthermore, intracerebroventricular administration of WVE-004 to mice with *C9orf72* repeat expansions reduced Poly-GP levels in the spinal cord and cerebral cortex and the amount of variant 3 (V3), a transcript of the *C9orf72* gene. Based on these results, WVE-004 was considered a potential therapeutic candidate for ALS/FTD and a clinical trial (FOCUS-C9 study) was conducted in ALS patients with *C9orf72* repeat expansions. Results showed that although there was a reduction in Poly-GP levels in the WVE-004 group at 24 weeks post-dose, there was no inhibition of progression of clinical symptoms and no correlation between changes in clinical symptoms and Poly-GP levels [44].

Prior to the FOCUS-C9 study, a clinical study of BIIB078, a compound with the same mechanism of action as WVE-004, also showed a reduction in Poly-GP levels in the BIIB078-treated group, but no reduction in the progression of clinical symptoms, as in WVE-004 [44, 45].

11.9 Drug Development for Parkinson's Disease (PD)

For PD, results of drug screening using disease-specific iPSCs have also been reported. Yamaguchi et al. established an imaging-based, semi-automatic, high-throughput system for the quantitative detection of disease-specific phenotypes in dopaminergic neurons differentiated from iPSCs derived from patients with familial Parkinson's disease (PARK2 and PARK6), who have abnormal function in impaired mitochondria clearance, and conducted screening with 320 drugs. The drug screening identified four compounds (MRS1220, tranylcypromine, flunarizine, and bromocriptine) with pathophysiological improvements, which were also confirmed in a PD drosophila model and in cells derived from some idiopathic PD cases [22].

The efficacy of several compounds has also been reported in neurons derived from PD patients with *PINK1* or *LRRK2* mutations, using LDH release under chemical stress as a marker for drug efficacy [46, 47]. In another report, dopamine cells derived from PD patients with *SNCA* mutations have been used to assess drug efficacy using neurite length as a marker [48].

11.10 Other Drug Discoveries and Related Research

Although not exhaustive screening, a number of studies have been reported using disease-specific iPSCs to evaluate the drug efficacy of specific compounds. In AD, the efficacy of drugs whose efficacy has been previously suggested using animal models was re-evaluated using disease-specific iPSCs. In the reports of NSAIDs and γ-secretase inhibitors, these drugs were less effective in disease-specific iPSC-derived neurons than in existing animal models or overexpression models [49, 50], raising hopes for a disease-specific iPSC-based evaluation system.

More recently, we have developed a machine learning system to predict hits in compound screening [51]. This screening system uses a heat-diffusion equation (HDE) model developed by applying partial differential equations describing the diffusion of heat; using iPSCs from ALS patients, we implemented a two million compound screening using the HDE model to predict and extract hit compounds. As a result, compounds that inhibited motor neuronal cell death were identified in many ALS patient iPSCs. The HDE model is expected to be used as a next-generation AI drug discovery tool.

We generated motor neurons from iPSCs from 15 healthy controls and 15 ALS patients and trained an artificial intelligence to learn images of the motor neurons to generate a model to identify ALS patient motor neurons. The produced model was able to distinguish ALS patient motor neurons from healthy controls with high accuracy and is expected to be used for future diagnostic support [52].

We also generated iPSCs from patients with spinocerebellar ataxia type 36 (SCA36) and differentiated them into neurons. The number of RNA foci-positive cells was increased in patient iPSCs and iPSC-derived neurons. Treatment with

2′-O, 4′-C-ethylene-bridged nucleic acid antisense oligonucleotides (ASOs) targeting the disease causative gene pre-mRNA reduced RNA foci-positive cells in patient iPSCs and iPSC-derived neurons, suggesting that disease-specific iPSCs may be useful for evaluating the effects of ASOs [53]. Disease-specific iPSCs can be used for predictive toxicology. We have generated neurons from iPSCs derived from patients with Charcot-Marie-Tooth disease (CMT). Some CMT patients are sensitive to anticancer drugs and experience neuropathy as an adverse reaction. We developed cellular assays and identified mitochondria-related assays in human iPSC-based models that were used to predict drug-induced neurotoxicity [54].

In reviewing previous studies, drug screening results have varied from study to study, despite studies using iPSCs derived from the same neurodegenerative disease; potential causes stemming from the iPSC evaluation system include the possible lack of uniformity in the characteristics of iPSC lines, and induction protocols to the target neuronal cell type. The results of drug screening differ from study to study due to the different lines, different induction protocols for the targeted neuronal cell types, different endpoints in screening, the presence or absence of stress factor administration, and the presence or absence of blinding [5].

11.11 Evaluation Measures for Disease-Specific iPSCs

One challenge in drug discovery research using disease-specific iPSCs is the selection of the phenotype to be evaluated for drug efficacy. First, it is necessary to establish strict negative controls. If healthy human-derived iPSCs are used, the genetic background other than the mutated gene will be different. Therefore, for diseases with known genetic mutations, isogenic controls with identical genetic backgrounds other than the mutated gene may be generated using genome editing technology, represented by CRISPR-Cas9.

As an example of CRISPR/Cas9 correction, we generated disease-specific iPSCs from patients with tetrahydrobiopterin (BH4) metabolism disorder, corrected the gene variant in the iPSCs using the CRISPR/Cas9 system, and differentiated the BH4 metabolism disorder patient- and isogenic control iPSCs into midbrain dopamine (DA) neurons. We found that gene correction restored BH4 levels, tyrosine hydroxylase protein levels and extracellular DA levels in DA neuronal culture using disease-specific iPSCs [55]. Another example is iPSCs from patients with hereditary motor and sensory neuropathy with proximal dominant involvement (HMSN-P). We differentiated the iPSCs into neural cells with spinal motor neurons (iPS-MNs). We found that HMSN-P patient iPS-MNs exhibited ubiquitin proteasome system (UPS) impairment, and HMSN-P patient iPS-MNs were vulnerable to UPS inhibitory stress. Genetic correction of the causal mutation of HMSN-P using the CRISPR-Cas9 system rescued the cellular phenotypes of HMSN-P patient iPS-MNs [56].

In the neuronal death of neurodegenerative diseases, non-cell-autonomous cell death associated with a microenvironment composed of multiple cell types, such as microglia and astrocytes, is involved, as well as abnormal molecular disease mechanisms within the neuronal cells. Flat culture methods on culture dishes have the problem of reproducing the extracellular environment, but by utilising the self-assembly potential of stem cells, it is possible to produce spheroids, which are assemblies of nervous system cells, and organoids with three-dimensional structures resembling neural tissue such as the cerebral cortex and hippocampus. Long-term culture of these has made it possible to recapitulate extracellular pathologies [57–61]. In spinocerebellar ataxia (SCA), which causes cerebellar ataxia, brain organoids were generated from iPSCs derived from SCA6 patients with *CACNA1A* mutations, and cerebellar Purkinje cells were induced to differentiate, resulting in neuronal cell death under thyroid hormone-deprived condition. They also reported that riluzole, a drug for ALS, was found to prevent this neuronal death [62]. Co-culture of brain organoids and microglia has also been reported to reproduce neuroinflammation associated with microglia [63]. 3D culture techniques using hydrogels such as Matrigel have made it possible to co-culture multiple cell types, such as neurons and glial cells, with greater reproducibility [64–68].

Furthermore, as a method for producing organoids that more closely resemble living neural tissues, an assembloid has been reported in which spheroid masses differentiated into the cerebral cortex, spinal cord, and muscle tissue were co-cultured in close proximity to each other [69]. It may be possible to reproduce functional connections as if they were biological tissues.

Kawada et al. successfully produced spontaneously forming neural organoids using iPSC-derived neurons [70]. Normally, in vitro neurons form axons in a random direction and do not form neural bundles. In vivo, during early development, along a preceding axon, called a 'pioneer axon', another axon elongates in the same direction as the pioneer axon due to axon-axon interaction; Kawada et al. used this principle to generate nerve bundle-forming nerve organoids in vitro by inducing axon elongation in a certain direction using microchambers that can facilitate axon-axon interactions. The axons of the neuro-organoids produced in this report have been shown to have the same physiological function as normal axons and to undergo axonal degeneration induced by hydrogen peroxide, which may be used as a model for neurodegenerative diseases with axonal degeneration for therapeutic drug screening. Other groups reported improved methods for making neural organoids for screening and models of axonal degeneration [71, 72], and it is hoped that future screening techniques will be established.

Li et al. reported on a method to produce vascularised neural organoids, showing the possibility of overcoming the problem of neural organoids' inadequate uptake of nerve growth factors, oxygen and other substances [73]. Sabate-Soler et al. also established a mesencephalic organoid model reflecting the extra-neuronal brain environment by co-culturing astrocytes and microglia with mesencephalic organoids [74]. Furthermore, Osaki et al. constructed a disease-specific iPSC-derived ALS model utilising organ-on-a-chip (culture technology in a fluid environment close to the living organism) and indicated the possibility of combining neural organoids to produce a model reflecting an environment closer to the living organism [75]. We investigated the effect of vertical mixing effect on brain organoid structures using bioreactors and established inverted brain organoids. The organoids generated by vertical mixing showed neurons that migrated from the outer periphery to the inner core of the organoids, in contrast to orbital mixing. Single-cell RNA sequencing revealed that neurons of inverted brain organoids exhibited a GABAergic character of the ventral forebrain, which is applicable to Alzheimer's disease modelling [76].

11.12 Conclusion

One of the advantages of drug discovery using iPSCs would be the possibility of keeping the cost of drug discovery low. In conventional drug development, the generation of an animal model of a disease and the evaluation of efficacy in an animal model are essential for the identification of candidate drugs; in iPSC-based drug discovery, once disease-specific iPSCs have been established and improvement of the disease condition can be measured appropriately, verification using animal models may be partially omitted and the results may be transferred to clinical trials on humans in some cases.

In addition, the screening of candidate drugs using disease-specific iPSCs is expected to increase the success rate in clinical trials, even for diseases for which efficacy has been recognised in animal experiments but not in clinical trials. Disease-specific iPSCs may become the potent infrastructure for drug discovery in the development of drugs for rare diseases and diseases for which the causative genes are unknown.

As a future challenge for drug discovery using disease-specific iPSCs, further development is expected in the direction of reproducing clinical symptoms, such as recapitulating disease mechanisms similar to living organisms and establishing an efficient way to produce mature, differentiated cells. It is also important to establish reliable quantitative indices suitable for high-throughput drug screening regarding the evaluation used in the phenotypic analysis of iPSCs to assess drug efficacy and toxicity.

It is hoped that the development of therapies using disease-specific iPSCs will be applied to more neurodegenerative diseases in the future and contribute to the establishment of new treatment methods.

Acknowledgments This work was supported by grants from MEXT/JSPS KAKENHI (JP23H03886 and JP20H05800) and Japan Agency for Medical Research and Development (AMED) (JP22bm0804034, JP22ek0109613, JP23bm1423014, JP23bm1423012, JP23bm1223013, JP24bk0104169, JP24wm0625501, JP24fm0208101, JP23bm1323001, JP23dk0207066, JP24wm0625201 and JP24bm1123047). This study was supported by the Canon Foundation Grant Program. The authors thank Mr. Masahiro Adachi for some text.

Competing Interests All authors declare they have no conflict of interest.

Ethics Approval All studies involving human participants or animals, which were conducted by any of the authors, were approved by the relevant institutional review board and/or ethics committee.

References

1. Takahashi K, Yamanaka S (2006) Induction of pluripotent stem cells from mouse embryonic and adult fibroblast cultures by defined factors. Cell 126:663–676. https://doi.org/10.1016/j.cell.2006.07.024
2. Takahashi K, Tanabe K, Ohnuki M, Narita M, Ichisaka T, Tomoda K et al (2007) Induction of pluripotent stem cells from adult human fibroblasts by defined factors. Cell 131:861–872. https://doi.org/10.1016/j.cell.2007.11.019
3. Inoue H, Nagata N, Kurokawa H, Yamanaka S (2014) iPS cells: a game changer for future medicine. EMBO J 33:409–417. https://doi.org/10.1002/embj.201387098
4. Breschi A, Gingeras TR, Guigó R (2017) Comparative transcriptomics in human and mouse. Nat Rev Genet 18:425–440. https://doi.org/10.1038/nrg.2017.19
5. Hawrot J, Imhof S, Wainger BJ (2020) Modeling cell-autonomous motor neuron phenotypes in ALS using iPSCs. Neurobiol Dis 134:104680. https://doi.org/10.1016/j.nbd.2019.104680
6. Ryan SD, Dolatabadi N, Chan SF, Zhang X, Akhtar MW, Parker J et al (2013) Isogenic human iPSC Parkinson's model shows nitrosative stress-induced dysfunction in MEF2-PGC1α transcription. Cell 155:1351–1364. https://doi.org/10.1016/j.cell.2013.11.009
7. Ashburn TT, Thor KB (2004) Drug repositioning: identifying and developing new uses for existing drugs. Nat Rev Drug Discov 3:673–683. https://doi.org/10.1038/nrd1468
8. Nosengo N (2016) Can you teach old drugs new tricks? Nature 534:314–316. https://doi.org/10.1038/534314a
9. Pushpakom S, Iorio F, Eyers PA, Escott KJ, Hopper S, Wells A et al (2019) Drug repurposing: progress, challenges and recommendations. Nat Rev Drug Discov 18:41–58. https://doi.org/10.1038/nrd.2018.168
10. Egawa N, Kitaoka S, Tsukita K, Naitoh M, Takahashi K, Yamamoto T et al (2012) Drug screening for ALS using patient-specific induced pluripotent stem cells. Sci Transl Med 4:145ra104. https://doi.org/10.1126/scitranslmed.3004052
11. Burkhardt MF, Martinez FJ, Wright S, Ramos C, Volfson D, Mason M et al (2013) A cellular model for sporadic ALS using patient-derived induced pluripotent stem. Mol Cell Neurosci 56:355–364. https://doi.org/10.1016/j.mcn.2013.07.007
12. Imamura K, Izumi Y, Watanabe A, Tsukita K, Woltjen K, Yamamoto T et al (2017) The Src/c-Abl pathway is a potential therapeutic target in amyotrophic lateral sclerosis. Sci Transl Med 9:eaaf3962. https://doi.org/10.1126/scitranslmed.aaf3962
13. Fujimori K, Ishikawa M, Otomo A, Atsuta N, Nakamura R, Akiyama T et al (2018) Modeling sporadic ALS in iPSC-derived motor neurons identifies a potential therapeutic agent. Nat Med 24:1579–1589. https://doi.org/10.1038/s41591-018-0140-5
14. Shi Y, Lin S, Staats KA, Li Y, Chang WH, Hung ST et al (2018) Haploinsufficiency leads to neurodegeneration in C9ORF72 ALS/FTD human induced motor neurons. Nat Med 24:313–325. https://doi.org/10.1038/nm.4490
15. Marrone L, Poser I, Casci I, Japtok J, Reinhardt P, Janosch A et al (2018) Isogenic FUS-eGFP iPSC reporter lines enable quantification of FUS stress granule pathology that is rescued by drugs inducing autophagy. Stem Cell Reports 10:375–389. https://doi.org/10.1016/j.stemcr.2017.12.018
16. Czuppa M, Dhingra A, Zhou Q, Schludi C, König L, Scharf E et al (2022) Drug screen in iPSC-neurons identifies nucleoside analogs as inhibitors of (G4C2)n expression in C9orf72 ALS/FTD. Cell Rep 39:110913. https://doi.org/10.1016/j.celrep.2022.110913

17. Casiraghi V, Sorce MN, Santangelo S, Invernizzi S, Bossolasco P, Lattuada C et al (2025) Modeling of TDP-43 proteinopathy by chronic oxidative stress identifies rapamycin as beneficial in ALS patient-derived 2D and 3D iPSC models. Exp Neurol 383:115057. https://doi.org/10.1016/j.expneurol.2024.115057
18. Kondo T, Imamura K, Funayama M, Tsukita K, Miyake M, Ohta A et al (2017) iPSC-based compound screening and in vitro trials identify a synergistic anti-amyloid β combination for Alzheimer's disease. Cell Rep 21:2304–2312. https://doi.org/10.1016/j.celrep.2017.10.109
19. van der Kant R, Langness VF, Herrera CM, Williams DA, Fong LK, Leestemaker Y et al (2019) Cholesterol metabolism is a druggable axis that independently regulates tau and amyloid-β in iPSC-derived Alzheimer's disease neurons. Cell Stem Cell 24:363–375.e9. https://doi.org/10.1016/j.stem.2018.12.013
20. Park JC, Jang SY, Lee D, Lee J, Kang U, Chang H et al (2021) A logical network-based drug-screening platform for Alzheimer's disease representing pathological features of human brain organoids. Nat Commun 12:280. https://doi.org/10.1038/s41467-020-20440-5
21. Nguyen LD, Wei Z, Silva MC, Barberán-Soler S, Zhang J, Rabinovsky R et al (2023) Small molecule regulators of microRNAs identified by high-throughput screen coupled with high-throughput sequencing. Nat Commun 14:7575. https://doi.org/10.1038/s41467-023-43293-0
22. Yamaguchi A, Ishikawa KI, Inoshita T, Shiba-Fukushima K, Saiki S, Hatano T et al (2020) Identifying therapeutic agents for amelioration of mitochondrial clearance disorder in neurons of familial Parkinson disease. Stem Cell Reports 14:1060–1075. https://doi.org/10.1016/j.stemcr.2020.04.011
23. Lee G, Ramirez CN, Kim H, Zeltner N, Liu B, Radu C et al (2012) Large-scale screening using familial dysautonomia induced pluripotent stem cells identifies compounds that rescue IKBKAP expression. Nat Biotechnol 30:1244–1248. https://doi.org/10.1038/nbt.2435
24. Clinical trial of Ezogabine (Retigabine) in ALS subjects. https://clinicaltrials.gov/study/NCT02450552. Accessed 10 Dec 2024
25. A study to assess the safety, tolerability, and pharmacokinetics of BIIB078 in adults with C9ORF72-associated amyotrophic lateral sclerosis. https://clinicaltrials.gov/study/NCT03626012. Accessed 10 Dec 2024
26. Phase 1/2a, double-blind, placebo-controlled study with an open-label extension of ropinirole hydrochloride extended-release tablets—explorative assessment of the safety, tolerability, and efficacy after oral treatment in patients with amyotrophic lateral sclerosis (ALS). https://center6.umin.ac.jp/cgi-open-bin/ctr_e/ctr_view.cgi?recptno=R000039856. Accessed 10 Dec 2024
27. Phase 1 dose escalation study of Bosutinib in patients with amyotrophic lateral sclerosis (ALS). https://center6.umin.ac.jp/cgi-open-bin/ctr_e/ctr_view.cgi?recptno=R000041294. Accessed 10 Dec 2024
28. Study of WVE-004 in patients with C9orf72-associated amyotrophic lateral sclerosis (ALS) or frontotemporal dementia (FTD) (FOCUS-C9). https://clinicaltrials.gov/study/NCT04931862. Accessed 10 Dec 2024
29. Phase 1/2 study of Bosutinib in patients with amyotrophic lateral sclerosis (ALS). https://jrct.niph.go.jp/en-latest-detail/jRCT2051220002. Accessed 10 Dec 2024
30. Double-blind comparative trial and open-label extension trial to investigate the safety and efficacy of TW-012R in Alzheimer's disease with presenilin 1 mutations (REBRAnD). https://jrct.niph.go.jp/en-latest-detail/jRCT2041200008. Accessed 10 Dec 2024
31. Repurposing bromocriptine for abeta metabolism in Alzheimer's disease (REBRAnD). https://clinicaltrials.gov/study/NCT04413344. Accessed 10 Dec 2024
32. Neurodegenerative Alzheimer's disease and amyotrophic lateral sclerosis (NADALS) Basket trial. https://clinicaltrials.gov/study/NCT05189106. Accessed 10 Dec 2024
33. Kondo T, Asai M, Tsukita K, Kutoku Y, Ohsawa Y, Sunada Y et al (2013) Modeling Alzheimer's disease with iPSCs reveals stress phenotypes associated with. Cell Stem Cell 12:487–496. https://doi.org/10.1016/j.stem.2013.01.009

34. Kondo T, Banno H, Okunomiya T, Amino Y, Endo K, Nakakura A et al (2021) Repurposing bromocriptine for aβ metabolism in Alzheimer's disease (REBRAnD) study: randomised placebo-controlled double-blind comparative trial and open-label extension trial to investigate the safety and efficacy of bromocriptine in Alzheimer's disease with presenilin 1 (PSEN1) mutations. BMJ Open 11:e051343. https://doi.org/10.1136/bmjopen-2021-051343
35. Rodriguez S, Hug C, Todorov P, Moret N, Boswell SA, Evans K et al (2021) Machine learning identifies candidates for drug repurposing in Alzheimer's disease. Nat Commun 12:1033. https://doi.org/10.1038/s41467-021-21330-0
36. Imamura K, Izumi Y, Nagai M, Nishiyama K, Watanabe Y, Hanajima R et al (2022) Safety and tolerability of bosutinib in patients with amyotrophic lateral sclerosis (iDReAM study): a multicentre, open-label, dose-escalation phase 1 trial. EClinicalMedicine 53:101707. https://doi.org/10.1016/j.eclinm.2022.101707
37. Morimoto S, Takahashi S, Fukushima K, Saya H, Suzuki N, Aoki M et al (2019) Ropinirole hydrochloride remedy for amyotrophic lateral sclerosis—protocol for a randomized, double-blind, placebo-controlled, single-Centre, and open-label continuation phase I/IIa clinical trial (ROPALS trial). Regen Ther 11:143–166. https://doi.org/10.1016/j.reth.2019.07.002
38. Wainger BJ, Kiskinis E, Mellin C, Wiskow O, Han SS, Sandoe J et al (2014) Intrinsic membrane hyperexcitability of amyotrophic lateral sclerosis patient-derived motor neurons. Cell Rep 10(7):1–11. https://doi.org/10.1016/j.celrep.2014.03.019
39. Wainger BJ, Macklin EA, Vucic S, McIlduff CE, Paganoni S, Maragakis NJ et al (2021) Effect of Ezogabine on cortical and spinal motor neuron excitability in amyotrophic lateral sclerosis: a randomised clinical trial. JAMA Neurol 78:186–196. https://doi.org/10.1001/jamaneurol.2020.4300
40. Imamura K, Sahara N, Kanaan NM, Tsukita K, Kondo T, Kutoku Y et al (2016) Calcium dysregulation contributes to neurodegeneration in FTLD patients iPSC-derived neurons. Sci Rep 6:34904. https://doi.org/10.1038/srep34904
41. Donnelly CJ, Zhang PW, Pham JT, Haeusler AR, Mistry NA, Vidensky S et al (2013) RNA toxicity from the ALS/FTD C9ORF72 expansion is mitigated by antisense intervention. Neuron 80:415–428. https://doi.org/10.1016/j.neuron.2013.10.015
42. Sareen D, O'Rourke JG, Meera P, Muhammad AK, Grant S, Simpkinson M et al (2013) Targeting RNA foci in iPSC-derived motor neurons from ALS patients with a C9ORF 72 repeat expansion. Sci Transl Med 5:208ra149. https://doi.org/10.1126/scitranslmed.3007529
43. Liu Y, Andreucci A, Iwamoto N, Yin Y, Yang H, Liu F et al (2022) Preclinical evaluation of WVE-004, an investigational stereo pure oligonucleotide for the treatment of C9orf72-associated ALS or FTD. Mol Ther Nucleic Acids 28:558–570. https://doi.org/10.1016/j.omtn.2022.04.007
44. Cammack AJ, Balendra R, Isaacs AM (2024) Failure of C9orf72 sense repeat-targeting antisense oligonucleotides: lessons learned and the path forward. Brain 147:2607–2609. https://doi.org/10.1093/brain/awae168
45. van den Berg LH, Rothstein JD, Shaw PJ, Babu S, Benatar M, Bucelli RC et al (2024) Safe-ty, tolerability, and pharmacokinetics of antisense oligonucleotide BIIB078 in adults with C9orf72-associated amyotrophic lateral sclerosis: a phase 1, randomised, double blinded, placebo-controlled, multiple ascending dose study. Lancet Neurol 23:901–912. https://doi.org/10.1016/S1474-4422(24)00216-3
46. Cooper O, Seo H, Andrabi S, Guardia-Laguarta C, Graziotto J, Sundberg M et al (2012) Pharmacological rescue of mitochondrial deficits in iPSC-derived neural cells from patients with familial Parkinson's disease. Sci Transl Med 4:141ra90. https://doi.org/10.1126/scitranslmed.3003985
47. Reinhardt P, Schmid B, Burbulla LF, Schöndorf DC, Wagner L, Glatza M et al (2013) Genetic correction of a LRRK2 mutation in human iPSCs links parkinsonian neurodegeneration to ERK-dependent changes in gene expression. Cell Stem Cell 12:354–367. https://doi.org/10.1016/j.stem.2013.01.008

48. Kouroupi G, Taoufik E, Vlachos IS, Tsioras K, Antoniou N, Papastefanaki F et al (2017) Defective synaptic connectivity and axonal neuropathology in a human iPSC-based model of familial Parkinson's disease. Proc Natl Acad Sci USA 114:E3679–E3688. https://doi.org/10.1073/pnas.1617259114
49. Mertens J, Stüber K, Wunderlich P, Ladewig J, Kesavan JC, Vandenberghe R et al (2013) APP processing in human pluripotent stem cell-derived neurons is resistant to NSAID-based γ-secretase modulation. Stem Cell Reports 1:491–498. https://doi.org/10.1016/j.stemcr.2013.10.011
50. Liu Q, Waltz S, Woodruff G, Ouyang J, Israel MA, Herrera C et al (2014) Effect of potent γ-secretase modulator in human neurons derived from multiple presenilin 1-induced pluripotent stem cell mutant carriers. JAMA Neurol 71:1481–1489. https://doi.org/10.1001/jamaneurol.2014.2482
51. Hidaka T, Imamura K, Hioki T, Takagi T, Giga Y, Giga MH et al (2020) Prediction of compound bioactivities using heat-diffusion equation. Patterns (N Y) 1(9):100140. https://doi.org/10.1016/j.patter.2020.100140
52. Imamura K, Yada Y, Izumi Y, Morita M, Kawata A, Arisato T et al (2021) Prediction model of amyotrophic lateral sclerosis by deep learning with patient induced. Ann Neurol 89:1226–1233. https://doi.org/10.1002/ana.26047
53. Matsuzono K, Imamura K, Murakami N, Tsukita K, Yamamoto T, Izumi Y et al (2017) Antisense oligonucleotides reduce RNA foci in spinocerebellar ataxia 36 patient iPSCs. Mol Ther Nucleic Acids 8:211–219. https://doi.org/10.1016/j.omtn.2017.06.017
54. Ohara R, Imamura K, Morii F, Egawa N, Tsukita K, Enami T et al (2017) Modeling drug-induced neuropathy using human iPSCs for predictive toxicology. Clin Pharmacol Ther 101:754–762. https://doi.org/10.1002/cpt.562
55. Ishikawa T, Imamura K, Kondo T, Koshiba Y, Hara S, Ichinose H et al (2016) Genetic and pharmacological correction of aberrant dopamine synthesis using patient iPSCs with BH4 metabolism disorders. Hum Mol Genet 25:5188–5197. https://doi.org/10.1093/hmg/ddw339
56. Murakami N, Imamura K, Izumi Y, Egawa N, Tsukita K, Enami T et al (2017) Proteasome impairment in neural cells derived from HMSN-P patient iPSCs. Mol Brain 1:7. https://doi.org/10.1186/s13041-017-0286-y
57. Lee HK, Velazquez Sanchez C, Chen M, Morin PJ, Wells JM, Hanlon EB et al (2016) Three dimensional human neuro-spheroid model of Alzheimer's disease based on differentiated induced pluripotent stem cells. PLoS One 11:e0163072. https://doi.org/10.1371/journal.pone.0163072
58. Raja WK, Mungenast AE, Lin YT, Ko T, Abdurrob F, Seo J et al (2016) Self-organizing 3D human neural tissue derived from induced pluripotent stem cells recapitulate Alzheimer's disease phenotypes. PLoS One 11:e0161969. https://doi.org/10.1371/journal.pone.0161969
59. Gonzalez C, Armijo E, Bravo-Alegria J, Becerra-Calixto A, Mays CE, Soto C (2018) Modeling amyloid beta and tau pathology in human cerebral organoids. Mol Psychiatry 23:2363–2374. https://doi.org/10.1038/s41380-018-0229-8
60. Yan Y, Song L, Bejoy J, Zhao J, Kanekiyo T, Bu G et al (2018) Modeling neurodegenerative microenvironment using cortical organoids derived from human stem cells. Tissue Eng Part A 24:1125–1137. https://doi.org/10.1089/ten.TEA.2017.0423
61. Pomeshchik Y, Klementieva O, Gil J, Martinsson I, Hansen MG, de Vries T et al (2020) Human iPSC-derived hippocampal spheroids: an innovative tool for stratifying Alzheimer disease patient-specific cellular phenotypes and developing therapies. Stem Cell Reports 15:256–273. https://doi.org/10.1016/j.stemcr.2020.06.001
62. Ishida Y, Kawakami H, Kitajima H, Nishiyama A, Sasai Y, Inoue H et al (2016) Vulnerability of Purkinje cells generated from spinocerebellar ataxia type 6 patient-derived iPSCs. Cell Rep 17:1482–1490. https://doi.org/10.1016/j.celrep.2016.10.026
63. Lin YT, Seo J, Gao F, Feldman HM, Wen HL, Penney J et al (2018) APOE4 causes widespread molecular and cellular alterations associated with Alzheimer's disease phenotypes in human iPSC-derived brain cell types. Neuron 98:1141–1154.e7. https://doi.org/10.1016/j.neuron.2018.05.008

64. Oksanen M, Petersen AJ, Naumenko N, Puttonen K, Lehtonen Š, Gubert Olivé M et al (2017) PSEN1 mutant iPSC-derived model reveals severe astrocyte pathology in Alzheimer's disease. Stem Cell Reports 9:1885–1897. https://doi.org/10.1016/j.stemcr.2017.10.016
65. Choi SH, Kim YH, Hebisch M, Sliwinski C, Lee S, D'Avanzo C et al (2014) A three-dimensional human neural cell culture model of Alzheimer's disease. Nature 515:274–278. https://doi.org/10.1038/nature13800
66. Zhang D, Pekkanen-Mattila M, Shahsavani M, Falk A, Teixeira AI, Herland A (2014) A 3D Alzheimer's disease culture model and the induction of P21-activated kinase mediated sensing in iPSC derived neurons. Biomaterials 35:1420–1428. https://doi.org/10.1016/j.biomaterials.2013.11.028
67. Park J, Wetzel I, Marriott I, Dréau D, D'Avanzo C, Kim DY et al (2018) A 3D human triculture system modelling neurodegeneration and neuroinflammation in Alzheimer's disease. Nat Neurosci 21:941–951. https://doi.org/10.1038/s41593-018-0175-4
68. Kwak SS, Washicosky KJ, Brand E, von Maydell D, Aronson J, Kim S et al (2020) Amyloid-β42/40 ratio drives tau pathology in 3D human neural cell culture models of Alzheimer's disease. Nat Commun 11:1377. https://doi.org/10.1038/s41467-020-15120-3
69. Andersen J, Revah O, Miura Y, Thom N, Amin ND, Kelley KW et al (2020) Generation of functional human 3D cortico-motor assembloids. Cell 183:1913–1929.e26. https://doi.org/10.1016/j.cell.2020.11.017
70. Kawada J, Kaneda S, Kirihara T, Maroof A, Levi T, Eggan K et al (2017) Generation of a motor nerve organoid with human stem cell-derived neurons. Stem Cell Reports 9:1441–1449. https://doi.org/10.1016/j.stemcr.2017.09.021
71. Osaki T, Chow SYA, Nakanishi Y, Hernández J, Kawada J, Fujii T et al (2020) Three-dimensional motor nerve organoid generation. J Vis Exp 163. https://doi.org/10.3791/61544
72. Seo WM, Yoon J, Lee JH, Lee Y, Lee H, Geum D et al (2022) Modeling axonal regeneration by changing cytoskeletal dynamics in stem cell-derived motor nerve organoids. Sci Rep 12:2082. https://doi.org/10.1038/s41598-022-05645-6
73. Li M, Gao L, Zhao L, Zou T, Xu H (2023) Toward the next generation of vascularized human neural organoids. Med Res Rev 43:31–54. https://doi.org/10.1002/med.21922
74. Sabate-Soler S, Nickels SL, Saraiva C, Berger E, Dubonyte U, Barmpa K et al (2022) Microglia integration into human midbrain organoids leads to increased neuronal maturation and functionality. Glia 70:1267–1288. https://doi.org/10.1002/glia.24167
75. Osaki T, Uzel SGM, Kamm RD (2018) Microphysiological 3D model of amyotrophic lateral sclerosis (ALS) from human iPS-derived muscle cells and optogenetic motor neurons. Sci Adv 4:eaat5847. https://doi.org/10.1126/sciadv.aat5847
76. Suong DNA, Imamura K, Inoue I, Kabai R, Sakamoto S, Okumura T et al (2021) Induction of inverted morphology in brain organoids by vertical-mixing bioreactors. Commun Biol 4:1213. https://doi.org/10.1038/s42003-021-02719-5

Chapter 12
Cellular Reprogramming to Stimulate Retina Regeneration

Daniel Urrutia-Cabrera, Luozixian Wang, and Raymond C. B. Wong

Abstract Retina degeneration is one of the major causes of blindness and visual impairment in the world, with millions of patients affected by diseases such as age-related macular degeneration (AMD), glaucoma, diabetic retinopathy (DR) and inherited retinal diseases (IRD). Although the recent progress in gene therapy has made treatments to prevent and slow retina degeneration more accessible, there are currently no effective therapies to restore vision once the retina has degenerated. Regenerative strategies such as cell-based therapies using stem cell-derived retinal neurons represent a promising avenue to regenerate the retina and restore vision. However, despite the promising results in pre-clinical research, there are currently no approved therapies to regenerate the retina. Cell-based treatments still have some hurdles to overcome before progressing into the clinics. For instance, it remains unclear whether the transplanted cells can integrate into the retinal circuit and achieve functional improvements in the long-term. Additionally, material exchange between transplanted and host cells such as photoreceptors has represented a big challenge for therapies using stem cells to regenerate the retina.

Recently, in vivo direct reprogramming of Müller glia into retinal neurons has emerged as an alternative approach to overcome the limitations associated with transplantation. Müller cells are an attractive target for regenerative therapies, as these cells have shown different degrees of regenerative capacity in vertebrates such as fish, chicken and mice. Notably, the Müller glia of teleost fish can dedifferentiate into retinal progenitors and regenerate the retina upon injury. Therefore, numerous studies have focused on the identification of genes and molecules that can potentiate the regenerative capacity of Müller glia in mammals. This chapter reviews the progress of pre-clinical research using cellular reprogramming to regenerate the retina,

D. Urrutia-Cabrera · L. Wang · R. C. B. Wong (✉)
Centre for Eye Research Australia, Royal Victorian Eye and Ear Hospital,
East Melbourne, VIC, Australia

Ophthalmology, Department of Surgery, University of Melbourne,
East Melbourne, VIC, Australia
e-mail: d.urrutiacabrera@unimelb.edu.au; ching.wong@unimelb.edu.au

R. C. B. Wong and Y. Hayashi (eds.), *Advances in Using Stem Cells to Treat Neural Diseases*, Stem Cell Biology and Regenerative Medicine 77,
https://doi.org/10.1007/978-3-031-94101-6_12

and it highlights the potential of using cellular reprogramming technologies to treat the major retinal degenerative diseases.

Keywords Cellular reprogramming · Regeneration · Müller glia · Cell therapy · Gene therapy · Partial reprogramming · Retinal progenitors · Vision loss · Retinal degeneration · Retinitis pigmentosa · Age-related macular degeneration · Inherited retinal diseases · In vivo reprogramming

12.1 Retina Degeneration Is a Major Cause of Vision Loss

Retina degeneration is an irreversible process that has become one of the main causes of visual impairment and blindness worldwide. Vision is a coordinated process that largely relies on the collaboration of different subtypes of neurons, involving multiple structures such as the retina, the thalamus and the visual cortex. Specifically, the retina is responsible for the detection of light and visual cues, which are subsequently processed, and ultimately relayed to the visual cortex to form vision. The retina is organised as a multilayered structure integrated by five major neuronal subtypes, photoreceptors (PR), bipolar (BP), amacrine, horizontal and retinal ganglion cells (RGC), which are supported by at least three types of glial cells, Müller glia, microglia and retinal astrocytes [1].

Photoreceptors and RGCs are commonly affected in degenerative diseases that result in the loss of vision, such as AMD, glaucoma and IRDs. Mammals have two types of photoreceptors, rods and cones. Rods regulate vision in dim light conditions, whereas cones mediate vision in daylight, which also includes the detection of colour. Additionally, the centre of highest visual acuity of the human retina, the fovea, is densely packed with cone photoreceptors [2]. Thus, cone photoreceptors are also responsible for high acuity vision, and damage to these cells results in severe visual impairment.

Photoreceptor degeneration is a major hallmark of diseases that cause visual impairment such as AMD and IRDs. AMD is one of the leading causes of blindness in the world, affecting around 196 million people [3]. Early AMD is characterised by the presence of lipid deposits called drusen between the basal lamina of the retinal pigment epithelium (RPE) and Bruch membrane, changes in pigmentation at the macula region and moderate vision loss [4]. There are two forms of advanced AMD: (1) dry AMD, also known as geographic atrophy, which is characterised by the degeneration of RPE in the macula and the death of the overlying photoreceptors, resulting in progressive loss of vision; (2) wet or neovascular AMD, which is caused by the uncontrolled growth of new blood vessels, usually invading the neuroretina from the choroid, leading to damage to the macula and severe loss of central vision [5].

IRDs are a group of diseases characterised by progressive loss of vision caused by retina degeneration [6]. IRDs are phenotypically and genotypically heterogeneous, with over 300 genes associated with the aetiology of IRDs (https://retnet.org). Some of these diseases primarily affect photoreceptors, such as retinitis pigmentosa, which is the most common form of IRD, estimated to occur in more than 1 in 4000 individuals [7]. Retinitis pigmentosa is characterised by a progressive degeneration of rod photoreceptors at the onset of the disease, followed by a gradual loss of cones at a later stage [7]. Generally, patients first experience night blindness and loss of peripheral vision caused by the degeneration of rods. In advanced retinitis pigmentosa, the death of cones severely impacts the visual field, resulting in the loss of central vision and colour detection, progressively leading to blindness.

On the other hand, glaucoma affects around 76 million people worldwide, thus representing one of the leading causes of blindness [8]. Glaucoma is characterised by progressive degeneration of RGCs, which are the neurons that form the optic nerve and are responsible for relaying the visual stimuli from the retina to the brain. The exact mechanisms that lead to optic nerve degeneration in glaucoma are not completely understood. However, high intraocular pressure (IOP) is one of the major risk factors in glaucoma, as it causes mechanical stress and strain on the posterior structures of the eye [9]. Imbalance between the secretion of aqueous humour by the ciliary body and its outflow through the trabecular and uveoscleral pathways can lead to an increase in the IOP [10]. Glaucoma can be broadly categorised into two groups: (1) open-angle glaucoma, where there is an increased resistance to the aqueous outflow through the trabecular meshwork, and the increase in IOP is often slow; (2) angle-closure glaucoma, where the drainage pathway for the outflow is partially or fully obstructed, which leads to a sharp increase in IOP [9].

The degeneration of retinal neurons is an irreversible process that takes place in numerous diseases that lead to vision loss. Currently, there are no effective treatments to restore vision once the retina has degenerated. Although the retina is a complex tissue formed by multiple cell types, photoreceptors and RGCs are one of the main neuronal subtypes affected in retinal diseases that lead to blindness. Thus, these cells are an attractive target for regenerative therapies that aim to restore loss of vision in patients with many of the major retinal diseases.

12.2 Approaches to Treat Retina Degeneration

Significant progress in clinical research has produced strategies to prevent and stop retina degeneration in some cases of IRDs, AMD and glaucoma. For instance, neuroprotective agents, such as vitamin A and nicotinamide (Fig. 12.1a), have been explored to prevent photoreceptor and RGC damage with varying degrees of success in slowing the degeneration rate [11–13].

Gene therapies have emerged as a promising strategy to develop treatments for numerous retinal diseases that currently have no cure (Fig. 12.1b). The viral-mediated delivery of the neurotrophic factor BDNF has shown promising protective

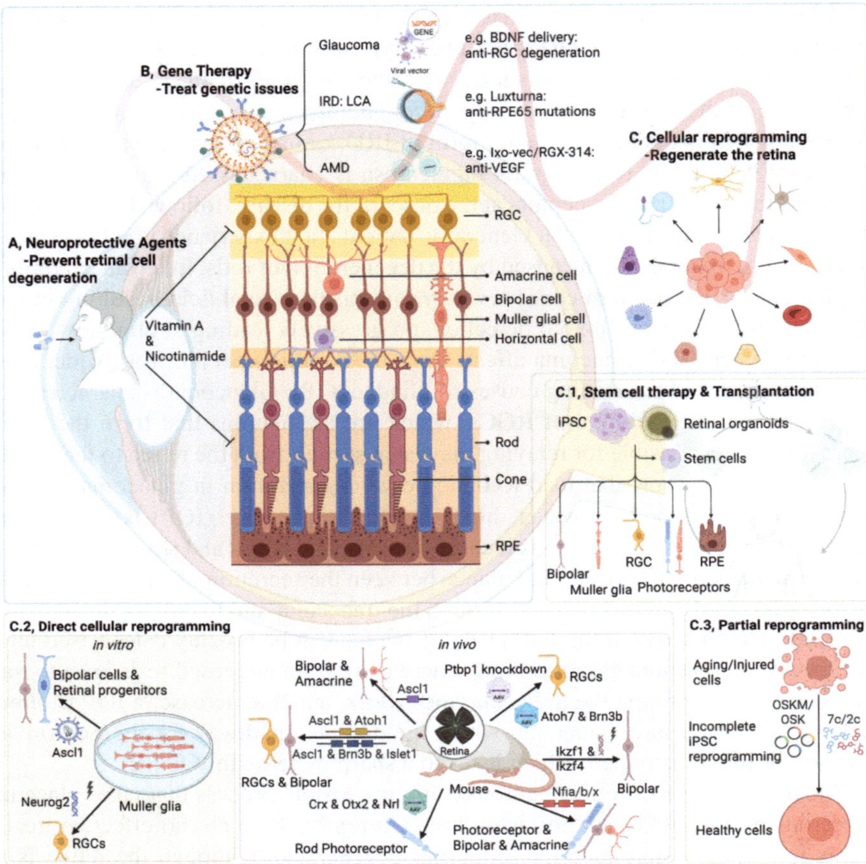

Fig. 12.1 Treatments for retina degeneration. (**a**) Strategies to prevent retina degeneration using neuroprotective agents. (**b**) Gene therapy approaches to treat the major retinal diseases that cause irreversible loss of vision (Glaucoma, AMD, and IRD). (**c**) Cellular reprogramming technology as a promising approach to regenerate the retina. (**c.1**) Cell-based therapies using stem cell-derived retinal neurons. (**c.2**) Reprogramming strategies to differentiate Müller glia into retinal neurons in vitro and in vivo reprogramming of Müller glia to promote retina regeneration. (**c.3**) Partial reprogramming technology to induce cell rejuvenation and tissue repair

effects against RGC degeneration caused by high IOP [14]. Notably, Luxturna is the first gene therapy approved by the FDA to treat retinal diseases [15]. Luxturna is a gene replacement therapy that uses adeno-associated viruses (AAV) to treat the biallelic mutation of *RPE65* that occurs in some cases of the IRD called Leber's congenital amaurosis [16]. For the treatment of AMD, anti-VEGF therapies have been the main focus for treating and managing the neovascularization that takes place in wet AMD [17]. Ixo-vec (formerly ADVM-022) and RGX-314 are anti-VEGF gene therapies using AAVs carrying aflibercept [18] or anti-VEGF-A-fab coding sequences, respectively [19]. Both treatments have shown promising results in phase 1 and 2 (Ixo-vec) and 1/2 (RGX-314) clinical trials; and they have received a Fast rack designation from the FDA. Similarly, GT005 is a gene therapy for dry AMD that uses AAVs to deliver the Complement Factor I, which allows management of an overactive complement system, thus representing a treatment for geographic atrophy in dry AMD patients [20]. GT005 also received a Fast Track designation from the FDA, but the phase 2 clinical trial "EXPLORE" (NCT04437368) was advised to be terminated early due to lack of treatment efficacy. In contrast, gene therapies for glaucoma are still in the pre-clinical phase with most treatments focusing on IOP reduction or RGC neuroprotection [21].

Although these treatments have shown encouraging results in preventing further neurodegeneration and managing retinal diseases, they are unable to regenerate retinal neurons and repair vision loss. To date, there are no effective treatments to restore vision following retina degeneration. Cell-based therapies using stem cell-derived retinal cells are an interesting approach to regenerate the retina and restore vision [22, 23]. Pre-clinical research has shown the therapeutic potential of transplanting stem cell-derived photoreceptors, retinal progenitors or products of retinal organoids to treat retinal degeneration (Fig. 12.1c.1) [24]. Notably, a clinical trial investigated the safety and survival rate of retinal organoid sheets transplanted in two retinitis pigmentosa patients with severe retina degeneration [25]. The study found that the graft could survive for 2 years without causing significant adverse events, but little improvement in visual function was detected. It has been challenging to demonstrate that the transplanted cells can successfully integrate into the retinal circuit, despite the identification of functional improvement using electrophysiology and behavioural assays [26, 27]. Notably, the discovery of bidirectional material exchange between the grafted and host cells, such as photoreceptors, has created controversy about the integration of grafted cells and the efficacy of these approaches [28]. Additionally, the manufacturing costs of cell therapy, risk of tumorigenicity and immune rejection remain key issues that need to be resolved for these cell therapies to progress into a treatment for retinal degeneration.

An alternative approach that has emerged as an encouraging avenue to regenerate the retina is the in vivo reprogramming of Müller glia into retinal neurons (Fig. 12.1c.2). Direct reprogramming uses transcription factors to induce the conversion of one somatic cell type into another; further expanding on the widely established induced pluripotent stem cell (iPSC) reprogramming technology, which demonstrated that cell fate is not definitive and it can be manipulated with the right signals [29]. In neuronal reprogramming, transcription factors that act as master regulators of the neuronal lineage during development can be used to induce neuronal fate [30]. A pivotal study by Vierbuchen et al. showed that mouse fibroblast could be reprogrammed into functional neurons through the overexpression of *Brn2*, *Ascl1* and *Myt1l* [31]. Subsequent research has further demonstrated that fine-tuning the combination of reprogramming genes allows the production of specific neuronal subtypes, such as dopaminergic neurons [32], sensory neurons [33] and retinal neurons [34].

Interestingly, in cellular reprogramming approaches, the transcriptional and epigenetic landscape of the starting cell type play a critical role in the differentiation process. For instance, the overexpression of the proneural gene *Ascl1* in dorsal midbrain astrocytes promotes reprogramming into glutamatergic and GABAergic neurons [35], whereas *Ascl1* overexpression in Müller glia induces reprogramming predominantly into bipolar and amacrine retinal neurons [36]. Thus, Müller glia cells represent an attractive target for regenerative approaches using cellular reprogramming technology, offering the opportunity to repair the retina in vivo and bypass the limitations associated with cell transplantation therapies.

12.3 The Regenerative Ability of Müller Glia in Vertebrates

In humans and mammals in general, retina degeneration is irreversible, and it causes permanent loss of vision. However, some lower vertebrates like teleost fish can regenerate their retina and restore their vision through a Müller glia-mediated repair. For instance, the Müller glia of zebrafish have the remarkable ability to dedifferentiate into multipotent progenitors after injury and divide asymmetrically to replenish all the damaged retinal cell types [37]. Notably, retina regeneration has been identified in other vertebrates such as chicken and amphibians, but this process has been largely studied in zebrafish and developing chicken [38].

In zebrafish, retinal injury first results in the inhibition of genes associated with Notch, pSmad, and GABA signalling pathways [39–41], which generally restrict proliferation and Müller glia reprogramming while they are active. Conversely, the signalling pathways of Stat3, Wnt/β-catenin, Sonic Hedgehog and the hippo pathway effector Yap become activated during Müller glia reprogramming in response to injury [42–44]. In particular, the proneural gene a*scl1a* plays a major role in the dedifferentiation of Müller glia into progenitors and the subsequent retina regeneration [45]. Notably, the microRNA let-7 is involved in maintaining quiescent Müller glia differentiated, as it suppresses the expression of progenitor-associated genes such as *pax6b*, a*scl1a*, *hspd1*, *c-myc*, *oct4* and *lin-28* [46, 47]. Induction of *ascl1a* expression upon injury upregulates *lin-28*, which in turn represses let-7 inhibitory regulation, thus promoting the activation of progenitor genes and Müller glia dedifferentiation [46].

Upon injury, the Müller glia from zebrafish dedifferentiate and re-enter the cell cycle to undergo one asymmetric cell division to produce a multipotent progenitor cell [48]. This process allows Müller cells to self-renew and retain the highly patterned multilayer structure of the retina, which in turn later facilitates the integration of the newly generated neurons into the retinal circuit. Multipotent progenitors migrate to the inner nuclear layer and proliferate to amplify the progenitor pool before undergoing differentiation into retinal cells [48]. During differentiation, Müller glia-derived progenitors exit the cell cycle and express specification genes such as *neurod1a*, *atoh7a* and *onecut1a* to give rise to the major retinal cells [41].

In contrast, the Müller glia of mammals have limited regenerative capacity, and display variable potential to re-enter the cell cycle depending on the species [49, 50]. Upon injury, Müller cells become reactive and initiate a gliotic response with the purpose of protecting retinal neurons from further damage [51]. Morphological, biochemical and physiological changes during gliosis allow Müller glia to release neurotrophic factors and protect the neurons from free radicals and glutamate neurotoxicity [51]. However, this type of response can represent a physical barrier for retina regeneration, and instead it promotes scar formation, which can result in permanent reduction of vision [52].

12.4 Cellular Reprogramming to Regenerate the Retina in Mammals

In recent years, significant efforts have been made to induce the reprogramming of mammalian Müller glia into retinal neurons, with the focus of harnessing the regenerative potential that these cells have shown in other vertebrates. Mouse Müller glia typically do not proliferate after photoreceptor injury despite the upregulation of *Pax6* and markers involved in cell cycle progression (cyclins) [53]. Indeed, mouse Müller glia require stimulation with growth factors such as EGF and FGF1 following retinal damage to proliferate and generate new neurons [49]. Notably, the forced expression of *Ascl1* reprogrammed the Müller glia of young mice in vitro into cells expressing markers characteristic of retinal progenitors and bipolar cells that could also elicit neuronal physiological responses [54]. In young mice, *Ascl1* overexpression can stimulate the neuronal reprogramming of Müller glia in vivo following retinal injury [36]. The reprogrammed neurons expressed markers of bipolar (*Otx2* and *Cabp5*+) and amacrine (*HuC/D* and *Pax6*+) neurons, and the presence of the synaptic marker CtBP2 suggested integration into the retinal circuit [36]. Notably, *Ascl1* overexpression was not sufficient to induce neuronal differentiation without retinal injury in Müller glia of adult mice, and the regenerative effect was limited when compared to young mice. Müller glia of adult mice have reduced chromatin accessibility for *Ascl1* target genes, hindering the capacity of *Ascl1* to induce reprogramming [55]. Therefore, treatment with epigenetic modulators such as trichostatin A (TSA) can enhance the efficiency of Müller glia reprogramming in adult mice following injury [55]. Additionally, it was identified that the STAT signalling prevents neurogenesis in the injured retina of adult mice through the upregulation of ID proteins [56]. These ID proteins can dimerise with basic helix-loop-helix (bHLH) proteins such as Ascl1 to inhibit their DNA binding and transcriptional activation activity [57]. Thus, treatment with the STAT pathway inhibitor (SH-4-54) can further improve the neurogenesis induced by Ascl1 and TSA in adult mice [56].

The overexpression of *Ascl1* induces the reprogramming of mouse Müller glia primarily into interneurons such as bipolar and amacrine cells [36, 56]. However, other proneural factors can also induce neuronal differentiation in mouse Müller glia, and the forced expression of a combination of genes can direct the reprogramming to induce other retinal neurons. For instance, overexpression of *Neurog2*, which is another proneural bHLH factor, has been reported to promote the conversion of mouse Müller glia into RGCs in vitro [58].

Notably, AAV-mediated delivery of *Atoh7* and *Brn3b* transgenes to mouse Müller glia promoted in vivo reprogramming into RGCs [59]. In this study, the authors included fluorescent reporter proteins (tdTomato and GFP) within an AAV polycistronic expression cassette in order to identify the transduced Müller glia. The reprogrammed cells projected their axons along the optic nerve to integrate into the visual circuit and contributed to the improvement of visual function in models with optic nerve crush and *Brn3b* null mutation [59]. Similarly, the overexpression of *Ascl1* and *Atoh1* enhanced the efficacy of Müller glia neuronal reprogramming, promoting an increase in the proportion of induced RGCs compared to *Ascl1* individual overexpression [60]. Remarkably, the overexpression of this transcription factor combination could induce reprogramming into functional neurons without the need for retinal injury. Conversely, retinal damage was still needed to stimulate the reprogramming of mouse Müller glia into RGC-like cells through the overexpression of *Ascl1*, *Pou4f2* and *Islet1* [61]. In this study, some of the reprogrammed RGCs also expressed markers of retinal progenitors, which were downregulated with the overexpression of *Atoh1*, suggesting that *Atoh1* also plays a role in the maturation process during reprogramming [61].

Boudreau-Pinsonneault et al. showed that electroporation of the temporal identity factors *Ikzf1* and *Ikzf4* promoted the conversion of Müller glia from mouse retinal explants into cells that expressed markers of bipolar cells [62]. *Ikzf1* is the mammalian homologue of Drosophila's gap gene hunchback, and it collaborates with *Ikzf4* in the developing retina to regulate the production of retinal cells [63]. Notably, the reprogramming induced by *Ikzf1/4* occurs without the proliferation of Müller glia and it is primed by changes in chromatin accessibility [62].

A reprogramming study targeting the regeneration of rod photoreceptors in mice, used the combination of *Crx*, *Otx2* and *Nrl* to promote conversion of Müller cells into rods in vivo [34]. The authors used AAVs to deliver the transgenes to the mouse retinatargeting the Müller cells. Overexpression of β-catenin was utilised to activate the proliferation of Müller glia in the absence of injury to facilitate neuronal differentiation. Notably, the induced photoreceptors could improve the visual response in a mouse model with impaired photoreceptor function [34].

These reports demonstrate that the overexpression of genes involved in the development of the neuroretina can stimulate the reprogramming of mouse Müller glia into retinal neurons. This encouraging research can be harnessed to develop regenerative therapies to treat vision loss. However, many of these studies have relied on the activation of Müller cells through retinal injury or epigenetic modulators, suggesting that more research is required to increase the efficacy of reprogramming. Additionally, the gene overexpression method plays a critical role in determining the specificity and efficacy of Müller glia reprogramming treatments.

12.5 Strategies to Support Müller Glia Reprogramming in Mammals

Regenerative research has identified molecules that can facilitate the reprogramming of mouse Müller glia in the absence of retinal injury, which includes growth factors, small molecules, epigenetic modulators and microRNAs (Table 12.1). Mitogenic agents like EGF, FGF1, FGF2 and β-catenin can activate mouse Müller glia proliferation and enhance their potential to differentiate into retinal neurons [34, 49, 58]. Thus, these molecules can be used in combination with transcription factors to improve the efficacy of reprogramming treatments. Yao et al. showed that in the absence of injury, stimulation with β-catenin is required to reprogram mouse Müller glia into rod photoreceptors in combination with the transcription factors Crx, Otx2 and Nrl [34]. Similarly, the histone deacetylase inhibitor TSA has been extensively used in combination with the proneural gene *Ascl1* to aid the reprogramming of Müller glia in adult mice [55].

An interesting approach to facilitate Müller glia reprogramming is the modulation of the REST complex, which contributes to the maintenance of glial fate through the inhibition of neuronal-specific genes [64]. Indeed, the overexpression of the microRNAs miR-124-9-9* can potentiate the neuronal reprogramming activity of *Ascl1* in primary cultures of mouse Müller glia [65]. The study showed that these microRNAs promoted the downregulation of REST complex components *Ctdsp1* and *Ptbp1* in targeted Müller glia cells, which in turn facilitated the acquisition of neuronal fate [65].

Notably, the nuclear factors Nfia, Nfib and Nfix have also been identified to regulate Müller glia regenerative capacity. The activity of Nfia/b/x in the mouse retina is critical for returning reactive Müller glia into a resting state following injury [66]. In this study, the authors demonstrated that the knockout of *Nfia/b/x* promoted an increase in the population of proliferative Müller glia after retinal damage, which led to the differentiation of a fraction of these cells into photoreceptors, bipolar and amacrine neurons [66].

Chemical reprogramming is an interesting alternative to identify molecules that could assist the neuronal reprogramming of Müller glia. Mahato et al. used a cocktail of small molecules including, valproic acid, CHIR99021, RepSox, forskolin, IWR1, Sonic hedgehog, taurine and retinoic acid, to reprogram mouse embryonic fibroblasts into photoreceptor-like cells [67]. The study showed that treatment with this cocktail caused the accumulation of mitochondrial reactive oxygen species, which led to NF-κB activation and the subsequent upregulation of *Ascl1*. Although the effect of this cocktail was not evaluated in Müller glia, these molecules could be paired with other reprogramming strategies to potentiate neuronal differentiation, which warrant future studies.

Overall, these studies suggest that supplementation of transcription factor overexpression with molecules that can potentiate Müller glia reprogramming is a promising strategy to enhance retina regeneration.

Table 12.1 Strategies to reprogram Müller glia into retinal neurons

Reprogramming genes	Delivery method	Treatment	Model	Cell type induction	References
Ascl1	Lentivirus	EGF	Mouse Müller glia (in vitro)	Retinal progenitors and bipolar cells	[54]
			Mouse retinal explants (in vitro)		
Ascl1	Transgenic mice	NMDA	Young mouse retina (in vivo)	Bipolar and amacrine cells	[36]
Ascl1	Transgenic mice	NMDA + TSA	Mouse retina (in vivo)	Bipolar and amacrine cells	[55]
Ascl1	Transgenic mice	NMDA + TSA + SH-4-54	Mouse retina (in vivo)	Bipolar cells	[56]
Neurog2	Electroporation	EGF + FGF2	Mouse Müller glia (in vitro)	Retinal ganglion cells	[58]
Ascl1 and Atoh1	Transgenic mice		Mouse retina (in vivo)	Retinal ganglion and bipolar cells	[60]
Atoh7 and Brn3b	AAVs		Mouse retina (in vivo)	Retinal ganglion cells	[59]
Ascl1, Brn3b and Islet1	Transgenic mice	NMDA + TSA	Mouse retina (in vivo)	Retinal ganglion and bipolar cells	[61]
Ikzf1 and Ikzf4	Electroporation		Young mouse retinal explant (in vitro)	Bipolar cells	[62]
			Young mouse retina (in vivo)		
Crx, Otx2 and Nrl	AAVs	β-Catenin	Mouse retina (in vivo)	Rod photoreceptors	[34]
Ascl1	Lentivirus	EGF	Mouse Müller glia (in vitro)	Retinal progenitors and retinal neurons	[65]
Nfia/b/x	Transgenic mice	NMDA + FGF2 + insulin	Mouse retina (in vivo)	Photoreceptors, bipolar and amacrine cells	[66]

12.6 Cell Rejuvenation Through Partial Reprogramming

Another interesting approach to repair retinal degeneration is partial reprogramming. Partial reprogramming aims to achieve molecular and cellular rejuvenation in ageing or injured cells while maintaining cellular identity (Fig. 12.1c.3). The technology of iPSC reprogramming has widely employed the Yamanaka factors Oct4, Sox2, Klf4 and c-Myc (OSKM) to dedifferentiate somatic cells back into a pluripotent stem cell state [29]. Interestingly, it has been reported that the transient expression of the OSKM gene combination leads to an incomplete iPSC reprogramming process, which can revert cellular and physiological hallmarks of ageing without the loss of the original cell identity. Studies performed in fibroblasts isolated from a premature ageing mouse model demonstrated that short-term OSKM expression (up to 12 days) promoted the reprogramming of ageing markers, such as the reduction of the DNA double-strand break marker y-H2AX and the restoration of heterochromatin maintenance markers H3K9me3 and H4K20me3 [68]. Similarly, partial reprogramming exerted a robust reversal of ageing markers in aged human muscle stem cells, including the reset of epigenetic clock and reduction of inflammation levels without losing the original identity [69]. Notably, the transient expression of OSKM for 6 days in the myofibers of 2-month-old mice induced phenotypic rejuvenation and accelerated muscle regeneration in vivo [70]. Therefore, the partial reprogramming technology represents an attractive approach to promote cellular rejuvenation and tissue regeneration, while it reduces the risk of modifying cell identity and teratoma formation.

Ageing is associated with several molecular markers including DNA damage, epigenetic modifications, transcriptome alterations, and proteostasis disruption [71, 72]. Changes in the epigenetic landscape take place during iPSC reprogramming, which can be used to track the progress and efficiency of the reprogramming process. These markers include the restoration of histone marks H3K9me3 and H4K20me3 levels, reset of telomere length, decrease in methylation clock age, and the reduction of progerin accumulation [73–75]. Considering that partial reprogramming is achieved through an incomplete iPSC reprogramming process, these molecular changes also can be used as markers to evaluate the partial reprogramming efficacy.

The reprogramming cocktail and the exposure to the reprogramming stimulus are key factors that determine partial reprogramming efficiency. Many studies have focused on optimising the length of OSKM transient expression, achieving different levels of rejuvenation. For instance, a study optimised the period of OSKM expression in fibroblasts to develop a method called maturation phase transient reprogramming (MPTR) to induce robust and substantial rejuvenation without erasing the original cell identity [76].

A complete iPSC reprogramming process includes three phases: (1) initiation phase, where cells undergo a mesenchymal-to-epithelial transition; (2) maturation phase, characterised by the activation of pluripotency-associated genes; (3) stabilisation phase, where a complete pluripotency programme is achieved and cell identity is reprogrammed [77]. Notably, transient OSKM expression up to the maturation phase during iPSC reprogramming resulted in the temporally loss of cell identity with the expression of early pluripotency markers [78]. However, cells in this intermediate state reacquired their initial cell identity after the reprogramming stimulus was removed, achieving the recovery of a youthful epigenetic landscape and a rejuvenated phenotype. The incomplete reprogramming process at this phase prevents the stable activation of pluripotency genes such as Nanog and Oct4 which are critical for the induction of iPSC identity. Similarly, 1 week of OSKM expression in mice was found to reverse age-related transcriptome alterations and partially restored a youthful DNA methylation pattern in the pancreas [79]. Thus, in vitro or in vivo experiments suggest that short-term OSKM expression can achieve reversal of a youthful cell signature without loss of cell identity. Furthermore, strategies using cyclic or pulse OSKM expression have also promoted the reversal of several hallmarks of ageing [70].

Similar to the complete iPSC reprogramming technology, partial reprogramming still faces serious challenges such as the activation of pluripotency and differentiation genes, risk of teratoma formation and difficulty in determining the optimal strategy for reprogramming [80]. For example, 6 days of OSKM expression in a mouse model with myocardial infarction can promote heart regeneration, while the extended expression of OSKM (up to 12 days) leads to tumour formation in the heart [81]. Therefore, efforts have focused on the development of novel reprogramming methods to improve the safety of partial reprogramming strategies. For example, c-Myc is a proto-oncogene that has been associated with teratoma formation during in vivo reprogramming. Thus, researchers have excluded this gene from the partial reprogramming cocktail to reduce the risk of teratoma formation [82]. Interestingly, the overexpression of OSK genes in an optic nerve crush-injury mouse model stimulated axon regeneration even after 15 months of continuous OSK expression without inducing tumours or structural changes in the mouse retina. This study showed that the expression of OSK induce partial reprogramming to restore a youthful epigenetic profile and transcriptomic landscape without erasing cell identity or inducing complete pluripotency [83]. Notably, mice with optic nerve injury treated with OSK exhibited recovered axon density, increased visual acuity and improved pattern electroretinogram waves. Future studies to validate these findings would be important to support the use of partial reprogramming as strategy to treat optic neuropathy.

Chemical reprogramming is an alternative way to induce the partial reprogramming process. For example, iPSC can be chemically reprogrammed using a cocktail of seven small molecules (7c) CHIR99021, DZNep, Forskolin, TTNPB, Valproic acid (VPA), Tranylcypromine (TCP), and RepSox [84]. In contrast, 6 days of 7c treatment in human fibroblasts isolated from aged dermal tissue appeared to ameliorate epigenetic hallmarks of ageing including the reduction of y-H2AX and the restoration of H3K9me3 levels [85]. Moreover, a selected cocktail of TCP and RepSox provided a similar efficacy in improving these molecular hallmarks of ageing. Notably, both chemical reprogramming methods were able to rejuvenate mouse fibroblasts at the multi-omics scale [86].

In summary, partial reprogramming is a promising technology to restore the molecular changes that occur during ageing, which often increase the risk for degenerative diseases. Notably, studies on glaucoma models suggest that partial reprogramming can be also used to regenerate the retina. Future studies to validate these findings would be important to investigate the effects of partial reprogramming on retinal degenerative diseases.

12.7 Challenges for In Vivo Reprogramming of Müller Glia

In vivo cellular reprogramming to regenerate retinal cells is a promising strategy to treat vision loss. However, delivery methods play a critical role in determining the efficiency of Müller glia reprogramming approaches. AAVs have emerged as one of the most popular tools to deliver transgenes to Müller glia, as these vectors have demonstrated cell type specificity and a safety profile in the retina. Nevertheless, a limitation of AAV vectors is their small packaging capacity (~4.3 kb), which currently restricts the size of the cargo and the promoter sequences that can be used to regulate tissue-specific expression. For this reason, the GFAP promoter has been widely used in glia-to-neuron reprogramming studies in combination with AAV vectors [34, 59, 87]. Notably, the use of reliable lineage tracing systems has become an important factor to consider when assessing the efficacy of Müller glia reprogramming studies. Lineage tracing mice that can genetically label Müller cells have helped to identify occasional unspecific transgene expression in neurons following the injection of AAV vectors with the GFAP promoter [88, 89]. Hoang et al. used a GlastCreERT2 mouse line to simultaneously induce the genetic labelling and disruption of *Ptbp1* expression selectively in Müller glia of adult mice [90]. This system demonstrated that the loss of function of *Ptbp1* is not sufficient to promote the conversion of Müller glia into retinal neurons. In contrast, this study suggests that previous reports require careful examination as the potential ectopic expression of GFAP vectors could have led to variable results.

These and other studies have highlighted the importance of using genetic lineage tracing to validate cellular reprogramming approaches [89–91]. Notably, transgenic mice lines have shown that the inducible overexpression of *Ascl1* alone or in combination with other transgenes in Müller glia, promotes their reprogramming into retinal neurons [55, 60, 61]. It would be interesting to use a similar approach to validate if AAV-mediated gene transfer of *Ascl1* can generate comparable results in vivo. Moreover, the identification of reliable promoters and therapy delivery methods for cellular reprogramming is critical for the development of regenerative therapies for the retina. AAV vectors have enabled the targetting of retinal cells with unprecedented specificity, which has allowed AAVs to become a widely used tool for gene therapy delivery in clinical trials [92]. Therefore, the optimisation of tissue-specific regulatory sequences has the potential to accelerate regenerative research in the retina.

12.8 Perspectives: In Vivo Reprogramming and Cell Rejuvenation

The use of cellular reprogramming to stimulate retina regeneration is an innovative approach to treat degenerative retinal diseases and restore vision loss. In recent years, significant progress has been made in the identification of genes that can help overcome the limited regenerative capacity of Müller glia in mammals. High-throughput techniques such as single-cell RNA sequencing and chromatin accessibility assays have been instrumental in understanding the complex processes of Müller glia and partial reprogramming. Nevertheless, more research is needed to improve the efficacy of reprogramming treatments as well as the specificity of delivery methods. The use of lineage tracing lines is essential for the optimisation of reprogramming cocktails and to demonstrate the efficacy of these promising regenerative strategies. Future studies will continue to identify gene combinations that can generate specific retinal subtypes, which is critical for the development of treatments for diseases such as glaucoma, AMD and IRDs.

Finally, it is still unclear if findings in pre-clinical animal models such as rodents can be translated to the human retina. Notably, Wohlschlegel et al. explored if Ascl1 could reprogram cultures of human Müller glia dissociated from retinal organoids and foetal retinas [93]. Overexpression of *ASCL1* in these cells through lentiviral delivery led to their reprogramming into retinal neurons that expressed markers of RGCs and amacrine cells, which also demonstrated electrophysiological properties. However, it remains to be studied if *ASCL1* or other genes can reprogram adult human Müller glia using clinically approved gene delivery methods such as AAVs. More studies to test Müller glia reprogramming in human cell models are needed to further develop this approach as an effective gene therapy treatment.

Acknowledgements This work is supported by the RIKEN-University of Melbourne Collaborative project scheme and the JSPS open partnership joint research projects co-funding schemes (RCBW). RCBW is supported by the University of Melbourne, the Centre for Eye Research Australia, National Health and Medical Research Council (GCT1184076) and Medical Future Research Fund (MRF2024365). DU is supported by the University of Melbourne Early Career Researcher grant. LW is supported by a CERA postgraduate scholarship. The Centre for Eye Research Australia receives operational infrastructure support from the Victorian Government.

References

1. Hoon M, Okawa H, Della Santina L, Wong RO (2014) Functional architecture of the retina: development and disease. Prog Retin Eye Res 42:44–84
2. Bringmann A, Syrbe S, Gorner K, Kacza J, Francke M, Wiedemann P et al (2018) The primate fovea: Structure, function and development. Prog Retin Eye Res 66:49–84
3. Wong WL, Su X, Li X, Cheung CM, Klein R, Cheng CY et al (2014) Global prevalence of age-related macular degeneration and disease burden projection for 2020 and 2040: a systematic review and meta-analysis. Lancet Glob Health 2(2):e106–e116

4. Hernandez-Zimbron LF, Zamora-Alvarado R, Ochoa-De la Paz L, Velez-Montoya R, Zenteno E, Gulias-Canizo R et al (2018) Age-related macular degeneration: new paradigms for treatment and management of AMD. Oxid Med Cell Longev 2018:8374647
5. Velez-Montoya R, Oliver SC, Olson JL, Fine SL, Quiroz-Mercado H, Mandava N (2014) Current knowledge and trends in age-related macular degeneration: genetics, epidemiology, and prevention. Retina 34(3):423–441
6. Olivares-Gonzalez L, Velasco S, Campillo I, Rodrigo R (2021) Retinal inflammation, cell death and inherited retinal dystrophies. Int J Mol Sci 22(4):2096
7. Hartong DT, Berson EL, Dryja TP (2006) Retinitis pigmentosa. Lancet 368(9549):1795–1809
8. Tham YC, Li X, Wong TY, Quigley HA, Aung T, Cheng CY (2014) Global prevalence of glaucoma and projections of glaucoma burden through 2040: a systematic review and meta-analysis. Ophthalmology 121(11):2081–2090
9. Weinreb RN, Aung T, Medeiros FA (2014) The pathophysiology and treatment of glaucoma: a review. JAMA 311(18):1901–1911
10. Fautsch MP, Johnson DH (2006) Aqueous humor outflow: what do we know? Where will it lead us? Invest Ophthalmol Vis Sci 47(10):4181–4187
11. Berson EL, Rosner B, Sandberg MA, Hayes KC, Nicholson BW, Weigel-DiFranco C et al (1993) A randomized trial of vitamin A and vitamin E supplementation for retinitis pigmentosa. Arch Ophthalmol 111(6):761–772
12. Chen X, Amorim JA, Moustafa GA, Lee JJ, Yu Z, Ishihara K et al (2020) Neuroprotective effects and mechanisms of action of nicotinamide mononucleotide (NMN) in a photoreceptor degenerative model of retinal detachment. Aging (Albany NY) 12(24):24504–24521
13. Tribble JR, Otmani A, Sun S, Ellis SA, Cimaglia G, Vohra R et al (2021) Nicotinamide provides neuroprotection in glaucoma by protecting against mitochondrial and metabolic dysfunction. Redox Biol 43:101988
14. Martin KR, Quigley HA, Zack DJ, Levkovitch-Verbin H, Kielczewski J, Valenta D et al (2003) Gene therapy with brain-derived neurotrophic factor as a protection: retinal ganglion cells in a rat glaucoma model. Invest Ophthalmol Vis Sci 44(10):4357–4365
15. Russell S, Bennett J, Wellman JA, Chung DC, Yu ZF, Tillman A et al (2017) Efficacy and safety of voretigene neparvovec (AAV2-hRPE65v2) in patients with RPE65-mediated inherited retinal dystrophy: a randomised, controlled, open-label, phase 3 trial. Lancet 390(10097):849–860
16. Cideciyan AV (2010) Leber congenital amaurosis due to RPE65 mutations and its treatment with gene therapy. Prog Retin Eye Res 29(5):398–427
17. Khanani AM, Thomas MJ, Aziz AA, Weng CY, Danzig CJ, Yiu G et al (2022) Review of gene therapies for age-related macular degeneration. Eye (London) 36(2):303–311
18. Khanani AM, Boyer DS, Wykoff CC, Regillo CD, Busbee BG, Pieramici D et al (2024) Safety and efficacy of ixoberogene soroparvovec in neovascular age-related macular degeneration in the United States (OPTIC): a prospective, two-year, multicentre phase 1 study. EClinical Medicine 67:102394
19. Campochiaro PA, Avery R, Brown DM, Heier JS, Ho AC, Huddleston SM et al (2024) Gene therapy for neovascular age-related macular degeneration by subretinal delivery of RGX-314: a phase 1/2a dose-escalation study. Lancet 403(10436):1563–1573
20. Dreismann AK, McClements ME, Barnard AR, Orhan E, Hughes JP, Lachmann PJ et al (2021) Functional expression of complement factor I following AAV-mediated gene delivery in the retina of mice and human cells. Gene Ther 28(5):265–276
21. Sulak R, Liu X, Smedowski A (2024) The concept of gene therapy for glaucoma: the dream that has not come true yet. Neural Regen Res 19(1):92–99
22. Hinkle JW, Mahmoudzadeh R, Kuriyan AE (2021) Cell-based therapies for retinal diseases: a review of clinical trials and direct to consumer “cell therapy” clinics. Stem Cell Res Ther 12(1):538
23. Klymenko V, Gonzalez Martinez OG, Zarbin MA (2024) Recent progress in photoreceptor cell-based therapy for degenerative retinal disease. Stem Cells Transl Med 13(4):332–345

24. West EL, Ribeiro J, Ali RR (2020) Development of stem cell therapies for retinal degeneration. Cold Spring Harb Perspect Biol 12(8):a035683
25. Hirami Y, Mandai M, Sugita S, Maeda A, Maeda T, Yamamoto M et al (2023) Safety and stable survival of stem-cell-derived retinal organoid for 2 years in patients with retinitis pigmentosa. Cell Stem Cell 30(12):1585–96.e6
26. Ribeiro J, Procyk CA, West EL, O'Hara-Wright M, Martins MF, Khorasani MM et al (2021) Restoration of visual function in advanced disease after transplantation of purified human pluripotent stem cell-derived cone photoreceptors. Cell Rep 35(3):109022
27. Mandai M, Fujii M, Hashiguchi T, Sunagawa GA, Ito SI, Sun J et al (2017) iPSC-derived retina transplants improve vision in rd1 end-stage retinal-degeneration mice. Stem Cell Rep 8(1):69–83
28. Nickerson PEB, Ortin-Martinez A, Wallace VA (2018) Material exchange in photoreceptor transplantation: updating our understanding of donor/host communication and the future of cell engraftment science. Front Neural Circuits 12:17
29. Takahashi K, Yamanaka S (2006) Induction of pluripotent stem cells from mouse embryonic and adult fibroblast cultures by defined factors. Cell 126(4):663–676
30. El Wazan L, Urrutia-Cabrera D, Wong RC (2019) Using transcription factors for direct reprogramming of neurons in vitro. World J Stem Cells 11(7):431–444
31. Vierbuchen T, Ostermeier A, Pang ZP, Kokubu Y, Sudhof TC, Wernig M (2010) Direct conversion of fibroblasts to functional neurons by defined factors. Nature 463(7284):1035–1041
32. Caiazzo M, Dell'Anno MT, Dvoretskova E, Lazarevic D, Taverna S, Leo D et al (2011) Direct generation of functional dopaminergic neurons from mouse and human fibroblasts. Nature 476(7359):224–227
33. Blanchard JW, Eade KT, Szucs A, Lo Sardo V, Tsunemoto RK, Williams D et al (2015) Selective conversion of fibroblasts into peripheral sensory neurons. Nat Neurosci 18(1):25–35
34. Yao K, Qiu S, Wang YV, Park SJH, Mohns EJ, Mehta B et al (2018) Restoration of vision after de novo genesis of rod photoreceptors in mammalian retinas. Nature 560(7719):484–488
35. Liu Y, Miao Q, Yuan J, Han S, Zhang P, Li S et al (2015) Ascl1 converts dorsal midbrain astrocytes into functional neurons in vivo. J Neurosci 35(25):9336–9355
36. Ueki Y, Wilken MS, Cox KE, Chipman L, Jorstad N, Sternhagen K et al (2015) Transgenic expression of the proneural transcription factor Ascl1 in Muller glia stimulates retinal regeneration in young mice. Proc Natl Acad Sci U S A 112(44):13717–13722
37. Sherpa T, Fimbel SM, Mallory DE, Maaswinkel H, Spritzer SD, Sand JA et al (2008) Ganglion cell regeneration following whole-retina destruction in zebrafish. Dev Neurobiol 68(2):166–181
38. Woodworth MB, Greig LC, Goldberg JL (2023) Intrinsic and induced neuronal regeneration in the mammalian retina. Antioxid Redox Signal 39(16–18):1039–1052
39. Lee MS, Wan J, Goldman D (2020) Tgfb3 collaborates with PP2A and notch signaling pathways to inhibit retina regeneration. Elife 9:e55137
40. Rao MB, Didiano D, Patton JG (2017) Neurotransmitter-regulated regeneration in the zebrafish retina. Stem Cell Rep 8(4):831–842
41. Jui J, Goldman D (2024) Muller glial cell-dependent regeneration of the retina in zebrafish and mice. Annu Rev Genet 58(1):67–90
42. Ramachandran R, Zhao XF, Goldman D (2011) Ascl1a/Dkk/beta-catenin signaling pathway is necessary and glycogen synthase kinase-3beta inhibition is sufficient for zebrafish retina regeneration. Proc Natl Acad Sci U S A 108(38):15858–15863
43. Wan J, Ramachandran R, Goldman D (2012) HB-EGF is necessary and sufficient for muller glia dedifferentiation and retina regeneration. Dev Cell 22(2):334–347
44. Lourenco R, Brandao AS, Borbinha J, Gorgulho R, Jacinto A (2021) Yap regulates muller glia reprogramming in damaged zebrafish retinas. Front Cell Dev Biol 9:667796
45. Fausett BV, Gumerson JD, Goldman D (2008) The proneural basic helix-loop-helix gene ascl1a is required for retina regeneration. J Neurosci 28(5):1109–1117

46. Ramachandran R, Fausett BV, Goldman D (2010) Ascl1a regulates muller glia dedifferentiation and retinal regeneration through a Lin-28-dependent, let-7 microRNA signalling pathway. Nat Cell Biol 12(11):1101–1107
47. Gao H, Luodan A, Huang X, Chen X, Xu H (2021) Muller glia-mediated retinal regeneration. Mol Neurobiol 58(5):2342–2361
48. Nagashima M, Barthel LK, Raymond PA (2013) A self-renewing division of zebrafish muller glial cells generates neuronal progenitors that require N-cadherin to regenerate retinal neurons. Development 140(22):4510–4521
49. Karl MO, Hayes S, Nelson BR, Tan K, Buckingham B, Reh TA (2008) Stimulation of neural regeneration in the mouse retina. Proc Natl Acad Sci U S A 105(49):19508–19513
50. Lahne M, Nagashima M, Hyde DR, Hitchcock PF (2020) Reprogramming muller glia to regenerate retinal neurons. Annu Rev Vis Sci 6:171–193
51. Bringmann A, Iandiev I, Pannicke T, Wurm A, Hollborn M, Wiedemann P et al (2009) Cellular signaling and factors involved in muller cell gliosis: neuroprotective and detrimental effects. Prog Retin Eye Res 28(6):423–451
52. Bringmann A, Pannicke T, Grosche J, Francke M, Wiedemann P, Skatchkov SN et al (2006) Muller cells in the healthy and diseased retina. Prog Retin Eye Res 25(4):397–424
53. Joly S, Pernet V, Samardzija M, Grimm C (2011) Pax6-positive muller glia cells express cell cycle markers but do not proliferate after photoreceptor injury in the mouse retina. Glia 59(7):1033–1046
54. Pollak J, Wilken MS, Ueki Y, Cox KE, Sullivan JM, Taylor RJ et al (2013) ASCL1 reprograms mouse muller glia into neurogenic retinal progenitors. Development 140(12):2619–2631
55. Jorstad NL, Wilken MS, Grimes WN, Wohl SG, VandenBosch LS, Yoshimatsu T et al (2017) Stimulation of functional neuronal regeneration from Muller glia in adult mice. Nature 548(7665):103–107
56. Jorstad NL, Wilken MS, Todd L, Finkbeiner C, Nakamura P, Radulovich N et al (2020) STAT signaling modifies Ascl1 chromatin binding and limits neural regeneration from muller glia in adult mouse retina. Cell Rep 30(7):2195–208.e5
57. Benezra R, Davis RL, Lockshon D, Turner DL, Weintraub H (1990) The protein Id: a negative regulator of helix-loop-helix DNA binding proteins. Cell 61(1):49–59
58. Guimaraes RPM, Landeira BS, Coelho DM, Golbert DCF, Silveira MS, Linden R et al (2018) Evidence of muller glia conversion into retina ganglion cells using neurogenin2. Front Cell Neurosci 12:410
59. Xiao D, Jin K, Qiu S, Lei Q, Huang W, Chen H et al (2021) In vivo regeneration of ganglion cells for vision restoration in mammalian retinas. Front Cell Dev Biol 9:755544
60. Todd L, Hooper MJ, Haugan AK, Finkbeiner C, Jorstad N, Radulovich N et al (2021) Efficient stimulation of retinal regeneration from muller glia in adult mice using combinations of proneural bHLH transcription factors. Cell Rep 37(3):109857
61. Todd L, Jenkins W, Finkbeiner C, Hooper MJ, Donaldson PC, Pavlou M et al (2022) Reprogramming muller glia to regenerate ganglion-like cells in adult mouse retina with developmental transcription factors. Sci Adv 8(47):eabq7219
62. Boudreau-Pinsonneault C, David LA, Lourenco Fernandes JA, Javed A, Fries M, Mattar P et al (2023) Direct neuronal reprogramming by temporal identity factors. Proc Natl Acad Sci U S A 120(19):e2122168120
63. Javed A, Santos-Franca PL, Mattar P, Cui A, Kassem F, Cayouette M (2023) Ikaros family proteins redundantly regulate temporal patterning in the developing mouse retina. Development 150(2):dev200436
64. Conaco C, Otto S, Han JJ, Mandel G (2006) Reciprocal actions of REST and a microRNA promote neuronal identity. Proc Natl Acad Sci U S A 103(7):2422–2427
65. Wohl SG, Reh TA (2016) miR-124-9-9* potentiates Ascl1-induced reprogramming of cultured muller glia. Glia 64(5):743–762
66. Hoang T, Wang J, Boyd P, Wang F, Santiago C, Jiang L et al (2020) Gene regulatory networks controlling vertebrate retinal regeneration. Science 370(6519):eabb8598

67. Mahato B, Kaya KD, Fan Y, Sumien N, Shetty RA, Zhang W et al (2020) Pharmacologic fibroblast reprogramming into photoreceptors restores vision. Nature 581(7806):83–88
68. Ocampo A, Reddy P, Martinez-Redondo P, Platero-Luengo A, Hatanaka F, Hishida T et al (2016) In vivo amelioration of age-associated hallmarks by partial reprogramming. Cell 167(7):1719–1733.e12
69. Sarkar TJ, Quarta M, Mukherjee S, Colville A, Paine P, Doan L et al (2020) Transient non-integrative expression of nuclear reprogramming factors promotes multifaceted amelioration of aging in human cells. Nat Commun 11(1):1545
70. Browder KC, Reddy P, Yamamoto M, Haghani A, Guillen IG, Sahu S et al (2022) In vivo partial reprogramming alters age-associated molecular changes during physiological aging in mice. Nat Aging 2(3):243–253
71. Schumacher B, Pothof J, Vijg J, Hoeijmakers JHJ (2021) The central role of DNA damage in the ageing process. Nature 592(7856):695–703
72. Kane AE, Sinclair DA (2019) Epigenetic changes during aging and their reprogramming potential. Crit Rev Biochem Mol Biol 54(1):61–83
73. Marion RM, Strati K, Li H, Tejera A, Schoeftner S, Ortega S et al (2009) Telomeres acquire embryonic stem cell characteristics in induced pluripotent stem cells. Cell Stem Cell 4(2):141–154
74. Wahlestedt M, Norddahl GL, Sten G, Ugale A, Frisk MA, Mattsson R et al (2013) An epigenetic component of hematopoietic stem cell aging amenable to reprogramming into a young state. Blood 121(21):4257–4264
75. Horvath S (2013) DNA methylation age of human tissues and cell types. Genome Biol 14(10):R115
76. Gill D, Parry A, Santos F, Okkenhaug H, Todd CD, Hernando-Herraez I et al (2022) Multi-omic rejuvenation of human cells by maturation phase transient reprogramming. Elife 11:e71624
77. Samavarchi-Tehrani P, Golipour A, David L, Sung HK, Beyer TA, Datti A et al (2010) Functional genomics reveals a BMP-driven mesenchymal-to-epithelial transition in the initiation of somatic cell reprogramming. Cell Stem Cell 7(1):64–77
78. Loh YH, Wu Q, Chew JL, Vega VB, Zhang W, Chen X et al (2006) The Oct4 and Nanog transcription network regulates pluripotency in mouse embryonic stem cells. Nat Genet 38(4):431–440
79. Chondronasiou D, Gill D, Mosteiro L, Urdinguio RG, Berenguer-Llergo A, Aguilera M et al (2022) Multi-omic rejuvenation of naturally aged tissues by a single cycle of transient reprogramming. Aging Cell 21(3):e13578
80. Roux AE, Zhang C, Paw J, Zavala-Solorio J, Malahias E, Vijay T et al (2022) Diverse partial reprogramming strategies restore youthful gene expression and transiently suppress cell identity. Cell Syst 13(7):574–87.e11
81. Chen Y, Luttmann FF, Schoger E, Scholer HR, Zelarayan LC, Kim KP et al (2021) Reversible reprogramming of cardiomyocytes to a fetal state drives heart regeneration in mice. Science 373(6562):1537–1540
82. Nakagawa M, Takizawa N, Narita M, Ichisaka T, Yamanaka S (2010) Promotion of direct reprogramming by transformation-deficient Myc. Proc Natl Acad Sci U S A 107(32):14152–14157
83. Lu Y, Brommer B, Tian X, Krishnan A, Meer M, Wang C et al (2020) Reprogramming to recover youthful epigenetic information and restore vision. Nature 588(7836):124–129
84. Hou P, Li Y, Zhang X, Liu C, Guan J, Li H et al (2013) Pluripotent stem cells induced from mouse somatic cells by small-molecule compounds. Science 341(6146):651–654
85. Yucel AD, Gladyshev VN (2024) The long and winding road of reprogramming-induced rejuvenation. Nat Commun 15(1):1941
86. Mitchell W, Goeminne LJE, Tyshkovskiy A, Zhang S, Chen JY, Paulo JA et al (2024) Multi-omics characterization of partial chemical reprogramming reveals evidence of cell rejuvenation. Elife 12:RP90579
87. Zhou H, Su J, Hu X, Zhou C, Li H, Chen Z et al (2020) Glia-to-neuron conversion by CRISPR-CasRx alleviates symptoms of neurological disease in mice. Cell 181(3):590–603.e16

88. Xie Y, Zhou J, Wang LL, Zhang CL, Chen B (2023) New AAV tools fail to detect Neurod1-mediated neuronal conversion of Muller glia and astrocytes in vivo. EBioMedicine 90:104531
89. Le N, Appel H, Pannullo N, Hoang T, Blackshaw S (2022) Ectopic insert-dependent neuronal expression of GFAP promoter-driven AAV constructs in adult mouse retina. Front Cell Dev Biol 10:914386
90. Hoang T, Kim DW, Appel H, Pannullo NA, Leavey P, Ozawa M et al (2022) Genetic loss of function of Ptbp1 does not induce glia-to-neuron conversion in retina. Cell Rep 39(11):110849
91. Wang LL, Serrano C, Zhong X, Ma S, Zou Y, Zhang CL (2021) Revisiting astrocyte to neuron conversion with lineage tracing in vivo. Cell 184(21):5465–81 e16
92. Yudaeva A, Kostyusheva A, Kachanov A, Brezgin S, Ponomareva N, Parodi A et al (2024) Clinical and translational landscape of viral gene therapies. Cells 13(22):1916
93. Wohlschlegel J, Finkbeiner C, Hoffer D, Kierney F, Prieve A, Murry AD et al (2023) ASCL1 induces neurogenesis in human muller glia. Stem Cell Rep 18(12):2400–2417

Index

R. C. B. Wong and Y. Hayashi (eds.), *Advances in Using Stem Cells to Treat Neural Diseases*, Stem Cell Biology and Regenerative Medicine 77, https://doi.org/10.1007/978-3-031-94101-6_12

MIX
Papier aus verantwortungsvollen Quellen
Paper from responsible sources
FSC® C105338

If you have any concerns about our products,
you can contact us on
ProductSafety@springernature.com

In case Publisher is established outside the EU,
the EU authorized representative is:
Springer Nature Customer Service Center GmbH
Europaplatz 3, 69115 Heidelberg, Germany

Printed by Libri Plureos GmbH
in Hamburg, Germany